Dynamical Systems

Proceedings of a University of Florida International Symposium

ACADEMIC PRESS RAPID MANUSCRIPT REPRODUCTION

Proceedings of a University of Florida International Symposium
on Dynamical Systems held at the University of Florida,
Gainesville, March 24–26, 1976

Dynamical Systems

Proceedings of a University of Florida
International Symposium

Edited by

A. R. Bednarek

University of Florida
Gainesville, Florida

L. Cesari

University of Michigan
Ann Arbor, Michigan

1977

ACADEMIC PRESS, INC.

New York San Francisco London

A Subsidiary of Harcourt Brace Jovanovich, Publishers

ACADEMIC PRESS, INC.
111 Fifth Avenue, New York, New York 10003

United Kingdom Edition published by
ACADEMIC PRESS, INC. (LONDON) LTD.
24/28 Oval Road, London NW1

Library of Congress Cataloging in Publication Data

International Symposium on Dynamical Systems, University of Florida, 1976.
Dynamical systems.

1. Differential equations–Congresses. 2. System analysis–Congresses. 3. Control theory–Congresses. 4. Stability–Congresses. I. Bednarek, A. R. II II. Cesari, Lamberto. III. Title.
QA371.I55 1976 515′.35 77-7283
ISBN 0–12–083750–1

PRINTED IN THE UNITED STATES OF AMERICA

Contents

CONTRIBUTED PAPERS

List of Contributors

SHOSHANA ABRAMOVICH, Department of Mathematics, University of Haifa, Mt. Carmel, Haifa, Israel 31999

S. R. BERNFELD, Department of Mathematics, University of Texas at Arlington, Arlington, Texas 76019

YURI BIBIKOV, Division of Applied Mathematics, Brown University, Providence, Rhode Island 02912

R. L. BOWDEN, Department of Physics, Virginia Polytechnic Institute and State University, Blacksburg, Virginia 24061

T. T. BOWMAN, Department of Mathematics, University of Florida, Gainesville, Florida 32611

LAMBERTO CESARI, Department of Mathematics, University of Michigan, Ann Arbor, Michigan 48104

ETHELBERT N. CHUKWU, Department of Mathematics, Cleveland State University, Cleveland, Ohio 44115

MICHAEL G. CRANDALL, Department of Mathematics, University of Wisconsin, Madison, Wisconsin 53706

C. M. DAFERMOS, Lefschetz Center for Dynamical Systems, Brown University, Providence, Rhode Island 02912

RICHARD DATKO, Department of Mathematics, Georgetown University, Washington, D.C. 20057

D. G. DEFIGUEIREDO, Department of Mathematics, University of Brasillia, Brasilia, Brasil

RICHARD I. DeVRIES, Department of Mathematics, University of Michigan, Ann Arbor, Michigan 48104

R. D. DRIVER, Department of Mathematics, University of Rhode Island, Kingston, Rhode Island 02881

ROBERT M. GOOR, Research Laboratories, General Motors Corporation, General Motors Technical Center, Warren, Michigan 48090

JOHN R. GRAEF, Department of Mathematics, Mississippi State University, Mississippi State, Mississippi 39762

JOHN GREGORY, Department of Mathematics, Southern Illinois University at Carbondale, Carbondale, Illinois 62901

JAN M. GRONSKI, Department of Mathematics, Cleveland State University, Cleveland, Ohio 44115

CHAITAN P. GUPTA, Department of Mathematics, Northern Illinois University, DeKalb, Illinois 60115

M. L. J. HAUTUS, Technological University Eindhoven, P. O. Box 513, Eindhoven, Netherlands

H. HERMES, Department of Mathematics, University of Colorado, Boulder, Colorado 80302

MICHAEL HEYMANN, Department of Electrical Engineering, Technion, Haifa, Israel

DEH-PHONE K. HSING, Department of Mathematics, University of Rhode Island, Kingston, Rhode Island 02881

J. INCIURA, Department of Applied Mathematics, University of Waterloo, Waterloo, Ontario, Canada

E. F. INFANTE, Division of Applied Mathematics, Brown University, Providence, Rhode Island 02912

GARY D. JONES, Department of Mathematics, Murray State University, Murray, Kentucky 42071

JOHN JONES, JR., Air Force Institute of Technology, Wright Patterson Air Force Base, Dayton, Ohio 45433

R. KANNAN, Department of Mathematics, University of Missouri at St. Louis, 8001 Natural Bridge Road, St. Louis, Missouri 63121

JUNJI KATO, Department of Mathematics, Michigan State University, East Lansing, Michigan 48824

GEORGE H. KNIGHTLY, Department of Mathematics, University of Massachusetts, Amherst, Massachusetts 01002

H. W. KNOBLOCH, Mathematisches Institut, Universität Würzburg, West Germany.

V. LAKSHMIKANTHAM, Department of Mathematics, Univ. of Texas at Arlington, Arlington, Texas 76019

J. P. LaSALLE, Lefschetz Center for Dynamical Systems, Brown University, Providence, Rhode Island 02912

JON LEE, Flight Dynamics Laboratory, Wright Patterson AFB, Dayton, Ohio 45433

W. S. LOUD, Department of Mathematics, University of Minnesota, Minneapolis, Minnesota 55455

ROGER C. McCANN, Department of Mathematics, Case Western Reserve Univeristy, Cleveland, Ohio 44106

P. J. McKENNA, Department of Mathematics, University of Michigan, Ann Arbor, Michigan 48104

JOHN MALLET-PARET, Division of Applied Mathematics, Brown University, Providence, Rhode Island 02912

A. MANITIUS, Centre de Recherches Mathematiques, Case Postal 6128, Montreal H3C 3J7, Canada

T. MATSUMOTO, Department of Electrical Engineering, Waseda University, Shinjuku, Tokio 160, Japan

JEAN MAWHIN, Institut Mathematique, Chemin du Cyclotron 2, B-1348 Louvain La Neuve, Belgium

PETER J. MELVIN, Liberal Arts and Sciences Administration, University of Illinois at Urbana-Champaign, Urbana, Illinois 61801

R. MENIKOFF, Department of Mathematics, John Hopkins University, Baltimore, Maryland 28218

ZBIGNIEW NITECKI, Department of Mathematics, Tufts University, Medford, Massachusetts 02155

ANNETT NOLD, Institute for Physical Science & Technology, Space Science Building, University of Maryland, College Park, Maryland 20742

ROGER D. NUSSBAUM, Department of Mathematics, Rutgers University, New Brunswick, New Jersey

V. M. POPOV, Department of Mathematics, University of Florida, Gainesville, Florida 32611

W. PRAGER, Tgess Tgampi 4/27 (7451) Savognin, Switzerland

PAUL H. RABINOWITZ, Department of Mathematics, University of Wisconsin, Madison, Wisconsin 53706

SAMUEL M. RANKIN III, Department of Mathematics, Murray State University, Murray, Kentucky 42071

Y. M. REDDY, Department of Mathematics, Paul Quinn College, Waco, Texas 76703

E. H. ROTHE, Department of Mathematics, University of Michigan, Ann Arbor, Michigan 48104

EMILIO O. ROXIN, Department of Mathematics, University of Rhode Island, Kingston, Rhode Island 02881

G. I. N. ROZVANY, Department of Civil Engineering, Monash University, Clayton, Victoria 3168, Australia

J. D. SCHUUR, Department of Mathematics, Michigan State University, East Lansing, Michigan 48823

GEORGE SEIFERT, Department of Mathematics, Iowa State University, Ames, Iowa 50010

YASUTAKA SIBUYA, Department of Mathematics, University of Minnesota, Minneapolis, Minnesota 55455

CARL P. SIMON, Department of Mathematics, University of Michigan, Ann Arbor, Michigan 48104

ALFREDO SOMOLINOS, Division of Applied Mathematics, Brown University, Providence, Rhode Island 02912

EDUARDO D. SONTAG, Center for Mathematical System Theory, University of Florida, Gainesville, Florida 32611

PAUL W. SPIKES, Department of Mathematics, Mississippi State University, Mississippi State, Miss. 39762

RONALD J. STERN, Department of Mathematics, McGill University, Montreal, Canada

M. B. SURYANARAYANA, Department of Mathematics, State University of New York at Albany, Albany, New York

RONALD SVERDLOVE, Department of Mathematics, Stanford University, Stanford, California

R. TRIGGIANI, Department of Mathematics, Iowa State University, Ames, Iowa 50010

J. A. WALKER, Department of Mechanical Engineering and Astronautical Sciences, Northwestern University, Evanston, Illinois 60201

Y. H. WAN, Department of Mathematics, SUNY at Buffalo, Amherst, New York 14226

H. F. WEINBERGER, School of Mathematics, University of Minnesota, Minneapolis, Minnesota 55455

YONG J. YOON, Department of Mathematics, University of Florida, Gainesville, Florida 32611

JAMES A. YORKE, Institute for Fluid Dynamics & Applied Mathematics, University of Maryland, College Park, Maryland 20742

P. F. ZWEIFEL, Department of Physics, Virginia Polytechnic Institute and State University, Blacksburg, Virginia 24061

List of Participants

Participants of the International Symposium on Dynamical Systems, Center for Applied Mathematics, University of Florida. March 24-26, 1976.

S. ABRAMOVICH (C), University of Haifa

J. ALBERT, Tufts University

P. BARRERA, National University of Mexico

A. R. BEDNAREK, University of Florida

S. R. BERNFELD (C), University of Texas

Y. N. BIBIKOV, USSR (C), Visiting Prof., Brown Univ.

L. BLOCK, University of Florida

T. T. BOWMAN (L), University of Florida

R. C. BROWN, University of Alabama

L. CESARI (L), University of Michigan

J. CHANDRA, U. S. Army Research Office

R. CHEN, University of Florida

S. CHOW (L), Michigan State University

E. N. CHUKWU (C), Cleveland State University

K. L. COOKE, Pomona College

C. DAFERMOS (L), Brown University

R. F. DATKO (C), Georgetown University

D. G. DEFIGUEIREDO (L), University of Brasilia

R. DEVRIES (C), University of Michigan

J. DRAPER, University of Florida

R. D. DRIVER (C), University of Rhode Island

P. DUNKELBERG, University of Florida

R. H. ELDERKIN, Pomona College

G. B. ERMENTROUT, Chicago University

R. E. FENNELL, Clemson University

R. GOOR (C), University of Delaware

J. R. GRAEF (C), Mississippi State University

J. GREGORY (C), Southern Illinois University at Carbondale

(L) = Lecture (C) = Contributed Paper

M. GROST, University of Michigan

C. P. GUPTA (C), Northern Illinois University

T. HAIGH, Rose Hulman Institute

J. K. HALE (L), Brown University

T. G. HALLAM, Florida State University

G. HEMP, University of Florida

G. P. HENDERSON, University of South Florida

F. HERMANO, University of Florida

H. HERMES (L), University of Colorado

M. HEYMANN (L), University of Florida

J. INCIURA (C), University of Waterloo

E. F. INFANTE (L), Brown University

A. F. IZE, Sao Paulo, Brazil

J. JONES (C), Air Force Inst. of Technology

R. KALMAN, Center for Mathematical System Theory, Univ. of Florida

R. KANNAN (L), University of Missouri

J. KATO (C), Michigan State University

J. KEESLING, University of Florida

K. W. KIRCHGASSNER (L), Stuttgart, Germany

G. H. KNIGHTLY (L), University of Massachusetts

H. W. KNOBLOCH (L), University of Wurzburg, W. Germany

D. KOLZOW (C), Erlangen, Germany

D. KU (C), University of Michigan

V. LAKSHMIKANTHAM, University of Texas

W. F. LANGFORD, McGill University

J. P. LaSALLE (L), Brown University

J. LEE (C), Wright-Patterson AFB

S. LIST (C), Brown University

O. F. LOPES (C), Campinas, Brazil

W. S. LOUD (L), University of Minnesota
D. L. LOVELADY, Florida State University
C. LUEHR, University of Florida
C. McCALLA, Harvard University
R. C. McCANN (C), Case Western Reserve University
P. J. McKENNA (L), University of Michigan
A. K. MAJUMDAR, University of Florida
J. MALLET-PARET (C), Brown University
A. MANITIUS (C), Université de Montreal
T. MATSUMOTO (C), Waseda University, Japan
J. MAWHIN (L), Belgium
P. J. MELVIN (C), University of Illinois
K. T. MILLSAPS, University of Florida
A. P. MORGAN, University of Miami
D. R. NAUGLER, Mt. St. Vincent University
D. NEUMANN, Bowling Green State University
Z. NITECKI (C), Tufts University
R. D. NUSSBAUM (L), Rutgers University
V. M. POPOV (L), University of Florida
Z. R. POP-STOJANOVIC, University of Florida
W. PRAGER (L), Savognin, Switzerland
T. G. PROCTOR, Clemson University
P. RABINOWITZ (L), University of Wisconsin
S. M. RANKIN (C), Murray State University
J. A. RENEKE, Clemson University
M. ROBERTSON, Auburn University
F. V. ROHDE, Mississippi State University
E. H. ROTHE (L), University of Michigan
E. O. ROXIN (L), University of Rhode Island
L. D. SABBAGH, Bowling Green State University
R. SAPIRO, University of Florida
J. SCHUUR (C), Michigan State University
G. SEIFERT (C), Iowa State University
CARL SIMON (L), University of Michigan
A. SOMOLINOS (C), Brown University
E. SONTAG (C), University of Florida
P. SPIKES, Mississippi State University
M. J. STECHER, Texas A & M University
R. J. STERN, McGill University
M. B. SURYANARAYANA (L), SUNY at Albany
R. SVERDLOVE (C), Stanford University
R. UNDERWOOD, University of South Carolina
J. A. WALKER, Northwestern University
Y. H. WAN (C), SUNY at Buffalo
H. G. WEINBERGER (L), University of Minnesota
G. WELLER
Y. YAMAMOTO, University of Florida
Y. J. YOON (C), University of Florida
J. YORKE (L), University of Maryland
E. C. YOUNG, Florida State University
P. ZWEIFEL (C), Virginia Polytechnic Institute

Preface

The present volume collects the lectures and the contributed papers read at the International Symposium on Dynamical Systems held at the University of Florida in Gainesville, Florida, March 24–26, 1976.

The field of science—mostly mathematics—that goes under the name Dynamical Systems is vast and complex. Its many parts are deeply intertwined and at present in a phase of intense development: ordinary and partial differential equations, dynamical systems proper, nonlinear analysis, optimal control, system theory, deterministic and stochastic problems, stability, limit cycles and invariant manifolds, nonlinear eigenvalues, bifurcation and alternative methods, functional analysis, topological methods.

This international symposium was sponsored by the Center for Applied Mathematics of the University of Florida and was made possible by grants from the University and from the U.S. Army Research Office.

The program committee for the symposium was composed of A. R. Bednarek, and K. T. Millsaps, Codirectors of the Center for Applied Mathematics at the University of Florida, J. P. LaSalle and J. K. Hale of Brown University, and L. Cesari of the University of Michigan.

The editors of the present volume wish to thank the colleagues of the program committee for their endeavors and Dr. T. T. Bowman for his help in editing these proceedings as well as for his very successful handling of the details of the symposium as chairman of the local committee.

Thanks should also go to Ms. Brenda Hobby for typing this volume.

Dynamical Systems

Proceedings of a University of Florida International Symposium

PERTURBATIONS OF LIÉNARD SYSTEMS WITH SYMMETRIES

T. T. Bowman
Department of Mathematics
University of Florida

1. INTRODUCTION

We consider here the existence of T-periodic solutions of the Liénard-type nonlinear differential system

(1) $x''+(d/dt)U_1(x,t)+h_1(x,t) = \varepsilon((d/dt)U_2(x,t)+h_2(x,t))$

where $U_i:R^{n+1} \to R^n$, $h_i:R^{n+1} \to R^n$, and $x = (x_1,x_2,\ldots,x_n)$, satisfying the symmetric conditions

$$U_i(-x,t+T/2) = -U_i(x,t)$$

and

$$h_i(-x,t+T/2) = -h_i(x,t).$$

The author in [1] examined equation (1) without the perturbation term. It was shown that under suitable growth conditions on $U_1(x,t)$ and $h_1(x,t)$, equation (1) has T-periodic solutions. Here we show that such a system will still have periodic solutions when symmetric perturbation terms are added with no growth restriction on the terms. We illustrate in sections 2 and 3 applications of the theorems to several classical nonlinear equations.

We shall use notation corresponding to [1]. Minor changes must be made to account for the fact that here we deal with T-periodic conditions instead of 2π-periodic conditions. In particular $S = (L_2[0,T])$, H^1 the Hilbert space of all absolutely continuous (AC) n-vector functions $x(t) =$

$(x_1, x_2, \ldots, x_n)$ with $x' \in S$, $x(0) = x(T)$, H^2 the Hilbert space of all AC n-vector functions x with $x' \in H^1$ and $x(0) = x(T)$, and $\mathcal{O}$ is the subspace of functions $x \in S$ with $x(t+T/2) = -x(t)$ for almost all t. Let $P: S \to S$ denote the projector operator defined by $Px = T^{-1} \int_0^T x(t)\,dt$, S_1 the kernel of P and S_0 the image of P. For $x \in S_1$, denote by $J_1 x$ the unique primitive of x with mean value zero; J_1 is an operator from $S_1 \to H^1 \cap S_1$. Define $J: S \to H^1 \cap S_1$ by $Jx = J_1(I-P)$ and define $J^* = -J$.

2. THE FIRST EXISTENCE THEOREM

2.1 Theorem. For the differential equation

$$x'' + (d/dt)\,\mathrm{grad}\,G(x) = h_1(x,t) + \varepsilon h_2(x,t) \tag{2}$$

where $G: R^n \to R$ of class C^2, $h_i: R^{n+1} \to R^n$, $i = 1,2$, is continuous, satisfying the symmetric conditions

(i) $G(x) = G(-x)$ for all $x \in R$;

(ii) $h_i(-x, t+\pi/w)) = -h_i(x,t)$, $i = 1,2$, for all $(x,t) \in R^{n+1}$ and w a fixed positive number; and

(iii) $h_1(x,t)x \geq -\delta|x| - \rho|x|^2$ for some constants $\delta \geq 0$, $0 \leq \rho < w^2$ and all $(x,t) \in R^{n+1}$;

then for every $r > 2\pi^{5/2}\delta(w+1)w^{-3/2}(w^2-\rho)^{-1}$ there exists a number $\varepsilon_r > 0$ such that for all $\varepsilon, |\varepsilon| < \varepsilon_r$, system (1) has a $2\pi/w$-periodic solution x with $x(t) = -x(t+\pi/w)$ and

$$\sup_t |x(t)| \leq r.$$

Before we prove the theorem we need the following lemma.

2.2 Lemma. If $T = 2\pi/w$ and $y \in \mathcal{O} \cap H^1$, then $\sup_t |y(t)| \leq (\pi/w)^{1/2} 2^{-1} ||y'||$.

Proof. The absolute continuity and the symmetric conditions imply there exists a $t_1 \in (t_0, t_0+\pi/w)$ such that $\sup_t |y(t)| = |y(t_1)|$. By the absolute continuity of y for

$t \in [t_0, t_0+\pi/w]$,

$$|y(t)| = \left|\int_{t_0}^{t} y'(\alpha)\,d\alpha\right| \le (\pi/w)^{1/2}\left(\int_{t_0}^{t_0+\pi/w} y'(\alpha)^2 d\alpha\right)^{1/2} =$$

$$= (\pi/w)^{1/2}\left(\int_{0}^{\pi/w} y'(\alpha)^2 d\alpha\right)^{1/2} \le (\pi/w)^{1/2} 2^{-1} ||y'||.$$

<u>Proof of Theorem 2.1.</u> In the proof of parts (a), (b) and (d) of theorem (2i), section 2 of [1], it was necessary and sufficient to solve the equation

(3) $$y + JNJ^* y = 0$$

for $y \in \mathcal{O}$ and

$$N = -(d/dt) \text{ grad } G(x) + h_1(x,t) + \varepsilon h_2(x,t).$$

For if $y \in \mathcal{O}$ is a solution of (3), then $x = J^* y \in \mathcal{O}$ is a solution of (2). Corresponding to [3], J^* from S to H^1 satisfies the inequality $||J^*x||_1 \le (1+w)w^{-1}||x||$.

Since h_2 is continuous and $2\pi/w$-periodic in t, h_2 is bounded on $S_r \times R$, say $|h_2(x,t)| \le M_r$, for each $r > 0$ and $S_r = \{x \in R^n : |x| \le r\}$. Define the map $\eta_r : R^n \to S_r$ by

$$\eta_r(x) = \begin{cases} x, & |x| \le r \\ (x/|x|)r, & |x| > r. \end{cases}$$

We now define the map $g_r : R^n \times R \to R$ by

$$g_r(x,t) = h_2(\eta_r x, t).$$

Note that $g_r(-x, t+\pi/w) = -g_r(x,t)$ and since it is the composition of continuous maps g_r is continuous. Also, note that

$$|g_r(x,t)x| = |h_2(\eta_r x, t)x| \le M_r\ |x|$$

which implies

$$g_r(x,t)x \ge -M_r\ |x|.$$

For $r > 2\pi^{5/2}\delta(w+1)w^{-3/2}(w^2-\rho)^{-1}$, define

$$\varepsilon_r = [r - 2\pi^{5/2}\delta(w+1)w^{-3/2}(w^2-\rho)^{-1}](w^2-\rho)w^{3/2}[2\pi^{5/2}(w+1)M_r]^{-1}.$$

For any ε with $|\varepsilon| < \varepsilon_r$, there exists an R such that $0 < R < 2rw^{3/2}\pi^{-1/2}(w+1)^{-1}$ and

$$(4) \qquad |\varepsilon| < R(w^2-\rho)(4\pi^2 M_r)^{-1} - \delta M_r^{-1}.$$

We now have

$$h_1(x,t) + \varepsilon g_r(x,t) \geq -\delta|x| - \rho|x|^2 - |\varepsilon|M_r|x|$$
$$= -(\delta+|\varepsilon|M_r)|x| - \rho|x|^2,$$

and solving (4) for R^{-1}, we have

$$(2\pi)^2(\delta+|\varepsilon|M_r)R^{-1} + \rho$$
$$< \frac{(2\pi)^2(\delta+|\varepsilon|M_r)\ (w^2-\rho)}{(|\varepsilon|+\delta M_r^{-1})\ 4\pi^2 M_r} + \rho = (w^2-\rho) + \rho = w^2.$$

By a slight generalization of the proof of part (d) of theorem 2(i), section 8 of [1], the equation

$$(3)' \qquad y + J\tilde{N}J^*y = 0$$

where $\tilde{N}x = -(d/dt)\ \mathrm{grad}\ G(x) + h_1(x,t) + \varepsilon g_r(x,t)$, has a solution $y \in \mathcal{O}$ with $||y|| \leq R < 2rw^{3/2}\pi^{-1/2}(w+1)^{-1}$. Then $x = J^*y$ is a solution of

$$x'' + (d/dt)\ \mathrm{grad}\ G(x) = h_1(x,t) + \varepsilon g_r(x,t)$$

and $||x||_1 = ||J^*y||_1 \leq (1+w)w^{-1}||y||$. By lemma 2.2 $\sup_t|x(t)| \leq (\pi/w)^{1/2}2^{-1}||x||_1 \leq (\pi/w)^{1/2}2^{-1}(1+w)w^{-1}||y||$ $= \pi^{1/2}2^{-1}(1+w)w^{-3/2}||y|| < r$. Since $g_r(x,t) = h_2(x,t)$ for $|x| \leq r$, x is a solution of (2).

We will illustrate the theorem with a few examples. Example (1), (3) and (4) are taken from [2].

Example 1. We consider the forced Hill equation

$$x'' + p(t)x = a \sin t$$

where $p(t)$ is a π-periodic continuous function. If $\sup_t p(t) < 1$, we may apply Theorem 2.1 by letting $G(x) = 0$, $h_2(x,t) = 0$ and $h_1(x,t) = -p(t)x + a \sin t$ to prove the existence of a 2π-periodic solution $x(t)$. If $b = \sup_t p(t)$ and $\rho = \max\{0,b\}$, then $\sup_t|x(t)| \leq 4|a|\pi^{5/2}(1-\rho)^{-1}$.

Example 2. We may use Theorem 2.1 to prove the existence of 2π-periodic solutions of

$$x'' + \varepsilon a \tan x = \varepsilon b \sin t$$

for ε sufficiently small. For any r, $0 < r < \pi/2$, let

$$g_r(x,t) = -a \tan \eta_r(x) + b \sin t$$

where η_r is defined as in the proof of Theorem 2.1. We now apply Theorem 2.1 by letting $G(x) = 0$, $h_1(x,t) = 0$ and $h_2(x,t) = g_r(x,t)$ to prove there exists a 2π-periodic solution $x(t)$ of the equation

$$x'' = \varepsilon g_r(x,t)$$

with $\sup_t|x(t)| < r$ for sufficiently small $\varepsilon > 0$. Since for $|x| < r$, $g_r(x,t) = -a \tan x + b \sin t$, $x(t)$ is a solution of the original equation.

Example 3. The equation

$$x'' + cx' + \sigma x = B \cos wt + \alpha \cos 2wt \cdot x + \varepsilon x^3$$

contains as particular cases the nonlinear Mathieu equation with a large forcing term ($\alpha \neq 0$, $c = 0$), the Duffing equation ($\alpha = 0$, $c = 0$), and the Duffing equation with damping ($\alpha = 0$, $c > 0$). We take $G(x) = c2^{-1}x^2$, $h_1(x,t) = -\sigma x + B \cos wt + \sigma \cos 2wt \cdot x$ and $h_2(x,t) = x^3$. With the condition that $|\alpha| < w^2 - \sigma$, Theorem 2.1 implies the existence of $2\pi/w$-periodic solutions for ε sufficiently small.

Example 4. We consider the Liénard type equation

$$x'' + p_1(x)x' + \varepsilon p_2(x) = a \cos(wt+\alpha)$$

where $p_i(x)$, $i = 1,2$, are polynomials with $p_1(x) = p_1(-x)$ and $p_2(x) = -p_2(-x)$. Solving

$$(d/dt) \text{ grad } G(x) = p_1(x)x',$$

we find $G(x)$ is a polynomial with $G(x) = G(-x)$. Letting $h_1(x,t) = a \cos(wt+\alpha)$ and $h_2(x,t) = -p_2(x)$, Theorem 2.1

implies the existence of $2\pi/w$-periodic solutions for sufficiently small ε. Note that the forced Van der Pol equation is a particular case of this type of equation.

Example 5. An n-dimensional generalization of example 4 is motivated by the work of Fermi, Pasta and Ulam on the nonlinear string [4]. For numerical calculations on a computer they replaced the continuous string by finite particles connected by nonlinear springs. One of the systems of equations they considered was of the form

$$x_i'' = a(x_{i+1}+x_{i-1}-2x_i) + \beta[(x_{i+1}-x_i)^3 - (x_i-x_{i-1})^3]$$
$$(i = 1,2,\ldots,64)$$

where x_i denotes the displacement of the i-th particle from its original position. If we now act upon the i-th particle with a forcing term such as

$$h_i(x,t) = b_i\cos(wt+\alpha_i),$$

Theorem 2.1 ($|a|<1/2$) guarantees the existence of $2\pi/w$-periodic solutions for sufficiently small β.

Fermi, Pasta and Ulam also considered the case where the restoring forces of the springs are piecewise linear. The equations of the system becomes

$$x_i'' = \delta_1(x_{i+1}-x_i) - \delta_2(x_i-x_{i-1}) + c$$

where the parameters δ_1,δ_2, c are not constants but assume different values depending on whether or not the quantities in parentheses are less than or greater than a certain value fixed in advance. Under mild restrictions, this prescription can be made to satisfy the conditions of Theorem 2.1, and again we see that adding forcing terms as before the system will have $2\pi/w$-periodic solutions.

3. THE SECOND EXISTENCE THEOREM.

The results of the following theorem will not be as sharp as those of Theorem 2.1. The explicit formulas become too complex. The proof will depend heavily on the proofs of [1].

3.1 Theorem. For the differential equation

(5) $$x'' + (d/dt)\,\mathrm{grad}\, G(x) + (d/dt)(V_1(x,t) + \varepsilon V_2(x,t)) = h_1(x,t) + \varepsilon h_2(x,t)$$

where $G: R^n \to R$ of class C^2, $V_i: R^{n+1} \to R$ of class C^1, $h_i: R^{n+1} \to R^n$ continuous, satisfying the symmetric conditions

(i) $G(x) = G(-x)$ for all $x \in R^n$

(ii) $V_i(-x,t+\pi) = -V_i(x,t)$ and $h_i(-x,t+\pi) = -h(x,t)$ for all $(x,t) \in R^{n+1}$;

then for all sufficiently small $\varepsilon > 0$ equation (5) has at least one 2π-periodic solution $x(t)$, $-\infty < t < \infty$, with $x(t+\pi) = -x(t)$ for all t, provided one of the following assumptions holds:

either (a) $G(x)$ is homogeneous in x of degree $2p$ of constant sign, satisfying $|G(x)| \geq c|x|^{2p}$ for some integer $p \geq 2$, constant $c > 0$, and all $x \in R^n$, $|V_1(x,t)| \leq C|x|^p + D$, $|h_1(x,t)| \leq C'|x|^p + D'$ for some constants $C,D,C',D' \geq 0$ and all $(x,t) \in R^{n+1}$;

or (b) the same as (a) with $p = 1$ and $2c > C + C'$;

or (c) the same as (a) with $p \geq 1$, with $V_i(x)$ depending on x only of the form $V_i(x) = \mathrm{grad}\, W_i(x)$, $V_i: R^n \to R$ of class C^2, $|\mathrm{grad}\, W_1(x)| \leq C|x|^{2p-2} + D$, and $|h_1(x,t)| \leq C'|x|^{2p-2} + D'$ for some constants $C,D,C',D' \geq 0$ and all $(x,t) \in R^{n+1}$.

Outline of proof. In the proof of (a), (b) and (c) of theorem 2.1 of [1], the problem of existence of 2π-periodic solutions of (5) with $\varepsilon = 0$ was reduced to finding solutions $y \in S$ of the equation

$$y + JNJ^*y = 0 \tag{6}$$

where $Nx = -(d/dt)\,\mathrm{grad}\,G(x(t)) - (d/dt)V_1(x,t) + h_1(x,t)$. Equation (6) was solved by applying an extended form of the Borsuk-Ulam theorem. The essence of the proof was the construction of a function $k(\xi) \geq 0$, $0 \leq \xi < \infty$, such that whenever $x + JNJ^*x = w$, then $||x|| \leq k(||w||)$. The function $k(\cdot)$ was constructed explicitly from the growth constants $C, C', D, D' \geq 0$ of (a), (b) or (c). Here we adopt a new viewpoint. For $w = 0$, $k(0)$ becomes a continuous function of $C > 0$, C', D, $D' \geq 0$; denote this function by $f(C,C',D,D')$. If we have any functions $h(x,t)$, $V(x,t)$ satisfying the conditions of (a), (b) or (c), $\varepsilon = 0$, for the constants $C > 0$, C', D, $D' \geq 0$ then any solution $y \in S$ of (6) satisfies $||y|| \leq f(C,C',D,D')$. It was also proved in [1] that (6) has a solution $y \in H^1 \cap \mathcal{O}$ with $||y|| \leq f(C,C',D,D')$.

Suppose $h_1(x,t)$, $h_2(x,t)$, $V_1(x,t)$ and $V_2(x,t)$ are given functions satisfying the conditions of the theorem with h_1 and V_1 satisfying the growth conditions of (a), (b) or (c) for growth constants $C > 0$, C', D, $D' \geq 0$. Let $r_1 = f(C,C',D,D')$. If $y \in S \cap \mathcal{O}$, then if $x = Jy$, $\sup_t |x(t)| \leq \pi^{1/2}2^{-1}||Jy||_1 \leq \pi^{1/2}||y||$. Let r be any fixed number greater than $\pi^{1/2}r_1$. By continuity of f there exists a $\delta > 0$ such that $f(C+\delta, C'+\delta, D+\delta, D'+\delta) < \pi^{-1/2}r$. Let $\nu(t)$ be a nondecreasing infinitely differentiable function on the nonnegative reals with the property that $\nu(t) = 1$, $0 \leq t \leq r$ and $\nu(t) = t/r$ for all $t > r'$ for some $r' > r$. Define $\eta: R^n \to R^n$ by $\eta x = x/\nu(|x|)$. Note that $|\eta x| \leq r'$ for all $x \in R^n$.

Define $h^*(x,t) = h_2(\eta x,t)$, $V^*(x,t) = V_2(\eta x,t)$ and in case (c) $W^*(x) = W_2(\eta x)$. By continuity and periodicity of h_2 and V_2, h^* and V^* are bounded. A straightforward verification will show that grad $W^*(x)$ is also bounded. For sufficiently small $\varepsilon_1 > 0$, the functions $\varepsilon_1 h^*$, $\varepsilon_1 V^*$ and $\varepsilon_1 W^*$ will satisfy the growth conditions of (a), (b) or (c) with all growth constants equal to δ. Then $h(x,t) = h_1(x,t) + \varepsilon_1 h^*(x,t)$, $V(x,t) = V_1(x,t) + \varepsilon_1 V^*(x,t)$ and $W(x) = W_1(x) + \varepsilon_1 W^*(x)$ will satisfy (a), (b) or (c) with growth constants $C + \delta$, $C' + \delta$, $D + \delta$, $D' + \delta$. Then all solutions $y \in S$ of (6) will satisfy $||y|| \leq f(C+\delta,C'+\delta,D+\delta,D'+\delta) < \pi^{-1/2}r$. Also in [1] theorem 2.1 and its proof guarantee the existence of a solution $y \in H^1 \cap \mathcal{O}$ of (6) where $Nx = -(d/dt)\text{grad } G(x(t)) - V(x,t) + h(x,t)$. It was shown in [1] that $x = Jy$ is a solution of $x'' + (d/dt)\text{grad } G(x) + (d/dt)V(x,t) = h(x,t)$ in $H^2 \cap \mathcal{O}$. By lemma 2.2,

$$\sup_t |x(t)| \leq \pi^{1/2} 2^{-1} ||x||_1 \leq \pi^{1/2} ||y|| \leq r.$$

Since $V^*(x,t) + V_2(x,t)$, $h^*(x,t) = h_2(x,t)$ and $W^*(x) = W_2(x)$ for $|x| \leq r$, then y is a solution of $x'' + (d/dt)\text{grad } G(x(t)) + (d/dt)(V_1(x,t) + \varepsilon_1 V_2(x,t)) = h_1(x,t) + \varepsilon_1 h_2(x,t)$.

Remark. Theorem 3.1 allows the introduction of very general forces depending on x' and t. Such forces may be disserpative such as friction or of a self-exciting nature as found in the Van der Pol equation. The forces even may undergo periodic disturbances of small intensity as demonstrated in the following example.

Example. We modify the Hill equation of example 1 as follows

$$x'' + (1-\varepsilon \sin 2t)x' + p(t)x = a \sin t.$$

We prove the existence of a 2π-periodic solution by applying part (b) of theorem 3.1 where $G(x) = x^2/2$, $V_1(x,t) = 0$, $V_2(x,t) = -(\sin 2t)x$, $h_1(x,t) = a \sin t - p(t)x$ and $h_2(x,t) = -2x \cos 2t$. We must have the condition that $\sup_t |p(t)| < 1$.

REFERENCES

1. Bowman, T. T., Periodic solutions of Liénard systems with symmetries, submitted.

2. Cesari, L., *Asymptotic Behavior and Stability Problems in Ordinary Differential Equations,* second edition, Springer-Verlag, New York, 1963.

3. Cesari, L. and Kannan, R., Periodic solutions in the large of nonlinear differential equations, *Rend. Mat.* *6*(No. 2, 1975), 633-654.

4. Fermi, E., Pasta, J. and Ulam, S., *Studies of Nonlinear Problems,* (Los Alamos report LA-1940, May 1955).

AN ABSTRACT EXISTENCE THEOREM ACROSS A POINT OF RESONANCE

Lamberto Cesari*
Department of Mathematics
University of Michigan

1. INTRODUCTION

Landesman and Lazer [18] proved in 1969, by the use of the alternative method, an existence theorem for selfadjoint elliptic Dirichlet nonlinear problems at resonance. A version of this theorem for periodic solutions of certain nonlinear second order ordinary differential equations had been proved before by Lazer and Leach [19] by essentially the same method. The theorem of Landesman and Lazer for nonlinear Dirichlet problems has been since repeatedly extended, again by the alternative method by Williams [27] and Necas [23] (and by other methods by Hess [15]; DeFigueiredo [9-11]; Fucik [12]; Fucik, Kucera, and Necas [13]).

We consider here an abstract problem <u>at resonance</u> of the form

(1) $$Ex = Nx, \qquad x \in S,$$

where $N:S \to S$ is a nonlinear continuous operator, S a real Hilbert space, and where $E:\mathcal{D}(E) \to S$, $\mathcal{D}(E) \subset S$, is a linear operator for which the set of solutions $x \in S$ of $Ex = 0$ is a nontrivial subspace $S_0 = \ker E$ of S of finite dimension, $1 \le m < \infty$.

In Section 3 we assume N uniformly bounded on S, and, in term of the alternative method, we state first an abstract existence theorem for problem (1) (Theorem I, existence at resonance) we proved in [8], from which one can derive the Landesman and Lazer theorem, the Lazer and Leach theorem, as well as a number of the aforementioned extensions (cfr. [5] for the actual derivation).

Next we consider the equation

$$Ex + \alpha Ax = Nx, \qquad x \in S, \tag{2}$$

where E and N are as above, α is a real parameter varying in a neighborhood of the origin, and thus including the value $\alpha = 0$ (the point of resonance since $S_0 = \ker E$ is not trivial), and $A:S \to S$ is any continuous nonnecessarily linear operator for which we assume only that A is bounded, that is, A maps bounded sets into bounded sets, or equivalently, $||Ax|| \leq \omega(||x||)$ for all $x \in S$ and some monotone function $\omega(\zeta) \geq 0$, $0 \leq \zeta < +\infty$.

We prove here (Theorem II, existence across a point of resonance), that, hypotheses on E and N only slightly stronger than those of Theorem 1, have a much stronger implication. Namely, they guarantee the existence and uniform boundedness of the solutions $x \in S$ to problem (2) for $|\alpha|$ sufficiently small. Yet the conditions on E and N of Theorem II are so weak that they are still implied by the specific hypotheses of the theorems of Landesman and Lazer, of Lazer and Leach, and of the analogous statements mentioned before. These theorems, therefore, hold under the same specific hypotheses and with a stronger conclusion. They have been already so restated in our papers [5] and [7].

In Section 4 we assume that Nx grows with $||x||$ at most of order k, $0 < k < 1$, and we prove both an existence theorem at resonance (Th.III) and an existence theorem across a point of resonance (Th.IV). Other extensions of the Landesman and Lazer theorem can be derived from these theorems as we show in [5].

In the present paper we limit ourselves to the self-adjoint case. We refer to the papers (Cesari [6,7]) for extensions of the present Theorems I, II, III, IV to the nonselfadjoint case.

For the alternative method as it developed, by the injection of ideas of functional analysis in the classical Poincaré, Lyapunov, Schmidt bifurcation method, we mention Cesari [1-5], Hale [14], and much concurrent work (see [1-7], [14] for references). (For the development of the alternative method in connection with topological degree arguments see Cesari [1], Williams [26], and Mawhin [20,21]).

2. NOTATIONS AND MAIN ASSUMPTIONS

Let S be a real separable Hilbert space with inner product (x,y) and norm $||x|| = (x,x)^{1/2}$. Let $E:\mathcal{D}(E) \to S$ be a linear operator with domain $\mathcal{D}(E) \subset S$ and finite dimensional null space $S_o = \ker E$. Let $P:S \to S$ denote the projection operator with range S_o and null space $S_1 = (I-P)S$. Furthermore we assume that S_1 is also the range of $E, S_1 = \mathcal{R}(E)$. Then, $E:\mathcal{D}(E) \cap S_1 \to S_1$ is one-one and onto, and $H:S_1 \to \mathcal{D}(E) \cap S_1$ as a partial inverse operator of E exists. We assume that H is a linear bounded <u>compact</u> operator, and that E,P,H, satisfy the usual axioms $(h_1) H(I-P)E = I-P$; $(h_2) EP = PE$; $(h_3) EH(I-P) = I-P$.

Let $N:S \to S$ be a continuous operator in S, not necessarily linear. The equation

(3) $$Ex = Nx, \qquad x \in \mathcal{D}(E) \subset S,$$

is equivalent to the system of auxiliary and bifurcation equations

(4) $$x = Px + H(I-P)Nx,$$

(5) $$P(Ex-Nx) = 0.$$

We refer to [3] for details, and we note that here, having assumed $\mathcal{R}(P) = S_o = \ker E$, the bifurcation equation reduces to $PNx = 0$. Also, for $x^* = Px$, the auxiliary equation has the form $x = x^* + H(I-P)Nx$. Let L denote the norm of the operator $H:S_1 \to S_1$, or $L = ||H||$.

Finally, We shall also consider (in Theorems II and IV) another continuous operator $A:S \to S$, not necessarily linear, satisfying $||Ax|| \leq \omega(||x||)$ for all $x \in S$ and some monotone function $\omega(\zeta)$, $0 \leq \zeta < +\infty$, (ω not necessarily continuous, nor zero at the origin). We shall then discuss the equation $Ex + \alpha Ax = Nx$, or $Ex = -\alpha Ax + Nx$, whose auxiliary and bifurcation equations are then of the form

$$x = x^* + H(I-P)(-\alpha Ax+Nx), \quad P(-\alpha Ax+Nx) = 0.$$

We have depicted above a particular case of what we denote "the selfadjoint case" in [3]. We refer to [3] for more general considerations for the selfadjoint case, and to ([3-7],[14]) for the nonselfadjoint case.

3. THE BOUNDED CASE

I. Theorem (existence of solutions at resonance). Under the assumptions above, if (B) there is a constant $J_o \geq 0$ such that $||Nx|| \leq J_o$ for all $x \in S$; and (N_o) there is a constant $R_o \geq 0$ such that $(Nx,x^*) \leq 0$. [or always $(Nx,x^*) \geq 0$] for all $x \in S$, $x^* \in S_o$ with $||x^*|| \geq R_o$, $Px = x^*$, $||x-x^*|| \leq LJ_o$, then equation $Ex = Nx$ has at least a solutions.

A first proof of this theorem was given by Cesari and Kannan [8] by the use of Schauder's fixed point theorem. Another proof has been later given by Kannan and McKenna [17] by the use of the Leray-Schauder topological degree argument.

Remark. Condition (B) is the usual assumption of the theorems of Lazer and Leach [19], Landesman and Lazer [18], Williams [27], of some work by DeFigueiredo [11], and others. Conditions (N_o), as well as the slightly stronger condition (N_ε) below, can be shown to be implied (cfr. [5]) by the specific hypotheses of the above-mentioned theorems.

II. Theorem (existence of solutions across a point of resonance). Under the assumptions of No. 2, if (B) there is a constant J_o such that $||Nx|| \le J_o$ for all $x \in X$; and (N_ε) there are constants $K > LJ_o$, $R_o \ge 0$, $\varepsilon > 0$ such that $(Nx,x^*) \le -\varepsilon||x^*||$ [or $(Nx,x^*) \ge \varepsilon||x^*||$] for all $x \in S$, $x^* \in S_o$, with $Px = x^*$, $||x^*|| \ge R_o$ and $||x-x^*|| \le K$; then there are also constants $\alpha_o > 0$, and $C > 0$ such that, for any $|\alpha| \le \alpha_o$, equation $Ex + \alpha Ax = Nx$ has at least a solution x with $||x|| \le C$.

Proof. First we need to determine constants R_1, R, R_2, α_o such that

$$R_o \le R_1 < R < R_2, \qquad \alpha_o > 0, \tag{6}$$

$$R + LJ_o + L\alpha_o\omega(R_2) \le R_2, \tag{7}$$

$$LJ_o + L\alpha_o\omega(R_2) \le K, \tag{8}$$

$$R_1 + \alpha_o\omega(R_2) + J_o \le R, \tag{9}$$

$$(\alpha_o\omega(R_2)+J_o)^2 + 2\alpha_o R\omega(R_2) \le 2\varepsilon R_1. \tag{10}$$

We can satisfy these relations by choosing in order R_1, R, R_2, α_o as follows

$$R_1 \geq \max[R_o, 2\varepsilon^{-1}J_o^2],\ R \geq (3/2)R_1 + J_o,\ R_2 = 2(R+LJ_o),$$

$$\alpha_o L\omega(R_2) \leq 2^{-1}R_2,\ \alpha_o L\omega(R_2) \leq K-LJ_o,\ \alpha_o\omega(R_2) \leq 2^{-1}R_1,$$

$$\alpha_o\omega(R_2) \leq 1,\ \alpha_o\omega(R_2) \leq (2J_o+1)^{-1}(\varepsilon R_1 - J_o^2),\ \alpha_o R\omega(R_2) \leq 2^{-1}\varepsilon R_1.$$

Now, relation (6) is trivially satisfied. Since $\alpha_o L\omega(R_2) \leq 2^{-1}R_2$ we have

$$R_2 - \alpha_o L\omega(R_2) - R - LJ_o \geq 2^{-1}R_2 - (R+LJ_o) = 0,$$

and (7) is satisfied. Since $\alpha_o L\omega(R_2) \leq K - LJ_o$, and $\alpha_o\omega(R_2) \leq 2^{-1}R_1$, we have

$$LJ_o + L\alpha_o\omega(R_2) \leq LJ_o + (K-LJ_o) = K,$$

$$R_1 + \alpha_o\omega(R_2) + J_o \leq (3/2)R_1 + J_o \leq R,$$

and (8) and (9) are satisfied. Finally

$$\begin{aligned}(\alpha_o\omega(R_2)+J_o)^2 + 2\alpha_o R\omega(R_2) &\leq J_o^2 + (2J_o+1)\alpha_o\omega(R_2) + 2\alpha_o R\omega(R_2)\\ &\leq J_o^2 + \varepsilon R_1 - J_o^2 + \varepsilon R_1 = 2\varepsilon R_1,\end{aligned}$$

and (10) is satisfied.

Let $m = \dim S_o = \dim \ker E$, $1 \leq m < \infty$, and let $w = (w_1,\dots,w_m)$ be an orthonormal basis of S_o. Then, every $x^* \in S_o$ can be written in the form $x^* = c_1w_1 + \dots + c_mw_m$, or briefly $x^* = cw$, $c = (c_1,\dots,c_m) \in R^m$. Again, for the sake of brevity, we shall write (z,w) with $z \in S_o$ for (z,w_i), $i = 1,\dots,m$. We shall assume that $(Nx,x^*) \leq -\varepsilon||x^*||$ for all x, x^* as stated under (N_ε). Now, we define the map $T:(x,c) \to (\bar{x},\bar{c})$ by taking

$$\text{(11)}\quad T:\bar{x} = cw + H(I-P)[-\alpha Ax+Nx],\ \bar{c} = c+(P[-\alpha A\bar{x}+N\bar{x}],w),$$

$$(x,c) \in C = [(x,c)|x\in S, c\in R^m, ||x|| \leq R_2, |c| \leq R].$$

Thus, for $x^* = cw$, $\bar{x}^* = \bar{c}w$, the transformation T can be also defined by taking

$$\bar{x} = x^* + H(I-P)[-\alpha Ax+Nx],\ \bar{x}^* = x^* + P[-\alpha A\bar{x}+N\bar{x}].$$

For $(x,c) \in C$ we have now, by using (7)

$$||\bar{x}|| \le ||cw|| + ||H(I-P)[-\alpha Ax+Nx]|| \le |c| + L(\alpha_o\omega(||x||)+J_o)$$
$$\le R + LJ_o + L\alpha_o\omega(R_2) \le R_2,$$

and by using (8) also, since $P\bar{x} = cw = x^*$,

$$||\bar{x}-P\bar{x}|| = ||\bar{x}-x^*|| \le ||H(I-P)[-\alpha Ax+Nx]|| \le LJ_o + L\alpha_o\omega(R_2) \le K.$$

Since $P\bar{x} = cw = x^*$, the second relation (11) becomes

$$\bar{c} = c - \alpha(PA\bar{x},w) + (PN\bar{x},w).$$

From here we derive first, for $|\alpha| \le \alpha_o$,

$$|\bar{c} - c| \le |\alpha|\ ||PA\bar{x}|| + ||PN\bar{x}|| \le \alpha_o\omega(R_2) + J_o.$$

Thus, for $|c| \le R_1$, by using (9) we have

$$|\bar{c}| \le |c| + \alpha_o\omega(R_2) + J_o \le R_1 + \alpha_o\omega(R_2) + J_o \le R.$$

For $R_o \le R_1 \le |c| \le R$, if d_i denote the numbers

$d_i = (P[-\alpha A\bar{x}+N\bar{x}],w_i)$, $i = 1,\ldots,m$, we have

$$\bar{c}-c = (P[-\alpha A\bar{x}+N\bar{x}],w) = (d_j, j=1,\ldots,m) = d,\quad c = (c_j, j=1,\ldots,m),$$
$$(\bar{c}-c)c = \textstyle\sum_1^m c_jd_j,$$
$$(P[-\alpha A\bar{x}+N\bar{x}],x^*) = (P[-\alpha A\bar{x}+N\bar{x}],\textstyle\sum_1^m c_jw_j) = \sum_1^m c_jd_j = cd = (\bar{c}-c)c.$$

Since $P\bar{x} = cw = x^*$, $||\bar{x}-P\bar{x}|| \le K$, $||x^*|| \ge R_1 \ge R_o$, we have

$$(PN\bar{x},x^*) = (N\bar{x},x^*) \le -\varepsilon||x^*|| = -\varepsilon|c| \le -\varepsilon R_1.$$

Now, by the identity

$$(\bar{c}-c,\ \bar{c}-c) = |\bar{c}|^2 - |c|^2 - 2(\bar{c}-c)c \tag{12}$$

and the relations

$$(\bar{c}-c)c = (N\bar{x},x^*) - \alpha(PA\bar{x},x^*) \le -\varepsilon R_1 + \alpha_o R\omega(R_2),$$

we derive, by using (10) and (12),

$$|\bar{c}|^2 - |c|^2 = |\bar{c}-c|^2 + 2(\bar{c}-c)c$$
$$\le (\alpha_o\omega(R_2)+J_o)^2 + 2\alpha_o R\omega(R_2) - 2\varepsilon R_1 \le 0.$$

Thus, $|\bar{c}|^2 - |c|^2 \le 0$, or $|\bar{c}| \le |c| \le R$. We have proved that $T: \mathcal{C} \to \mathcal{C}$ maps $\mathcal{C}$ into itself.

Let us prove that T is compact. For this we consider any (bounded) sequence (x_k,c_k), $k = 1,2,\ldots$, of elements of $\mathcal{C}$. Then the sequences $[x_k]$, $[Nx_k]$ are bounded. Namely,

$||x_k|| \le R_2$, $||Nx_k|| \le J_o$, and so is the sequence $y_k = H(I-P)[-\alpha Ax_k+Nx_k]$, with $||y_k|| \le L(\alpha_o\omega(R_2)+J_o)$. Since H is compact, there is a subsequence, say still [k], such that y_k is convergent in S. Analogously, the sequences $x_k^* = c_k w$, $z_k = P[-\alpha A\bar{x}_k+N\bar{x}_k] = d_k w$ are bounded, thus $[c_k]$, $[d_k]$ are bounded sequences in R^m, a finite dimensional space, and we can take a further subsequence, say still [k], such that c_k, d_k converge in R^m. Thus, $\bar{x}_k = x_k^* + y_k$, $\bar{c}_k = c_k + d_k$, are also convergent in S and R^m, respectively. We have proved that T is compact.

By Schauder's fixed point theorem T has at least a fixed point (x,c) in C, or (x,c) = T(x,c), and (11) yields

$$x = x^* + H(I-P)[-\alpha Ax+Nx], \quad P[-\alpha Ax+Nx] = 0.$$

These are the auxiliary and bifurcation equations for the problem $Ex = -\alpha Ax + Nx$, and thus x satisfies this equation, with $||x|| \le R_2 = C$. This holds for every α real with $|\alpha| \le \alpha_o$, and $C = R_2$ is independent of α.

Remark. The following particular case of the inequality $(Nx,x^*) \ge 0$ [or ≤ 0] is of interest and will be used in [7]. Assume that S is a space of real-valued functions x(t), $t \in G$, on a bounded domain G in R^ν, $\nu \ge 1$, and that $S_o = \ker E$ is a space of constants; hence m = 1 and $Px = (\text{meas } G)^{-1}\int_G x(t)dt$. Let $F:R \to R$ be a real-valued function with $sF(s) \ge 0$ for all s real [or $sF(s) \le 0$ for all s], and take $Nx = F[x(t)]$, $t \in G$. Then, for $x(t) \ge 0$ we have $Nx = F[x(t)] \ge 0$, $x^* = PNx \ge 0$. Analogously, for $x(t) \le 0$ we have $Nx = F[x(t)] \le 0$, $x^* = PNx \le 0$. In any case we have $(Nx,x^*) \ge 0$. We see that the condition $(Nx,x^*) \ge 0$ is satisfied by all functions x(t), $t \in G$, of constant sign on G. (Analogous conclusion if $sF(s) \le 0$ for all s.)

4. THE CASE OF SLOW GROWTH

In this section we consider the case where the nonlinear operator N has the property that $||Nx|| \le J_0 + J_1||x||^k$ for some k, $0 < k < 1$. For the existence theorem at resonance we need the following simple

1. Lemma. Given constants $L > 0$, $R_o \ge 0$, $\varepsilon > 0$, $0 < \lambda < 1$, and the function $\phi(r) = J_0 + J_1 r^k$, $0 \le r < +\infty$, $J_o \ge 0$, $J_1 > 0$, $0 < k < 1$, there are constants R_1, R, R_2 such that

(13) $R_0 \le R_1 < R < R_2$, $R_1 + \phi(R_2) = R$, $R + L\phi(R_2) = R_2$,

$$(\phi(R_2))^2 \le 2\varepsilon R_1^{1+k}, \quad R_2 \le (1+\lambda)R_1.$$

Proof. For $R > 0$, $R_2 = (1+\rho)R$, $0 \le \rho < +\infty$, the expression $\Delta = R+L\phi(R_2) - R_2$ becomes $\Delta(\rho) = L\phi((1+\rho)R) - \rho R = 0$ with $\Delta(0) \ge 0$, and $\Delta(\rho) < 0$ for $\rho > 0$ large, since $L\phi((1+\rho)R)$ grows as ρ^k, that is, slower than ρR. Thus, $R_2 = R_2(R)$ can be determined in such a way that $R + L\phi(R_2) = R_2$, and then $R_1 = R - \phi(R_2) = R - L^{-1}(R_2-R)$. Note that, given σ, $0 < \sigma < 1$, $\sigma < (L+1)^{-1}$, then for r sufficiently large, $\phi(r) < \sigma r$, hence $R < R_2 < R + L\sigma R_2$, or $R_2 < (1-L\sigma)^{-1}R$, and then $R_1 > (1-L\sigma)^{-1}[1-(1+L)\sigma]R$. In other words, given any η, $0 < \eta < 1$, then for $R > 0$ sufficiently large we certainly have

$$R(1-\eta) < R_1 < R < R_2 < R(1+\eta).$$

Thus, for η sufficiently small we also have

$$R_2 < R_1(1+\eta)(1-\eta)^{-1} < (1+\lambda)R_1.$$

We can now take R sufficiently large so that

$$R_0 \le R_1, \quad (\phi(R_2))^2 \le 2\varepsilon R_1^{1+k}.$$

This is possible since $R_1 \to +\infty$ as $R \to \infty$, and because, for R large we have, since $0 < k < 1$, hence $2k < 1 + k$,

$$(\phi(R_2))^2 < [J_0+J_1(1+\eta)^k R^k]^2 \le 2\varepsilon(1-\eta)^{1+k}R^{1+k} < 2\varepsilon R_1^{1+k}.$$

III. Theorem (an existence theorem at resonance). Under the main assumptions of Section 2, if (B_k) there are constants $J_o \geq 0$, $J_1 > 0$, $0 < k < 1$, such that $||Nx|| \leq J_o + J_1||x||^k$ for all $x \in S$; and if $(N_{\varepsilon k})$ there are constants $R_o \geq 0$, $\varepsilon > 0$, $K_o \geq LJ_o$, $K_1 > LJ_1$ such that $(Nx,x^*) \leq -\varepsilon||x^*||^{1+k}$ [or always $(Nx,x^*) \geq \varepsilon||x^*||^{1+k}$] for all $x \in S$, $x^* \in S_o$, with $Px = x^*$, $||x^*|| \geq R_o$, $||x-x^*|| \leq K_o + K_1||x||^k$; then equation $Ex = Nx$ has at least one solution $x \in S$.

Proof. Let λ, $0 < \lambda < 1$, denote any number such that $LJ_1(1+\lambda)^k \leq K_1$. Let $L = ||H||$, $R_o > 0$, $\varepsilon > 0$, $\phi(r) = J_o+J_1r^k$, $0 \leq r < +\infty$, and let R_1, R, R_2, be the numbers determined in Lemma 1 in relation to the number λ just chosen. Then $R_2 \leq (1+\lambda)R_1$. Let $T: (x,c) \rightarrow (\bar{x},\bar{c})$ denote the transformation defined by

$$T:\bar{x} = cw + H(I-P)Nx,\quad \bar{c} = c + (PN\bar{x},w),$$
$$(x,c) \in C = [(x,c)\,|\,x\in S, c\in R^m, ||x||\leq R_2, |c|\leq R].$$

We need to prove that T maps C into itself. First we have, for $(x,c) \in C$,

$$\bar{x} = cw + H(I-P)Nx,\quad ||x^*|| = ||cw|| = |c| \leq R,\quad ||x|| \leq R_2,$$
$$||\bar{x}|| \leq |c| + ||H(I-P)Nx|| \leq |c| + L(J_o+J_1R_2^k) \leq R + L\phi(R_2) = R_2.$$

For $|c| \leq R_1$ we have now $\bar{c} = c + (PN\bar{x},w)$, $||\bar{x}|| \leq R_2$, and

$$|\bar{c}| \leq |c| + ||PN\bar{x}|| \leq |c| + (J_o+J_1R_2^k) \leq R_1 + \phi(R_2) = R.$$

For $R_1 \leq |c| \leq R$ we have now

$$\begin{aligned} P\bar{x} = cw = x^*,\quad ||\bar{x}-P\bar{x}|| = ||\bar{x}-x^*|| &\leq L(J_o+J_1R_2^k) \\ &\leq LJ_o + LJ_1(1+\lambda)^kR_1^k \\ &\leq LJ_o + LJ_1(1+\lambda)^k|c|^k \\ &\leq K_o + K_1||\bar{x}||^k. \end{aligned}$$

By $(N_{\varepsilon k})$ we have

$$(N\bar{x},x^*) \leq -\varepsilon||x^*||^{1+k}.$$

We have now, as in the proof of Theorem II with $\alpha = 0$,

$$(\bar{c}-c)c = (N\bar{x},x^*) \le -\varepsilon|c|^{1+k},$$

and

$$\begin{aligned}(J_o+J_1||\bar{x}||^k)^2 &\ge ||PN\bar{x}||^2 = ||\bar{c}-c||^2 = (\bar{c}-c,\bar{c}-c)\\ &= |\bar{c}|^2 - |c|^2 - 2(\bar{c}-c)c \ge |\bar{c}|^2 - |c|^2 + 2\varepsilon|c|^{1+k}, \text{ or}\\ 0 &\ge (\phi(R_2))^2 - 2\varepsilon R_1^{1+k} \ge |\bar{c}|^2 - |c|^2.\end{aligned}$$

Thus, $|\bar{c}| \le |c| \le R$. The proof proceeds now as for Theorem II. Theorem III is thereby proved.

For the existence theorem across a point of resonance we need the following.

2. Lemma. Given constants $L > 0$, $R_o \ge 0$, $\varepsilon > 0$, $0 < \lambda < 1$, $K_o > LJ_o$, the function $\phi(r) = J_o + J_1 r^k$, $0 \le r < +\infty$, $J_o \ge 0$, $J_1 > 0$, $0 < k < 1$, and any monotone function $\omega(r) \ge 0$, $0 \le r < +\infty$, there are constants α_o, R_1, R, R_2, such that

(14) $$R_o \le R_1 < R < R_2, \quad \alpha_o > 0,$$

(15) $$R + L\alpha_o\omega(R_2) + L\phi(R_2) \le R_2,$$

(16) $$L\alpha_o\omega(R_2) + LJ_o \le K_o, \quad R_2 \le (1+\lambda)R_1,$$

(17) $$R_1 + \alpha_o\omega(R_2) + \phi(R_2) \le R,$$

(18) $$(\alpha_o\omega(R_2) + \phi(R_2))^2 + 2\alpha_o R\omega(R_2) \le 2\varepsilon R_1^{1+k}.$$

Proof. First, let us restrict α_o to satisfy $\alpha_o L\omega(R_2) \le K_o - LJ_o$. For any constant $\sigma > 0$ and r sufficiently large we have $\phi(r) \le \sigma r$. We shall take $\sigma < (2L)^{-1}$, $\sigma < 2^{-1}(L+1)^{-1}$. We shall also restrict α_o to satisfy $\alpha_o\omega(R_2) \le \sigma R_2$. Then, for $R_2 = (1-2L\sigma)^{-1}R$ and R sufficiently large we have $R+2L\sigma R_2 = R_2$, and $R + \alpha_o L\omega(R_2) + L\sigma R_2 \le R_2$, that is, (15) is satisfied. On the other hand, for $R_1 = R(1-2L\sigma)^{-1}(1-2L\sigma-2\sigma)$ we also have

$$\begin{aligned}R_1 + \alpha_o\omega(R_2) + \phi(R_2) &\le R(1-2L\sigma)^{-1}(1-2L\sigma-2\sigma) + \sigma R_2 + \sigma R_2\\ &= R_2(1-2L\sigma-2\sigma) + 2\sigma R_2\\ &= R_2(1-2L\sigma) = R,\end{aligned}$$

and (17) is also satisfied. We may note that by taking σ sufficiently small we shall have

$$R_2 = (1-2L\sigma)^{-1}R = (1-2L\sigma-2\sigma)^{-1}R_1 < (1+\lambda)R_1$$

and thus (16) is satisfied. Note that

$$(1-2L\sigma)^{-1}(1-2L\sigma-2\sigma)R = R_1 < R < R_2 = (1-2L\sigma)^{-1}R.$$

This shows that, given η, $0 < \eta < 1$, we can take $\sigma > 0$ sufficiently small so that

$$R(1-\eta) < R_1 < R < R_2 < R(1+\eta).$$

Thus, for η sufficiently small we also have

$$R_2 < R_1(1+\eta)(1-\eta)^{-1} < (1+\lambda)R_1.$$

We see that R_2 grows as fast as R and R_1, and thus $(\phi(R_2))^2$ grows as R_1^{2k} with $2k = k+k < 1+k$, that is, slower than $2\varepsilon R_1^{1+k}$. For R sufficiently large we have then $(\phi(R_2))^2 + 2 \le 2\varepsilon R_1^{1+k}$ as well as $R_o \le R_1$. For $\alpha_o\omega(R_2) \le 1$, $2\alpha_o R\omega(R_2) \le 1$, and $\alpha_o\omega(R_2)(2\phi(R_2)+1) \le 1$, we have then

$$\begin{aligned}(\alpha_o\omega(R_2) + \phi(R_2))^2 + 2\alpha_o R\omega(R_2) &\le \\ &\le (\phi(R_2))^2 + (2\phi(R_2)+1)\alpha_o\omega(R_2) + 2\alpha_o R\omega(R_2) \\ &\le (\phi(R_2))^2 + 2 \le 2\varepsilon R_1^{1+k}.\end{aligned}$$

This proves Lemma 2.

IV. Theorem (existence theorem across a point of resonance). Under the main assumptions of Section 2, if (B_k) there are constants $J_o \ge 0$, $J_1 > 0$, $0 < k < 1$, such that $||Nx|| \le J_o + J_1||x||^k$ for all $x \in S$; and if $(N'_{\varepsilon k})$ there are constants $R_o \ge 0$, $\varepsilon > 0$, such that $(Nx,x^*) \le -\varepsilon||x^*||^{1+k}$ [or always $(Nx,x^*) \ge \varepsilon||x^*||^{1+k}$] for all $x \in S$, $x^* \in S_o$, with $Px = x^*$, $||x^*|| \ge R_o$, $||x-x^*|| \le K_o + K_1||x||^k$ for constants $K_o > LJ_o$, $K_1 > LJ_1$, then there are constants $\alpha_o > 0$, $C > 0$ such that, for every real $|\alpha| \le \alpha_o$, the equation $Ex + \alpha Ax = Nx$ has at least a solution $x \in S$ with $||x|| \le C$.

Proof. Let λ, $0 < \lambda < 1$, be a number such that $LJ_1(1+\lambda)^k \le K_1$. Let $L = ||H||$, and let α_0, R_1, R, R_2 be the constants determined by Lemma 2 in relation to the constant λ just chosen. Then $R_2 \le (1+\lambda)R_1$. Let $T: (x,c) \to (\bar{x},\bar{c})$ be the map defined by

$$(19)\quad T:\ \bar{x} = cw + H(I-P)[-\alpha Ax+Nx],\quad \bar{c} = c + (P[-\alpha A\bar{x}+N\bar{x}],w),$$

$$(x,c) \in C = [(x,c)\,|\,x \in S, c \in R^m, ||x|| \le R_2, |c| \le R].$$

For $(x,c) \in C$ we have now, using (15),

$$\begin{aligned} ||\bar{x}|| &\le ||cw|| + ||H(I-P)[-\alpha Ax+Nx]|| \\ &\le |c| + L(\alpha_0\omega(||x||) + \phi(||x||)) \\ &\le R + L\alpha_0\omega(R_2) + L\phi(R_2) \le R_2, \end{aligned}$$

and also

$$\begin{aligned} ||\bar{x}-P\bar{x}|| = ||\bar{x}-x^*|| &\le ||H(I-P)[-\alpha Ax+Nx]|| \\ &\le L\alpha_0\omega(R_2) + L\phi(R_2), \end{aligned}$$

since $P\bar{x} = cw = x^*$. The second relation (19) becomes

$$\bar{c} = c - \alpha(PA\bar{x},w) + (PN\bar{x},w).$$

From here we derive first, for $|\alpha| \le \alpha_0$,

$$|\bar{c}-c| \le \alpha_0\omega(R_2) + \phi(R_2).$$

Thus, for $|c| \le R_1$, by using (17), we have

$$|\bar{c}| \le |c| + \alpha_0\omega(R_2) + \phi(R_2) \le R_1 + \alpha_0\omega(R_2) + \phi(R_2) \le R.$$

For $R_1 \le |c| \le R$ we have now

$$\begin{aligned} P\bar{x} = cw = x^*,\quad ||\bar{x}-P\bar{x}|| = ||\bar{x}-x^*|| &\le L\alpha_0\omega(R_2) + L(J_0+J_1R_2^k) \\ &\le (L\alpha_0\omega(R_2)+LJ_0) + LJ_1(1+\lambda)^kR_1^k \\ &\le K_0 + K_1|c|^k \le K_0 + K_1||\bar{x}||^k. \end{aligned}$$

By $(N'_{\varepsilon k})$ we have

$$(PN\bar{x},x^*) = (N\bar{x},x^*) \le -\varepsilon||x^*||^{1+k}.$$

We have now, as in the proof of Theorem II,

$$\bar{c} = c - \alpha(PA\bar{x},w) + (PN\bar{x},w),$$

$$(\bar{c}-c)c = -\alpha(PA\bar{x},x^*) + (PN\bar{x},x^*)$$
$$\leq -\varepsilon||x^*||^{1+k} + \alpha_o R\omega(R_2),$$

and by the identity

$$(\bar{c}-c,\ \bar{c}-c) = |\bar{c}|^2 - |c|^2 - 2(\bar{c}-c)c,$$

we derive, by using (18),

$$|\bar{c}|^2 - |c|^2 = |\bar{c}-c|^2 + 2(\bar{c}-c)c$$
$$\leq (\alpha_o\omega(R_2) + \phi(R_2))^2 - 2\varepsilon R_1^{1+k} + 2\alpha_o R\omega(R_2) \leq 0.$$

Thus, $|\bar{c}|^2 - |c|^2 \leq 0$, or $|\bar{c}| \leq |c| \leq R$. We have proved that $T:C \to C$ maps C into itself. The argument is now the same as for Theorem II. Theorem IV is thereby proved.

REFERENCES

1. Cesari, L., Functional analysis and periodic solutions of nonlinear differential equations, *Contributions to Differential Equations 1*, 149-187, Wiley, 1963.

2. Cesari, L., Functional analysis and Galerkin's method, *Mich. Math. J. 11*(1964), 385-414.

3. Cesari, L., Alternative methods in nonlinear analysis, *International Conference on Differential Equations*, (Los Angeles), Academic Press, (Antosiewicz, ed.), 1975, 95-148.

4. Cesari, L., Nonlinear oscillations in the frame of alternative methods, *Dynamical Systems, An International Symposium*, (Providence, R.I.), Academic Press, (Cesari, Hale, LaSalle, eds.), Vol. 1, 1976, 29-50.

5. Cesari, L., Functional analysis, nonlinear differential equations, and the alternative method, *Nonlinear Functional Analysis and Differential Equations*, (Michigan State University Conference), (Cesari, Kannan, Schuur, eds.), Marcel Dekker, New York, 1976, 1-197.

6. Cesari, L., Nonlinear oscillations across a point of resonance for nonselfadjoint systems, *J. Differential Equations*, to appear.

7. Cesari, L., Nonlinear problems across a point of resonance for nonselfadjoint problems, *Nonlinear Analysis*, A volume in honor of E. H. Rothe, Academic Press, to appear.

8. Cesari, L., and Kannan, R., An abstract existence theorem at resonance, *Proc. Am. Math. Soc.*, to appear.

9. DeFigueiredo, D. G., Some remarks on the Dirichlet problem for semilinear elliptic equation, Univ. de Brisilia, *Trabalho de Mat. No. 57*, (1974).

10. DeFigueiredo, D. G., On the range of nonlinear operators with linear asymptotes which are not invertible, Univ. de Brasilia, *Trabalho de Mat. No. 59*, (1974).

11. DeFigueiredo, D. G., The Dirichlet problem for nonlinear elliptic equations: A Hilbert space approach, *Partial Differential Equations and Related Topics* (Dold and Eckman, eds.), *Springer Verlag Lecture Notes Math., No. 446* (1975), 144-165.

12. Fucik, S., Further remarks on a theorem of E. M. Landesman and A. C. Lazer, *Comm. Math. Univ. Carolin. 15* (1974), 259-271.

13. Fucik, S., Kucera, M., and Necas, J., Ranges of nonlinear asymptotically operators, to appear.

14. Hale, J. K., Applications of alternative problems, Lecture Notes, Brown University, 1971.

15. Hess, P., On a theorem by Landesman and Lazer, *Indiana Univ. Math. J. 23* (1974), 827-829.

16. Kannan, R., and Locker, J., On a class of nonlinear boundary value problems, to appear.

17. Kannan, R., and McKenna, J., An existence theorem by alternative method for semilinear abstract equations at resonance, to appear.

18. Landesman, E. M., and Lazer, A. C., Nonlinear perturbations of linear elliptic boundary value problems at resonance, *J. Math. Mech. 19* (1970), 609-623.

19. Lazer, A. C., and Leach, D. E., Bounded perturbations of forced harmonic oscillations at resonance, *Annali di Matematica Pura Appl. 72* (1969), 49-68.

20. Mawhin, J., Equivalence theorem for nonlinear operator equations and coincidence degree theory for some mappings in locally convex topological vector spaces, *J. Differential Equations 12* (1972), 610-636.

21. Mawhin, J., Topology and nonlinear boundary value problems, *Dynamical Systems, An International Symposium*, (Providence, R.I.), Academic Press, (Cesari, Hale, LaSalle, eds.), Vol. 1, 1976, 51-82.

22. McKenna, J., Nonselfadjoint semilinear problems in the alternative method, a Ph.D. thesis at the University of Michigan, 1976.

23. Necas, J., On the range of nonlinear operators with linear asymptotes which are not invertible, *Comm. Math. Univ. Carolin. 14*(1973), 63-72.

24. Shaw, H., A nonlinear elliptic boundary value problem, *Nonlinear Functional Analysis and Differential Equations*, (Michigan State University Conference), (Cesari, Kannan, Schuur, eds.), Marcel Dekker, New York, 1976, 339-346.

25. Shaw, H., Nonlinear elliptic boundary value problems at resonance, *J. Differential Equations*, 1976, to appear.

26. Williams, S. A., A connection between the Cesari and Leray-Schauder methods, *Mich. Math. J. 15*(1968), 441-448.

27. Williams, S. A., A sharp sufficient condition for solutions of a nonlinear elliptic boundary value problem, *J. Differential Equations 8*(1970), 580-586.

---

*This research was partially supported by AFOSR Research Project 71-2122 at the University of Michigan.

THE PRINCIPLE OF EXCHANGE OF STABILITY

Michael G. Crandall
and Paul H. Rabinowitz*
Department of Mathematics
University of Wisconsin

"The principle of exchange of stability" is one of those catch-all phrases that is used in different senses by different authors. It is generally employed in the context of a family of evolution equations for which bifurcation occurs and it refers to a qualitative relationship between the shape of the bifurcating curve of solutions and their stability.

As a simple illustration, consider the ordinary differential equation

$$\frac{du}{dt} + f(\mu,u) = 0 \tag{0.1}$$

where $\mu \in \mathbb{R}$, $u \in \mathbb{R}^m$ and $f \in C^2(\mathbb{R}\times\mathbb{R}^m,\mathbb{R}^m)$. Suppose that $f(\mu,0)=0$ so that (0.1) possesses the family of equilibrium solutions $\{(\mu,0)\,|\,\mu\in\mathbb{R}\}$. For convenience, further assume $f(\mu,u) = g(u) - \mu u$. If in addition 0 is a simple eigenvalue of $g_u(0)$, a curve of equilibrium solutions $(\mu(s),u(s))$ with $u(s) \neq 0$ if $s \neq 0$ bifurcates from $(0,0)$. In such bifurcation situations one is generally interested in the stability of the resulting solutions of (0.1). Towards that end suppose that 0 is the smallest eigenvalue of $g_u(0)$. This implies that the trivial solution $(\mu,0)$ of (0.1) is stable for $\mu < 0$ and unstable for $\mu > 0$. The principle of exchange of stability then relates the stability of the nontrivial solution curve $\{(\mu(s),u(s))\}$

to its shape. E.g. if $s\mu'(s) > 0$ for $0 \neq s$ near 0, i.e., the curve "bends" to the right, the principle says the spectum of $f_u(\mu(s),u(s))$ lies in the interior of the right half plane and therefore the bifurcating solutions are stable while if $s\mu'(s) < 0$ for $0 \neq s$ near 0, they are unstable.

As is evident from a closer inspection of the above example, what we are calling the principle of exchange of stability consists of two parts: (i) a qualitative relationship between $\sigma(f_u(\mu,0))$, $\sigma(f_u(\mu(s),u(s)))$ and the shape of the bifurcating curve; and (ii) a qualitative criterion for the stability of the equilibrium solutions in terms of $\sigma(f_u(\mu,0))$ and $\sigma(f_u(\mu(s),u(s)))$. Our results in this paper will only concern (i). Thus more properly we should refer to the <u>principle of exchange of linearized stability</u>. However we will abuse language and omit the "linearized." (Concerning (ii), conditions on $\sigma(f_u(\mu,0))$ and $\sigma(f_u(\mu(s),u(s)))$ which guarantee stability or instability are well known for ordinary differential equations, both for the time independent and time periodic cases [1]. For abstract evolution equations, similar results have been carried out in fairly general situations. See e.g. [2].)

Our main goal in this paper is to present a version of the principle of exchange of stability in the context of what is often called Hopf bifurcation, i.e., the bifurcation of a family of time periodic solutions from a family of equilibrium solutions. This turns out to be the analogue of earlier work of the authors on the time independent case. Therefore we will describe the latter case first in §1 together with some preliminaries on bifurcation. Then the results for the time periodic case will be given in §2, first in the simpler context of Hopf's work and then for an evolution equation in a Banach space. Another version of these results due to H. Weinberger can be found in these proceedings [3].

1. THE EQUILIBRIUM CASE

We begin with a few preliminaries. Recall if X and Y are Banach spaces, $U \subset X$, $F: U \to Y$, and there is a simple curve $I \subset U$ such that $F(x) = 0$ for all $x \in I$, we say an interior point z of I is a bifurcation point (for F with respect to I) if every neighborhood of z contains zeroes of F other than those on I. In the sequel we will always take $X = \mathbb{R} \times E$, with E a real Banach space, and $z = (0,0)$.

Let B(E,Y) denote the set of bounded linear maps from E to Y. If $L, K \in B(E,Y)$ and $r \in \mathbb{R}$, we say r is a K-simple eigenvalue of L if $\dim N(L-rK) = 1 = \operatorname{codim} R(L-rK)$ and $Kv \notin R(L-rK)$ where v spans $N(L-rK)$. For the special case in which $E = Y$, $K = I$, the identity map, and L is compact, it is easy to see that μ is an I-simple eigenvalue of L if and only if it is a simple eigenvalue in the usual sense.

Next we give a useful sufficient condition for bifurcation to occur for the equation $F(\mu,u) = 0$.

1.1 Theorem. Let $U \subset \mathbb{R} \times E$ be a neighborhood of $(0,0)$, $F \in C^2(U,Y)$, and $F(\mu,0) = 0$ for μ near 0. If 0 is an $F_{\mu u}(0,0)$ simple eigenvalue of $F_u(0,0)$, then $(0,0)$ is a bifurcation point. Moreover if Z is any complement of $N(F_u(0,0)) \equiv \operatorname{span}\{v\}$ in E, then there is a neighborhood $\mathcal{O}$ of $(0,0)$ in $\mathbb{R} \times E$, a $\delta > 0$, and functions $\mu \in C^1((-\delta,\delta),\mathbb{R})$, $\psi \in C^1((-\delta,\delta),Z)$ such that $\mu(0) = 0$, $\psi(0) = 0$, and $F^{-1}(0) \cap \mathcal{O} = \{(\mu,0) \mid (\mu,0) \in \mathcal{O}\} \cup \{(\mu(s),u(s)) \equiv sv+s\psi(s)) \mid s \text{ near } 0\}$.

1.2 Remark. The proof of Theorem 1.1 can be found in [4] and is a consequence of the implicit function theorem and an additional estimate. If F is analytic, μ and u are analytic in s and can be determined by a series expansion procedure. The hypotheses of Theorem 1.1 can be verified in several bifurcation problems of interest, e.g. in the Bénard problem which involves a convecting fluid and the Taylor problem which concerns a rotating fluid [5].

In applications, one often has $X \subset Y$ with continuous injection I and the equation $F = 0$ is the equilibrium form of

$$\frac{du}{dt} + F(\mu,u) = 0 \tag{1.3}$$

Thus the linearized stability of the two families of solutions we have in Theorem 1.1 hinges on how the eigenvalue 0 of $F_u(0,0)$ changes along the two curves in Theorem 1.1. The following result which is a simple consequence of a perturbation theorem for K-simple eigenvalues in [6] gives a quantitative answer to the linearized stability question.

1.4 Theorem. Let the hypotheses of Theorem 1.1 be satisfied with $E \subset Y$ with continuous injection I. Suppose 0 is also an I simple eigenvalue of $F_u(0,0)$. Then there are functions β,γ continuously differentiable from a neighborhood A of 0 in $\mathbb{R}$ into $\mathbb{R}$ and functions x,w continuously differentiable from A to E such that

(i) $\beta(\mu)$ is an I-simple eigenvalue of $F_u(\mu,0)$ and $F_u(\mu,0)\,x(\mu) = \beta(\mu)\,x(\mu)$, $\mu \in A$

(ii) $\gamma(s)$ is an I simple eigenvalue of $F_u(\mu(s),u(s))$ and $F_u(\mu(s),u(s))\,w(s) = \gamma(s)\,w(s)$, $s \in A$

(iii) $\beta(0) = 0 = \gamma(0)$; $x(0) = v = w(0)$

(iv) $x(\mu) - v \in Z$; $w(s) - v \in Z$

(v) $\beta'(0) \neq 0$.

1.5 Remark. Theorem 1.4 is constructive in that the underlying existence tool is the implicit function theorem. However obtaining approximate solutions is often difficult or cumbersome and therefore as much qualitative information as is possible about the functions $\beta(\mu)$, $\gamma(s)$ is desirable. The following principle of exchange of stability for this case provides some such information.

1.6 Theorem. Under the hypotheses of Theorem 1.4, $\gamma(s)$ and $\mu'(s)$ have the same zeroes and whenever $\mu'(s) \neq 0$, $\gamma(s)$ and $-s\mu'(s)\beta'(0)$ have the same sign.

1.7 Remark. Actually Theorem 1.6 can be strengthened slightly in that the quotient of $\gamma(s)$ and $s\mu'(s)$ is continuous for s near 0. Theorem 1.6, which is due to Crandall and Rabinowitz [6], is an elementary consequence of Theorems 1.1 and 1.4 together with some simple estimates. Perhaps the first result of this nature was mentioned by Hopf [7] who treats a special case of (0.1). Sattinger [8] treated another special case involving compact operators using a topological argument and was the first to give an abstract version of the principle. Weinberger [3] gives a qualitative result related to Theorem 1.6 in this volume.

1.8 Example. Returning to (0.1), since $\beta'(0) = -1$ for this example, by Theorem 1.7, $\gamma'(s)$ and $s\mu'(s)$ have the same sign which justifies our earlier remarks on stability for this case. As a second example, for the Bénard problem [5], it is not difficult to show $\beta'(0) < 0$ and if we have suitably normalized the problem so that 0 is the smallest eigenvalue of $F_u(\mu,\cdot)$, then a uniqueness theorem for the problem shows $\mu(s) > 0$ for $0 \neq s$ near 0. Therefore since this problem is analytic, $s\mu'(s) > 0$ in a deleted neighborhood of $s = 0$ and

$\gamma(s) > 0$. Consequently the convection solution is stable in an appropriate class of periodic functions. It is not known whether the solution is stable when considered in larger function classes. For the Taylor problem, the I simplicity of the zero eigenvalue is not known.

1.9 Remark. There are analogues of Theorems 1.4 and 1.6 for some situations where bifurcation does not occur. See [6].

2. THE TIME PERIODIC CASE

The simplest result concerning the bifurcation of a family of time periodic solutions from a family of equilibrium solutions is due to Hopf [7]. He also obtained a version of the principle of exchange of stability for this case. Our goal in this section is to give an infinite dimensional version of the latter result. Since this is somewhat complicated, we begin with Hopf's case so as to have a better model for illustrative purposes. Again consider the system of ordinary differential equations:

$$\frac{du}{dt} + F(\mu,u) = 0 \tag{2.1}$$

with $F \in C^2(\mathbb{R} \times \mathbb{R}^m, \mathbb{R}^m)$ and $F(\mu,0) = 0$ for μ near 0. In seeking periodic solutions of (2.1) near $(0,0)$, since the period is unknown, it is convenient to make a change of variables $\tau = \rho^{-1}t$ so that (2.1) becomes

$$\frac{du}{d\tau} + \rho F(\mu,u) = 0 \tag{2.2}$$

and we want 2π periodic solutions u of (2.2) for μ near 0 and ρ near 1. (This last requirement on ρ is a normalization.) Let $L_0 = F_u(0,0)$. We require

$$(F_1)\quad \begin{cases} \text{(a)} & i \text{ is an I simple eigenvalue of } L_0 \\ \text{(b)} & ni \notin \sigma(L_0) \text{ for } n = 0,2,3,\ldots . \end{cases}$$

The requirement that $i \in \sigma(L_0)$ is a necessary condition for bifurcation to occur. The remaining conditions above are required for technical reasons. (F_1) (a) implies that there are continuously differentiable functions $\beta(\mu)$, $x(\mu)$ for μ near 0 such that

(2.3) $$F_u(\mu,0)x(\mu) = \beta(\mu)\, x(\mu)$$

with $\beta(0) = i$ and $x(0)$ spanning $N(L_0 - iI)$. Further assume

(F_2) $$\operatorname{Re} \beta'(0) \neq 0.$$

Let $C^1_{2\pi}(\mathbb{R},\mathbb{R}^m)$ denote the set of continuously differentiable functions on $\mathbb{R}$ which are 2π periodic and take values in $\mathbb{R}^m$. We put the natural maximum norm on $C^1_{2\pi}(\mathbb{R},\mathbb{R}^m)$. For later use we require

<u>2.4</u> <u>Lemma.</u> If F satisfies (F_1),

(i) There is a $\phi_0 \in C^1_{2\pi}(\mathbb{R},\mathbb{R}^m)\setminus\{0\}$ satisfying

$$\frac{d\phi_0}{dt} + L_0\phi_0 = 0$$

(ii) If $\phi_1 = \frac{d\phi_0}{d\tau}$, $\{\phi_0,\phi_1\}$ is a basis for $N(\frac{d}{d\tau} + L_0)$ in $C^1_{2\pi}(\mathbb{R},\mathbb{R}^m)$ and $\frac{d\phi_1}{d\tau} = -\phi_0$

(iii) There is a solution $\psi_0 \in C^1_{2\pi}(\mathbb{R},\mathbb{R}^m)$ of

$$-\frac{d\psi_0}{d\tau} + L_0^*\psi_0 = 0$$

and if $\psi_1 = \frac{d\psi_0}{d\tau}$, $(\phi_i,\psi_j) = \delta_{ij}$ for $i,j = 0,1$.

In Lemma 2.4, δ_{ij} denotes the Kronecker delta, L_0^* the adjoint of L_0, and

$$(\phi,\psi) = \int_0^{2\pi} (\phi(\tau),\psi(\tau))_{\mathbb{C}^m}\, d\tau.$$

We can now state the Hopf bifurcation theorem.

<u>2.5</u> <u>Theorem.</u> Let $\Omega \subset \mathbb{R} \times \mathbb{R}^m$ be an open neighborhood of $(0,0)$, $F \in C^2(\Omega,\mathbb{R}^m)$, and $F(\mu,0) = 0$ for $(\mu,0) \in \Omega$. If (F_1) and (F_2) hold, then there are positive numbers ε,η and continuously differentiable functions $(\rho,\mu,u):(-\eta,\eta) \to \mathbb{R} \times \mathbb{R} \times C^1_{2\pi}(\mathbb{R},\mathbb{R}^m)$

with the following properties:

(a) $(\rho(s), \mu(s), u(s))$ is a nontrivial solution of (2.2) for $s \in (-\eta,\eta)$ and $s \neq 0$,

(b) $\mu(0) = 0$, $u(0) = 0$, $\rho(0) = 1$

(c) If (μ_1, u_1) is a solution of (2.1) of period $2\pi\rho_1$, and $|\rho_1 - 1| < \varepsilon$, $|\mu_1| < \varepsilon$, and $|u_1(\tau)| < \varepsilon$ for $\tau \in [0, 2\pi\rho_1]$, then there is an $s \in [0,\eta)$ and a $\Theta \in [0,2\pi)$ such that $\rho_1 = \rho(s)$, $\mu_1 = \mu(s)$, and $u_1(\rho_1\tau) = u(s)(\tau+\Theta)$ for $\tau \in [0,2\pi]$.

2.6 Remark. Let $V = \{v \in C^1_{2\pi}(\mathbb{R},\mathbb{R}^m) \mid (v,\psi_i) = 0,\ i = 0,1\}$ where ψ_i is as in Lemma 2.4. The proof of Theorem 2.5 shows $u(s)$ can be chosen so that $u(s) = s(\phi_0 + v(s))$ where ϕ_0 is as in Lemma 2.4 and $v(s) \in V$ with $v(0) = 0$.

2.7 Remark. Hopf [7] actually assumed F was real analytic and $\pm\, i$ were the only purely imaginary eigenvalues of L_0. A proof of Theorem 2.5 in the spirit of that of Theorem 1.1 can be found in [9]. There have been several extensions of Hopf's result both in finite and infinite dimensions, e.g. [10-12] and we refer to [9] for more references.

To motivate the form the principle of exchange of stability takes for this situation, it is worth recalling some concepts from the theory of ordinary differential equations. Let $A(t)$ be a time dependent matrix which is periodic in t. The Floquet multipliers of the problem

$$\frac{dw}{dt} + A(t)\, w = 0 \tag{2.8}$$

are the eigenvalues of $U(p)$ where $w(t) = U(t)x$ is the solution of (2.8) satisfying $w(0) = x$. A number κ is a Floquet exponent of (2.8) if $e^{p\kappa}$ is a Floquet multiplier. It is easy to verify that $-\kappa$ is a Floquet exponent if and only if the problem

$$\frac{dz}{dt} + A(t)\ z = \kappa z,\ z(p) = z(0) \tag{2.9}$$

has a nontrivial solution. This characterization of the Floquet exponents has the advantage that it may be a more convenient means by which to calculate them than using the definition. For a nonlinear equation

$$\frac{du}{dt} + g(u) = 0, \tag{2.10}$$

the Floquet exponents and multipliers for a p periodic solution u are defined to be those of (2.8) with $A(t) = g_u(u(t))$. Since (2.10) is autonomous, if $\frac{du}{dt} \not\equiv 0$, differentiating (2.10) and using (2.9) shows that 0 is a Floquet exponent and 1 a Floquet multiplier for u. It has been shown in various contexts [1,2] that a periodic solution of (2.10) is stable if one multiplier is one and the remaining multipliers are less than one in modulus. Thus we focus our attention on these quantities.

Returning to (2.2), we have $p = 2\pi$ and $g(u) = \rho(s)F(\mu(s), u(s))$. Using (2.9) at s = 0, we see the exponents are $\{\sigma(L_0) \pm in \mid n=0,1,\dots\}$ and the multipliers are $e^{-2\pi\sigma(L_0)}$. In particular by (F_1)(a), $\kappa = 0$ is a double eigenvalue of (2.9). Therefore 1 is a multiplier for (2.2) at s = 0 of multiplicity 2 and as we noted above, it is also a multiplier of (2.2) for s near 0. Thus the stability of the periodic solution u(s) of (2.2) is determined by how the second exponent $\kappa(s)$ which is 0 at s = 0 changes. The following lemma answers this question. The functions ϕ_i, ψ_j, v are as in Lemma 2.4 and Remark 2.6. Let $\mathcal{F}(\rho(s),\mu(s),u(s)) = \frac{du(s)}{d\tau} + \rho(s)\ F(\mu(s),u(s))$ and $\mathcal{F}_u(s)$ denote its Frechet derivative with respect to u.

<u>2.11</u> <u>Lemma</u>. Under the hypotheses of Theorem 2.5, there are unique functions $\kappa(s), \eta(s), z(s)$ with values in $\mathbb{R}, \mathbb{R}$, and V respectively, which are defined and continuously differen-

tiable for s near 0, which vanish at s = 0, and which satisfy

(2.12) $$F_u(s)(\phi_0+z(s)) = \kappa(s)(\phi_0+z(s)) + \eta(s)(\phi_1+\frac{d}{d\tau}v(s)).$$

2.13 Remark. The function $-\kappa(s)$ given by Lemma 2.11 provides the continuation of the second zero exponent of (2.2) since if $\kappa(s) = 0$, by earlier remarks and (ii) of Lemma 2.4.

(2.14) $$F_u(s)(\phi_1+\frac{d}{d\tau}v(s)) = 0.$$

Combining (2.14) with (2.12) shows that 0 is a double eigenvalue of $F_u(s)$ for this case. On the other hand if $\kappa(s) \neq 0$, (2.12) can be written as

$$F_u(s)[\phi_0+z(s)+\eta(s)\kappa(s)^{-1}(\phi_1+\frac{d}{d\tau}v(s))] = \kappa(s)[\phi_0+z(s)+\eta(s)\kappa(s)^{-1}(\phi_1+\frac{d}{d\tau}v(s))]$$

which shows, by (2.9), that $-\kappa(s)$ is a Floquet exponent.

Lemma 2.11 is the analogue of Theorem 1.4 for this case and provides a quantitative answer to the stability question. The proof of Lemma 2.11 can be found in [9]. The corresponding qualitative result, i.e., the principle of exchange of stability for this case is

2.15 Theorem. Under the hypotheses of Theorem 2.5, $\kappa(s)$ and $\mu'(s)$ have the same zeroes for s near 0 and whenever $\mu'(s) \neq 0$, $\kappa(s)$ and $-(\mathrm{Re}\beta'(0))s\mu'(s)$ have the same sign.

2.16 Remark. The proof of Theorem 2.15 is similar to that of Theorem 1.6. See [9]. Hopf actually treated the special case of F analytic and $\mu''(0) \neq 0$ in which case $\kappa'(0) = -\mathrm{Re}\beta'(0)\mu''(0)$. Joseph and Nield [13] recently proved a version of Theorem 2.15 in the analytic case and their result partly motivated our work. Weinberger also gives a variant of this result in this volume [3].

To illustrate Theorem 2.15, suppose aside from $\pm i$, $\sigma(L_0) \subset \{\rho\in\mathbb{C} \mid \mathrm{Re}\,\rho>0\}$. If $\mathrm{Re}\,\beta'(0) < 0$ and $s\mu'(s) > 0$ for

$s \neq 0$, then $\kappa(s) > 0$ by Theorem 2.15 and standard results on ordinary differential equations imply the periodic solution is stable. Similarly Re $\beta'(0) < 0$ and $s\mu'(s) < 0$ for $s \neq 0$ implies $\kappa(s) < 0$ and the periodic solution is unstable.

We conclude with generalizations of Theorems 2.5 and 2.15 to an infinite dimensional situation. Naturally the technicalities of dealing with unbounded operators make this case considerably more complicated. However the same basic ideas as in the earlier cases are employed here. Only a statement of the results will be given below. For a more detailed discussion see [9].

Let X be a real Banach space with complexification $X_c = X + iX$. If L is a linear operator on X, L will also denote its extension to X_c. The spectium $\sigma(L)$ is computed relative to X_c so $\lambda \in \sigma(L)$ if and only if $\bar{\lambda} \in \sigma(L)$.

Consider the evolution equation

$$\frac{du}{dt} + L_0 u + f(\mu,u) = 0. \tag{2.17}$$

We assume L_0 is a densely defined linear operator on X satisfying:

(HL_0)

(i) $-L_0$ is the infinitesimal generator of a stongly continuous semigroup $T(t)$ on X,

(ii) $T(t)$ is a holomorphic semigroup on X_c,

(iii) $(\lambda I-L_0)^{-1}$ is compact for λ in the resolvent set of L_0

(iv) i is an I-simple eigenvalue of L_0

(v) $ni \notin \sigma(L_0)$ for $n = 0,2,3...$

If $r > -\text{Re}\,\lambda$ for all $\lambda \in \sigma(L_0)$, (HL_0) (i) and (ii) imply the fractional powers $(L_0+rI)^\alpha$ are defined for $\alpha \geq 0$ [14]. In a standard fashion, the Banach spaces $X_\alpha \subset X$ with norms $||\cdot||_\alpha$ are defined by

$$X_\alpha = D((L_0+rI)^\alpha)$$
$$||x||_\alpha = ||(L_0+rI)^\alpha x|| \text{ for } x \in X_\alpha.$$

We can now state the hypotheses on f:

(Hf) There is an $\alpha \in [0,1)$ and a neighborhood Ω of $(0,0)$ in $\mathbb{R} \times X_\alpha$ such that $f \in C^2(\Omega,X)$. Moreover $f(\mu,0) = 0$ and $f_x(0,0) = 0$ if $(\mu,0) \in \Omega$.

Conditions (Hf) and (HL_0) (iv) imply if $x_0 \in N(L_0-iI)\setminus\{0\}$, there are C^1 functions $\beta(\mu)$, $x(\mu)$ defined for μ near 0 such that

$$(2.18) \qquad \begin{aligned} (L_0+f_x(\mu,0))x(\mu) &= \beta(\mu)x(\mu) \\ \beta(0) = i,\ x(0) &= x_0 . \end{aligned}$$

The analogue of (F_2) now becomes

(H_β) $\qquad$ $\operatorname{Re} \beta'(0) \neq 0$

Set $\tau = \rho^{-1}t$. Then (2.17) transforms to

$$(2.19) \qquad \frac{du}{d\tau} + \rho(L_0u+f(\mu,u)) = 0$$

and we seek 2π periodic solutions of (2.19). The notion of solution must be made more precise here.

<u>2.20</u> <u>Lemma</u>. Let $b > 0$, (HL_0) and (Hf) hold, and $u \in C([0,b],X_\alpha)$. Then the following statements are equivalent

$$(2.21) \qquad \frac{du}{d\tau} \in C((0,b),X),\ u((0,b)) \subset D(L_0),$$

and (2.19) is satisfied on (0,b)

$$(2.22) \qquad u(\tau) = T(\rho\tau)\ u(0) + \rho\int_0^\tau T(\rho(\tau-\xi))f(\mu,u(\xi))d\xi = 0$$

for $0 \le \tau \le b$.

A proof of Lemma 2.20 can be found in [2]. Because of the lemma we can define u to be a solution of (2.19) if $u \in C([0,b],X_\alpha)$ and satisfies (2.22).

Let $C_{2\pi}(\mathbb{R},X_\alpha)$ denote the Banach space of continuous 2π periodic functions from $\mathbb{R}$ into X_α and $C_0([0,2\pi],X_\alpha)$ the Banach space of continuous $h : [0,2\pi] \to X_\alpha$ such that $h(0) = 0$. Finally define

(2.23) $$F(\rho,\mu,u)(\tau) = u(\tau) - T(\rho\tau)\,u(0) + \rho\int_0^{\rho} T(\rho(\tau-\xi))f(\mu,u(\xi))\,d\xi\ .$$

With these preliminaries in hand we can state our infinite dimensional version of the Hopf theorem:

2.24 Theorem. Let (HL_0), (Hf), and $(H\beta)$ be satisfied. Then there are positive numbers ε,η and continuously differentiable functions $(\rho,\mu,u):(-\eta,\eta)\to \mathbb{R}\times\mathbb{R}\times C_{2\pi}(\mathbb{R},X_\alpha)$ with the following properties:

(a) $F(\rho(s),\mu(s),u(s)) = 0$ for $|s| < \eta$

(b) $\mu(0) = 0, u(0) = 0, \rho(0) = 1$ and $u(s)\neq 0$ if $0<|s|<\eta$

(c) If $(\mu_1,u_1)\in C(\mathbb{R},X_\alpha)$ is a solution of (2.1) of period $2\pi\rho_1$, $|\rho_1-1|<\varepsilon$, $|\mu_1|<\varepsilon$, and $||u_1||_\alpha<\varepsilon$, then there is an $s\in[0,\eta)$ and a $\theta\in[0,2\pi)$ such that $\rho_1=\rho(s)$, $\mu_1=\mu(s)$, and $\mu_1(\rho_1\tau) = u(s)(\tau+\theta)$ for $\tau\in[0,2\pi]$.

To obtain the corresponding principle of exchange of stability, we first require:

2.25 Lemma. Let (HL_0) hold. If $x_0\in N(I-T(2\pi))\setminus\{0\}$, then $\{x_0, x_1 = -L_0x_0\}$ is a basis for $N(I-T(2\pi))$. Moreover there exists $x_0^*\in N(I^*-T(2\pi)^*)$ such that $(x_0^*,x_0) = (x_1^*,x_1) = 1$, $(x_1^*,x_0) = (x_0^*,x_1) = 0$ where $x_1^* = L_0^*x_0^*$ and $(\cdot,\cdot)$ denotes the pairing between X^* and X.

We can now define $V = \{v\in C_{2\pi}(\mathbb{R},X)\,|\,(x_i^*,\int_0^{2\pi} T(2\pi-\xi)v(\xi)d\xi)=0$ for $i = 0,1\}$

Next let $K(r)$ denote the bounded linear operator from $C_{2\pi}(\mathbb{R},X)$ into $C_0([0,2\pi],X_\alpha)$ defined by

(2.26) $$(K(r)u)(\tau) = \int_0^{\tau} T(r(\tau-\xi))u(\xi)\,d\xi$$

The generalization of Lemma 2.11 can now be given:

2.27 Lemma. Under the hypotheses of Theorem 2.24, there exist unique functions $\kappa(s),\eta(s),z(s)$ with values in $\mathbb{R},\mathbb{R},$

and V respectively, which are defined and continuously differentiable near s = 0, which vanish at s = 0, and which satisfy

$$F_u(s)(\phi_0+z(s)) = K(\rho(s))[\kappa(s)(\phi_0+z(s)) + \eta(s)(\phi_1+\frac{du}{d\tau}(s)] \tag{2.28}$$

where $F_u(s) = F_u(\rho(s),\mu(s),u(s))$.

Finally the principle of exchange of stability becomes:

2.29 Theorem. Let the hypotheses of Theorem 2.24 be satisfied. Then the conclusions of Theorem 2.15 obtain.

The proofs of the above results can be found in [9].

REFERENCES

1. Hartman, R., *Ordinary Differential Equations*, John Wiley, New York, 1964.

2. Henry, D., *Geometric Theory of Semilinear Parabolic Equations*, (University of Kentucky lecture notes), 1974.

3. Weinberger, H. F., The stability of solutions bifurcating from steady or periodic solutions, these *Proceedings*.

4. Crandall, M. G. and Rabinowitz, P. H., Bifurcation from simple eigenvalues, *J. Fctl. Anal.* *8* (1971), 321-340.

5. Kirchgässner, K. and Kielhöfer, H., Stability and bifurcation in fluid dynamics, *Rocky Mtn. Math. J.* *3* (1973), 275-318.

6. Crandall, M. G. and Rabinowitz, P. H., Bifurcation, perturbation of simple eigenvalues, and linearized stability, *Arch. Rat. Mech. Anal.* *52* (1973), 161-180.

7. Hopf, E., Abzweigung einer periodischen Lösung von einer stationären Lösung eines Differentialsystems, *Ber. Math.-Phys. Kl. Sächs. Akad. Wiss., Leipzig* *94* (1942), 3-22.

8. Sattinger, D. H., Stability of bifurcating solutions by Leray-Schauder degree, *Arch. Rat. Mech. Anal.* *43* (1971), 154-166.

9. Crandall, M. G. and Rabinowitz, P. H., *The Hopf Bifurcation Theorem*, (Math. Res. Center Tech. Rep. 1604, University of Wisconsin) April, 1976.

10. Iudovich, V. I., The onset of auto-oscillations in a fluid, *P.M.M.* *35* (1971), 638-655.

11. Joseph, D. D. and Sattinger, D. H., Bifurcating time periodic solutions and their stability, *Arch. Rat. Mech. Anal.* *45* (1972), 79-109.

12. Iooss, G., Existence et stabilite de la solution periodiques secondaire intervenant dans les problems d'evolution du type Navier-Stokes, *Arch. Rat. Mech. Anal. 47* (1972), 301-329.

13. Joseph, D. D. and Nield, D. A., Stability of bifurcating time-periodic and steady state solutions of arbitrary amplitude, *Arch. Rat. Mech. Anal. 58* (1975), 369-380.

14. Friedman, A., *Partial Differential Equations*, Holt, Rinehart and Winston, Inc., New York, 1969.

---

*This research was sponsored in part by the Office of Naval Research under Contract No. N00014-76-C-0300, in part by the United States Army under Contract No. DAAG 29-75-C0024, and in part by the National Science Foundation under grant No. MPS 73-8720. Any reproduction in part or in full for the purposes of the U.S. government is permitted.

ALMOST PERIODIC PROCESSES AND ALMOST PERIODIC SOLUTIONS OF EVOLUTION EQUATIONS

C. M. Dafermos*
Lefschetz Center for Dynamical Systems
Brown University

1. INTRODUCTION

The problem of existence of almost periodic solutions of nonlinear evolution equations has been studied extensively (e.g. [2,3,4,11,12]). A careful study of the literature reveals that in most cases the methodology is in the spirit of the classical treatment of ordinary differential equations by Amerio [1]. The details, however, vary considerably, from case to case, in consequence of the variety of forms of partial differential equations. Since the proofs involve translates of the equation, assumptions are usually made guaranteeing the compactness of the set of translates of derivatives of the solution which seem extraneous to the problem.

In an effort to bring forward the essential features of the methodology and, at the same time, avoid the imposition of extraneous assumptions, we attempt here a study of the problem at the level of the evolution operator generated by the equation rather than for the equation itself. The relevant concept of a <u>process</u> was introduced and studied in [6,7, 8]. Roughly speaking, a process is a nonautonomous dynamical system. Almost periodic evolution equations generate the class of almost periodic processes [6]. We establish here a complete analogy between these processes and almost periodic ordinary differential equations by showing that existence and uniqueness of complete trajectories with relatively compact range imply almost periodicity of these trajectories. We also give conditions under which a unique complete trajectory with relatively compact range exists.

For the convenience of the reader we have collected the statements of results in Section 2. The proofs are given in Section 3. The abstract theory can be employed to investigate almost periodicity in a broad class of evolution equations. As an illustration we present, in Sections 4 and 5, two examples of the type previously considered in the literature, for comparison.

A general remark regarding the methodology of proofs. The popular classical Bochner criterion of almost periodicity involves uniform convergence and is generally difficult to use here. An alternative, more recent, criterion, also discovered by Bochner [5], which involves only pointwise convergence, has proved quite effective in the applications [9,13]. We have adopted a similar viewpoint here by founding our proof structure on a simple proposition (Lemma 3.1) which characterizes uniform convergence, in the presence of compactness, via pointwise convergence. This makes the proofs particularly simple.

2. ALMOST PERIODIC TRAJECTORIES OF ALMOST PERIODIC PROCESSES

<u>2.1 Definition</u>. A <u>process</u> on a complete metric space X is a two-parameter family of maps $U(t,\tau): X \to X$, parameters $t \in R$, $\tau \in R^+$, with the following properties:

i) $U(t,0) = \text{identity}, \quad \text{for } t \in R;$ (2.1)

ii) $U(t,\sigma+\tau) = U(t+\tau,\sigma)U(t,\tau)$ for $t \in R$; $\sigma,\tau \in R^+$; (2.2)

iii) For any fixed $\tau \in R^+$, the one-parameter family of maps $U(\cdot,\tau):X \to X$, parameter $t \in R$, is equicontinuous.

The above definition was designed so that $U(t,\tau)x$ gives the state at $t + \tau$ of an evolving system whose state at t was x. With this in mind, we now state

2.2 Definition. Let U be a process of X. The positive trajectory through $(t,x) \in R \times X$ is the map $U(t,\cdot)x:R^+ \to X$. A complete trajectory through (t,x) is a map $u(\cdot):R \to X$ such that $u(0) = x$ and $u(s+\tau) = U(t+s,\tau)u(s)$ for all $s \in R$, $\tau \in R^+$.

We recall that a function $u \in C(R;X)$ is almost periodic if the set of translates $\{u(s+\sigma)|\sigma\in R\}$ is relatively compact in $C(R;X)$. The closure in $C(R;X)$ of the set of translates of u is called the hull of u. In analogy one defines almost periodic processes as follows:

2.3 Definition. Let U be a process on X and $\sigma \in R$. The σ-translate of U is the process U_σ defined by $U_\sigma(t,\tau) = U(t+\sigma,\tau), t \in R, \tau \in R^+$.

2.4 Definition. A process U on X is called almost periodic if the set of translates $\{U_\sigma|\sigma\in R\}$ is sequentially relatively compact in the following sense: For any sequence $\{\sigma_n\}$ in R there is a subsequence, denoted again by $\{\sigma_n\}$, and a map $V:R \times R^+ \times X \to X$ so that $U_{\sigma_n}(t,\tau)x \to V(t,\tau)x$, pointwise in $\tau \in R^+$ and $x \in X$, and uniformly in $t \in R$. The sequential closure of the set of translates of U in the topology induced by the above sense of convergence is called the hull of U and is denoted by $\mathcal{H}[U]$.

It will be shown (Corollary 3.5) that if U is an almost periodic process then every $V \in \mathcal{H}[U]$ is also an almost periodic process and $H[V] = H[U]$.

A function $v \in C(R^+;X)$ is called asymptotically almost periodic if there is an almost periodic function u such that $\rho(u(t),v(t)) \to 0$, as $t \to \infty$ (ρ denotes the metric of X). We say v is asymptotic to u. It can be shown that $v \in C(R^+;X)$ is asymptotically almost periodic if and only if the set of right translates $\{v(t+\sigma) | \sigma \in R^+\}$ is relatively compact in $C(R^+;X)$.

The following proposition is the analog, for almost periodic processes, of the classical result for almost periodic ordinary differential equations.

__2.5__ __Theorem__. Let U be an almost periodic process on X. Assume that each $V \in H[U]$ has precisely one complete trajectory, $u_V(\cdot)$, with relatively compact range in X and $u_V(\cdot)$ is continuous. Then

i) $u_U(\cdot)$ is almost periodic;

ii) the set $\{u_V(\cdot) | V \in H[U]\}$ is the hull of $u_U(\cdot)$;

iii) mod $u_U \subset$ mod U in the sense that if for some sequence $\{\sigma_n\}$ in R $\{U_{\sigma_n}\}$ is convergent, then the sequence of translates $\{u_U(s+\sigma_n)\}$ is also convergent;

iv) any continuous positive trajectory of U with relatively compact range in X is asymptotically almost periodic, asymptotic to $u_V(\cdot)$, for some $V \in H[U]$.

__2.6__ __Corollary__. Under the assumptions of Theorem 2.5, if U is periodic of period σ, i.e., $U_\sigma = U$, then $u_U(\cdot)$ is also periodic of period σ. In particular, if U is a dynamical system, i.e., $U_\sigma = U$ for all $\sigma \in R$, then $u_U(\cdot)$ is a rest point.

In the applications of Theorem 2.5 to establishing existence of almost periodic solutions to evolution equations, one has to show first that the equation generates an almost periodic process and then that each process in the hull of the generated process has precisely one complete trajectory with relatively compact range. The existence of complete trajectories with relatively compact range is usually established by means of the following proposition.

2.7 Theorem. Let U be an almost periodic process on X which has some positive trajectory with relatively compact range. Then every $V \in H[U]$ has at least one complete trajectory with relatively compact range.

Uniqueness of complete trajectories with relatively compact range is more delicate and is expected to hold only for restricted classes of almost periodic processes. As a preparation for the applications discussed in Sections 4 and 5, we state the following proposition.

2.8 Theorem. Let U be an almost periodic process on X which is contractive, i.e., $\rho(U(t,\tau)x,U(t,\tau)y) \leq \rho(x,y)$ for $t \in R$, $\tau \in R^+$, $x,y \in X$, and has two (distinct) complete trajectories with relatively compact range. Then each $V \in H[U]$ has two complete trajectories, $u_V(\cdot),v_V(\cdot)$, with relatively compact range, such that $\rho(u_V(t),v_V(t))$ = positive constant for $t \in R$.

2.9 Corollary. If U is an almost periodic, contractive process on X which has precisely one complete trajectory with relatively compact range, then every $V \in H[U]$ has precisely one complete trajectory with relatively compact range.

2.10 Corollary. If U is an almost periodic, strictly contractive process on X, i.e., $\rho(U(t,\tau)x,U(t,\tau)y) < \rho(x,y)$ for $t \in R$, $\tau > 0$, $x \neq y$, then U has at most one complete trajectory with relatively compact range.

Often in the applications one knows that a function is almost periodic in a certain space and has to establish almost periodicity in another space. To this end, the following proposition is helpful.

2.11 Theorem. Let Y, Z be complete metric spaces, continuously embedded in a Hausdorff space $\mathcal{C}$. Suppose $u(\cdot):R \to Y \cap Z$ is almost periodic in Y and its range is relatively compact in Z. Then $u(\cdot)$ is also almost periodic in Z.

3. PROOFS

In proving the propositions of Section 2, our strategy is to circumvent the difficulty of establishing directly equicontinuity and uniform convergence by having recourse to pointwise convergence. To this end we will use the following lemma whose proof is very simple and will thus be omitted.

3.1 Lemma. Let $\mathcal{R}$ be a set, X a complete metric space and $\{u_n\}$ a sequence in $X^{\mathcal{R}}$ which converges pointwise on $\mathcal{R}$ to a map u which has relatively compact range in X. The following are equivalent:

i) $\{u_n\}$ converges to u uniformly on $\mathcal{R}$;

ii) for any sequence $\{s_n\}$ in $\mathcal{R}$, there is a subsequence $\{s_k\}$ such that $\{u_k(s_k)\}$ and $\{u(s_k)\}$ are both convergent and have the same limit;

iii) for any sequence $\{s_n\}$ in $\mathcal{R}$ for which $\{u(s_n)\}$ is convergent, $\{u_n(s_n)\}$ is also convergent and has the same limit.

3.2 Corollary. (Bochner [5]). A function $u \in C(R;X)$ is almost periodic if and only if for any sequences $\{s_n\},\{\sigma_n\}$ in R there are common subsequences $\{s_k\},\{\sigma_k\}$ such that $\{u(t+s_k)\}$ converges pointwise to, say, $v(t)$ and $\{v(t+\sigma_k)\}$,

$\{u(t+s_k+\sigma_k)\}$ are both pointwise convergent and have the same limit.

3.3 Corollary. A function $u \in C(R^+;X)$ is asymptotically almost periodic if and only if for any sequences $\{s_n\},\{\sigma_n\}$ in R^+, $s_n \to \infty$, there are common subsequences $\{s_k\},\{\sigma_k\}$ such that $\{u(t+s_k)\}$ converges pointwise to, say, $v(t)$ and $\{v(t+\sigma_k)\},\{u(t+s_k+\sigma_k)\}$ are both pointwise convergent and have the same limit.

3.4 Corollary. A two-parameter family of maps $U(t,\tau):X \to X$, $t \in R$, $\tau \in R^+$, which satisfies (2.1), (2.2), is an almost periodic process on X if and only if for any sequences $\{s_n\}$, $\{\sigma_n\}$ in R there are common subsequences $\{s_k\},\{\sigma_k\}$ such that

i) $\{U_{s_k}\}$ converges pointwise on $R \times R^+ \times X$ to, say, V;

ii) $\{V_{\sigma_k}\}$ and $\{U_{s_k+\sigma_k}\}$ are both pointwise convergent and have the same limit;

iii) if $\{x_k\}$ is any convergent sequence in X with limit, say, y, then $\{U_{\sigma_k}(t,\tau)x_k\}$ converges to $V(t,\tau)y$ for any $t \in R$, $\tau \in R^+$.

3.5 Corollary. If U is an almost periodic process on X, then every $V \in H[U]$ is an almost periodic process and $H[V] = H[U]$.

3.6 Lemma. If U is an almost periodic process on X and $V \in H[U]$, there are sequences $\{s_n\},\{\sigma_n\}$, $s_n \to -\infty$, $\sigma_n \to \infty$, such that $\{U_{s_n}\}$ and $\{U_{\sigma_n}\}$ converge to V.

Proof. We start with any sequence $\{q_k\}$, $q_k \to \infty$, for which $\{U_{q_k}\}$ is convergent with limit, say, $W \in H[U]$. By Corollary 3.5, $V \in H[W]$ so that $\{W_{p_n}\}$ converges to V for some sequence $\{p_n\}$ in R. We now select a subsequence $\{q_n\}$ of $\{q_k\}$ such that $\sigma_n \overset{\text{def}}{=} p_n + q_n \to \infty$. Then, by Corollary 3.4, $\{U_{\sigma_n}\}$

converges to V. The existence of $\{s_n\}$ is established in an analogous fashion.

3.7 Lemma. Let U be an almost periodic process on X and assume that some positive trajectory $U(t,\cdot)x$ has relatively compact range in X. Suppose that $\{\sigma_n\}$ is a sequence in R^+, $\sigma_n \to \infty$, such that U_{σ_n} converges to $V \in H[U]$. Then there is a subsequence of $\{\sigma_n\}$, denoted again by $\{\sigma_n\}$, such that $\{U(t,s+\sigma_n)x\}$ converges for all $s \in R$ to a complete trajectory $v(s)$ of V with relatively compact range in X.

Proof. By Cantor's diagonal procedure we can find a subsequence of $\{\sigma_n\}$, denoted again by $\{\sigma_n\}$, so that $\{U(t,s+\sigma_n)x\}$ is convergent for $s = 0, -1, -2, \ldots$. For fixed $s \in R$, $\tau \in R^+$, and sufficiently large n so that $\sigma_n + s \geq 0$, we have $U(t,s+\tau+\sigma_n)x = U_{\sigma_n}(t+s,\tau)U(t,s+\sigma_n)x$ which implies, on account of Corollary 3.4, that $\{U(t,s+\sigma_n)x\}$ converges for all $s \in R$ to, say, $v(s)$ and $v(s+\tau) = V(t+s,\tau)v(s)$. Thus $v(\cdot)$ is a complete trajectory of V and its range is contained in the closure of the range of $U(t,\cdot)x$ and is, therefore, relatively compact.

By the same procedure one shows the following proposition.

3.8 Lemma. Let $u(\cdot)$ be a complete trajectory with relatively compact range of an almost periodic process U. Assume that for some sequence $\{\sigma_n\}$ in R $\{U_{\sigma_n}\}$ converges to $V \in H[U]$. Then there is a subsequence of $\{\sigma_n\}$, denoted again by $\{\sigma_n\}$, such that $u(s+\sigma_n)$ converges for all $s \in R$ to a complete trajectory $v(s)$ of V with compact range in X.

3.9 Lemma. Let U be an almost periodic process on X and assume that some positive trajectory $U(t,\cdot)x$ has relatively compact range in X and is asymptotically almost periodic, asymptotic to, say, $v \in C(R;X)$. Then $v(\cdot)$ is an almost periodic complete trajectory of some $V \in H[U]$.

Proof. Since $v(\cdot)$ is almost periodic, there is $\{\sigma_n\}, \sigma_n \to \infty$, such that $v(s+\sigma_n) \to v(s)$ for all $s \in R$. Since $\rho(U(t,s+\sigma_n)x, v(s+\sigma_n)) \to 0$, for any $s \in R$, it follows that $\{U(t,s+\sigma_n)x\}$ converges to $v(s)$. Then Lemma 3.7 implies that $v(\cdot)$ is a complete trajectory of some $V \in H[U]$.

Proof of Theorem 2.5. Let $\{s_n\}$, $\{\sigma_n\}$ be sequences in R. By Corollary 3.4, there are common subsequences $\{s_k\}$, $\{\sigma_k\}$ such that $U_{s_k} \to V, V_{\sigma_k} \to W$, $U_{s_k+\sigma_k} \to W$. On account of Lemma 3.8, $u_U(t+s_k) \to u_V(t), u_V(t+\sigma_k) \to u_W(t)$, $u_U(t+s_k+\sigma_k) \to u_W(t)$, for all $t \in R$. Then Corollary 3.2 implies that $u_U(\cdot)$ is almost periodic and mod $u_U \subset$ mod U. It is also clear that $\{u_V(\cdot) | V \in H[U]\}$ is the hull of $u_U(\cdot)$. By a similar argument one deduces from Lemmas 3.7, 3.9, and Corollary 3.3 that any continuous positive trajectory of U with relatively compact range is asymptotically almost periodic, asymptotic to $u_V(\cdot)$ for some $V \in H[U]$.

Proof of Theorem 2.7. It is an immediate corollary of Lemmas 3.6 and 3.7.

Proof of Theorem 2.8. Let $g(\cdot), h(\cdot)$ be two distinct trajectories of U with relatively compact range. We set $\delta \overset{\text{def}}{=} \lim_{t \to -\infty} \rho(g(t),h(t))$. For any $V \in H[U]$, there is by Lemma 3.6 a sequence $\{s_n\}, s_n \to -\infty$, such that $U_{s_n} \to V$. Then Lemma 3.8 implies that there is a subsequence of $\{s_n\}$, which will be denoted again by $\{s_n\}$, such that $\{g(t+s_n)\}$, $\{h(t+s_n)\}$ converge for all $t \in R$ to complete trajectories $u_V(t)$, $v_V(t)$, respectively, of V having relatively compact range in X. Furthermore, $\rho(u_V(t),v_V(t)) = \delta$, for all $t \in R$.

Proof of Theorem 2.11. It follows easily from Corollary 3.2.

4. EVOLUTION EQUATIONS WITH STRICTLY MONOTONE OPERATORS

Let H be a real Hilbert space with inner product $< , >$ and V a reflexive Banach space contained in H algebraically and topologically. Furthermore, let V be dense in H in which case H can be identified with a subspace of the dual V' of V and $< , >$ can be extended by continuity to $V' \times V$.

We consider the initial value problem

$$\dot{u}(t) + Au(t) = f(t) \tag{4.1}$$

$$u(0) = \chi \tag{4.2}$$

where $A:V \to V'$ is a (generally nonlinear) bounded,

$$||Av||_{V'} \le C||v||_V^{p-1} + K, \quad v \in V, \; p > 1, \tag{4.3}$$

coercive,

$$<Av,v> \ge \alpha||v||_V^p, \quad v \in V, \; \alpha > 0, \tag{4.4}$$

strictly monotone,

$$<Av-Aw,v-w> > 0, \quad v,w \in V, \; v \ne w, \tag{4.5}$$

hemicontinuous operator (see [10, Ch. 2]).

The "elliptic" operator

$$Au = -\sum_{i=1}^{n} \frac{\partial}{\partial x_i} \Phi\left(\frac{\partial u}{\partial x_i}\right) \quad \text{in } \Omega \subset R^n$$

$$u = 0 \quad \text{on } \partial\Omega,$$

where $\Phi(\cdot)$ is a strictly increasing function satisfying $c|\xi|^p < \xi\Phi(\xi) < C|\xi|^p$, provides an example with $H = L^2(\Omega)$, $V = W_0^{1,p}(\Omega)$, $V' = W^{-1,p'}(\Omega)$, $p' = p/p-1$.

If W is a Banach space and $q \ge 1$, we let $S^q(R;W)$ denote the Banach space of $w \in L^q_{loc}(R;W)$ with

$$||w||_{S^q(R;W)} \overset{\text{def}}{=} \sup_{t \in R}\left[\int_t^{t+1} ||w(s)||_W^q \, ds\right]^{1/q} < \infty. \tag{4.6}$$

In an analogous fashion we define $S^q(R^+;W)$.

The following result is established in [10, Ch. 2, §1 and Ch. 4 §8]. If $\chi \in H$ and $f \in S^{p'}(R^+;V')$, $p' = p/p-1$, then there exists a unique solution $u \in S^p(R^+;V) \cap C(R^+;H) \cap W^{1,p'}_{loc}(R^+;V')$ of (4.1), (4.2) and we have the estimate

$$||u||^2_{L^\infty(R^+;H)} + ||u||^p_{S^p(R^+;V)} \le ||\chi||^2_H + \beta||f||^{p'}_{S^{p'}(R^+;V')} + \gamma||f||^{2p'/p}_{S^{p'}(R^+;V')} \quad (4.7)$$

where β and γ are positive constants independent of χ and f.

We now turn to the problem of almost periodicity of solutions of (4.1).

4.1 Theorem. Assume that $f(\cdot)$ is almost periodic in $S^{p'}(R;V')$ and the injection of V into H is compact. Then there is a unique solution of (4.1) which is almost periodic in H as well as in $S^p_{weak}(R;V)$. Furthermore, every solution of (4.1), (4.2) is asymptotically almost periodic in the above spaces.

Proof. If $u(\cdot)$ is the solution of (4.1), (4.2) and $v(\cdot)$ the solution of $\dot{v}(t) + Av(t) = g(t)$, $v(0) = \psi$, $g \in S^{p'}(R^+;V')$, $\psi \in H$, we have the identity

$$||u(t)-v(t)||^2_H - ||\chi-\psi||^2_H + 2\int_0^t \langle Au(s)-Av(s),u(s)-v(s)\rangle \, ds = 2\int_0^t \langle f(s)-g(s),u(s)-v(s)\rangle \, ds \quad (4.8)$$

which implies, on account of (4.5) and (4.7), that the map which takes $(\chi,f) \in H \times S^{p'}(R^+;V')$ into $u(t) \in H$, the solution of (4.1), (4.2) at a fixed $t \in R^+$, is continuous. Therefore, if $f(\cdot)$ is almost periodic in $S^{p'}(R^+;V')$, (4.1) generates, by Corollary 3.4, an almost periodic process U on H which, by virtue of (4.8), (4.5), is strictly contractive.

We now prove that the positive trajectory $U(0,\cdot)\chi$ has relatively compact range in H. Let $\{\sigma_n\}$ be any sequence in R^+, $\sigma_n \to \infty$. On account of (4.7) and Fatou's lemma,

$$\int_0^1 \liminf ||U(0,\sigma_n-s)\chi||^p_V \, ds \le \liminf \int_0^1 ||U(0,\sigma_n-s)\chi||^p_V \, ds<\infty$$

so that we can find $s \in [0,1]$ and a subsequence of $\{\sigma_n\}$, which will be denoted again by $\{\sigma_n\}$, such that $\{U(0,\sigma_n-s)\chi\}$ is convergent in H. We may also select the subsequence $\{\sigma_n\}$ so that $U_{\sigma_n} \to V \in \mathcal{H}[U]$. Then, since $U(0,\sigma_n) = U_{\sigma_n}(-s,s)U(0,\sigma_n-s)$, it follows from Corollary 3.4 that $\{U(0,\sigma_n)\chi\}$ is convergent in H.

Using Theorems 2.7, 2.8, 2.5 we conclude that (4.1) has a unique almost periodic solution in H and every solution is asymptotically almost periodic in H. We now observe that a function which is almost periodic in H is also almost periodic in C(R;H). Thus, applying Theorem 2.11 with $Y = C(R;H)$, $Z = S^p_{weak}(R;V)$, $C = S^p_{weak}(R;H)$, we establish almost periodicity in $S^p_{weak}(R;V)$. The proof is complete.

5. EVOLUTION EQUATIONS WITH UNIFORMLY MONOTONE OPERATORS

In the example discussed in Section 4, the compactness of the injection of V into H was used in order to establish the relative compactness of the range of solutions. Here we will trade off this assumption with uniformity of monotonicity of the operator.

On a Hilbert space H, we consider the equation

$$\dot{u}(t) + Au(t) = f(t) \tag{5.1}$$

where $A:D(A) \to H$ is a (generally nonlinear) densely defined, maximal, uniformly monotone operator, i.e.,

$$\langle Av-Aw, v-w\rangle \geq \alpha||v-w||^p_H, \quad v,w \in D(A), \tag{5.2}$$

with $\alpha > 0$, $p > 1$. We also assume $0 \in D(A)$ and $A0 = 0$.

A typical example is provided by the damped wave equation

$$\frac{\partial^2 u}{\partial t^2} = \Delta u - \phi\left(\frac{\partial u}{\partial t}\right) + h, \tag{5.3}$$

in an open bounded $\Omega \subset R^n$, together with boundary conditions

$u = 0$ on $\partial\Omega$. Here ϕ is a smooth function satisfying $\phi(0) = 0$ and $0 < c \le \phi'(\xi) \le C$, $\xi \in R$. In order to cast (5.3) into the form (5.1), we rewrite it as a first order system,

$$\frac{\partial u}{\partial t} = v$$
$$\frac{\partial v}{\partial t} = \Delta u - \phi(v) + h, \tag{5.4}$$

and we select $H = W_0^{1,2}(\Omega) \times L^2(\Omega)$ with inner product

$$<(u,v),(u^*,v^*)> = \int_\Omega [vv^* + \nabla u \cdot \nabla u^* + \lambda uv^* + \lambda u^* v]dx$$

where λ is a small positive constant depending solely on c,C, and the constant in the Poincaré inequality for Ω. A simple computation shows that the operator defined by the right-hand side of (5.4) satisfies (5.2) for $p = 2$.

We now turn to the problem of almost periodicity of solutions of (5.1).

5.1 Theorem. If $f(\cdot)$ is almost periodic in $S^{p'}(R;H)$, $p' = p/p-1$, then there is a unique weak solution of (5.1) which is almost periodic in H and all weak solutions of (5.1) are asymptotically almost periodic in H.

As in Section 4, one shows that for $f(\cdot)$ almost periodic in $S^{p'}(R;H)$, (5.1) has weak solutions in $C(R^+;H)$ and generates an almost periodic, strictly contractive process on H. Thus, in order to prove Theorem 5.1, it suffices to show that every weak solution of (5.1) on R^+ has relatively compact range in H. To this end we first establish the following.

5.2 Lemma. Let $\eta(\cdot)$ be a non-negative function on R^+ which satisfies

$$\eta(t) - \eta(s) + \beta\int_s^t \eta(\tau)^{p/2}\, d\tau < \varepsilon, \quad s \in R^+,\ t \in [s,s+1] \tag{5.5}$$

where ε and β are positive constants. Then

$$\eta(t) < (2\varepsilon/\beta)^{2/p} + \varepsilon, \text{ for } t \ge t_0 \overset{\text{def}}{=} \frac{\eta(0)}{\varepsilon} + 1. \tag{5.6}$$

Proof. The proof is a variant of a familiar argument (e.g. [12]). From (5.5),

$$\beta\int_0^{t_0} \eta(\tau)^{p/2}\, d\tau \le (t_0+1)\varepsilon + \eta(0) = 2t_0\varepsilon.$$

Hence, there is $\tau_0 \varepsilon\ [0,t_0]$ with $\eta(\tau_0) \le (2\varepsilon/\beta)^{2/p}$. From (5.5) it follows $\eta(t) < (2\varepsilon/\beta)^{2/p} + \varepsilon$ for $t \in [\tau_0,\tau_0 + 1]$. Suppose now τ_1 is the smallest number greater than $\tau_0 + 1$ for which $\eta(\tau_1) = (2\varepsilon/\beta)^{2/p} + \varepsilon$. Then $\eta(\tau_1) > \eta(\tau_1-1)$ and $\eta(\tau) > (2\varepsilon/\beta)^{2/p}$, $\tau \in [\tau_1-1,\tau_1]$. Then $\eta(\tau_1)-\eta(\tau_1-1) + \beta\int_{\tau_1-1}^{\tau_1} \eta(\tau)^{p/2}\, d\tau > 2\varepsilon$. Contradiction.

5.3 Lemma. The range of any weak solution of (5.1) on R^+ is relatively compact in H.

Proof. If $u(\cdot)$ is a weak solution of (5.1) and $v(\cdot)$ a weak solution of $\dot{v}(t) + Av(t) = g(t)$, it is easy to obtain, with the help of (5.2), the estimate

$$||u(t)-v(t)||_H^2 - ||u(s)-v(s)||_H^2 + \beta\int_s^t ||u(\tau)-v(\tau)||_H^p\, d\tau \le \gamma||f-g||^{p'}_{S^{p'}(R;H)}, \quad s \in R^+,\ t \in [s,s+1], \quad (5.7)$$

where β,γ are positive constants depending solely on α and p.

Given any increasing sequence $\{\sigma_n\}$ in R^+, $\sigma_n \to \infty$, we find subsequence, denoted again by $\{\sigma_n\}$, so that $\{f(t+\sigma_n)\}$ is Cauchy in $S^{p'}(R;H)$. For $\varepsilon > 0$, we determine n_0 so that, if $m > n \ge n_0$, $\gamma||f(t) - f(t+\sigma_m-\sigma_n)||^{p'}_{S^{p'}(R;H)} = \gamma||f(t+\sigma_m) - f(t+\sigma_n)||^{p'}_{S^{p'}(R;H)} < \varepsilon$ and, at the same time, $\sigma_{n_0} > 2||u||_{L^\infty(R^+;H)}/\varepsilon + 1$. We now apply (5.7) with $g(\tau) = f(\tau+\sigma_m-\sigma_n)$, $v(\tau) = u(\tau+\sigma_m-\sigma_n)$, $t = \sigma_n$ and we use Lemma 5.2 to deduce

$$||u(\sigma_n) - u(\sigma_m)||_H^2 < (2\varepsilon/\beta)^{2/p} + \varepsilon, \quad m > n \ge n_0$$

which shows that $\{u(\sigma_n)\}$ is Cauchy in H. The proof is complete.

REFERENCES

1. Amerio, L., Soluzioni quasi-periodiche, o limitate, di sistemi differenziali non lineari quasi-periodici, o limitati, *Ann. Mat. Pura Appl.* *39*(1955), 97-119.

2. Amerio, L. and Prouse, G., *Almost Periodic Functions and Functional Equations*, Van Nostrand, New York, 1971.

3. Biroli, M., Sur les solutions bornées et presque périodiques des équations et inéquations d'évolution, *Ann. Mat. Pura Appl.* *93*(1972), 1-79.

4. Biroli, M., Sur les solutions bornées ou presque périodiques des équations d'évolution multivoques sur un espace de Hilbert, *Ric. Mat.* *21*(1972), 17-47.

5. Bochner, S., A new approach to almost periodicity, *Proc. Nat. Acad. Sci. U.S.A.* *48*(1962), 2039-2043.

6. Dafermos, C. M., An invariance principle for compact processes, *J. Diff. Equations* *9*(1971), 239-252.

7. Dafermos, C. M., Uniform processes and semicontinuous Liapunov functionals, *J. Diff. Equations* *11*(1972), 401-415.

8. Dafermos, C. M., Semiflows associated with compact and uniform processes, *Math. Syst. Theory* *8*(1974), 142-149.

9. Fink, A. M., *Almost Periodic Differential Equations* (Lecture Notes in Math. No. 377), Springer Verlag, Berlin, 1974.

10. Lions, J. L., *Quelques Méthodes de Résolution des Problèmes aux Limites non Linéaires*, Dunod, Paris, 1969.

11. Nakao, M., On boundedness, periodicity and almost periodicity of solutions of some nonlinear partial differential equations, *J. Diff. Equations* *19*(1975), 371-385.

12. Prouse, G., Periodic and almost periodic solutions of a nonlinear functional equation, *Rend. Accad. Naz. Lincei* *43*(1967), 161-167, 281-287, 448-452; *44*(1968), 3-10.

13. Sell, G. R., Almost periodic solutions of linear partial differential equations, *J. Math. Anal. Appl.* *42*(1973), 302-312.

---

*Research supported in part by the Office of Naval Research under NONR N00014-75-C-0278, National Science Foundation under MPS71-02923 A04, and United States Army under ARO-D-AAG 29-76-G-0052.

SEMILINEAR ELLIPTIC PROBLEMS WITH NONLINEARITIES NEAR THE FIRST EIGENVALUE

D. G. de Figueiredo
Department of Mathematics
University of Brasilia

1. INTRODUCTION

In this paper we investigate the existence of solutions of the Dirichlet problem for higher order semilinear elliptic equations of the type

$$(1) \qquad Lu \equiv \sum_{\substack{|\alpha|\le m \\ |\beta|\le m}} (-1)^{|\beta|} D^{\beta}(a_{\alpha\beta} D^{\alpha} u) = f(x, u)$$

when the nonlinearity is below the first eigenvalue of L in a sense to be made precise shortly.

The main point here is to show how the classical compact operator theory can be used to handle cases when the nonlinearity f is not monotone, which precludes the direct use of the monotone operator theory. Also the consideration of higher order elliptic operators contrasts with the second order case where the maximum principles available allow the use of ordered Banach spaces techniques yielding nice and strong results; see for example the recent paper of Kazdan Warner [1]. We have only partial results where the nonlinearity f has no growth condition $u \to \infty$. To minimize technical hypotheses we restrict ourselves to the case where f does not depend on derivatives of u. Results in this later case can be obtained in a similar way as we did in [2]. Other boundary value problems can be considered, and this will be object of a later paper. We also believe that the restrictions imposed on the growth of $f(x,s)$ as a function of s can be lifted somehow by the eventual use of L^p estimates.

In Section 4 we treat the so-called non-resonant case and in the final Section 5 we discuss problems with resonance.

2. ON THE LINEAR PART OF (1)

We suppose that the linear operator

$$L = \sum_{\substack{|\alpha|\le m \\ |\beta|\le m}} (-1)^{|\beta|} D^{\beta}(a_{\alpha\beta}(x)D^{\alpha}) \tag{2}$$

is uniformly strongly elliptic, that is, there is a constant c_0 such that

$$\sum_{\substack{|\alpha|=m \\ |\beta|=m}} a_{\alpha\beta}(x)\xi^{\alpha}\xi^{\beta} \ge c_0|\xi|^{2m} \tag{3}$$

for all $x \in \Omega$ and $\xi \in R^N$. Ω is a bounded domain in R^N with a smooth boundary. The coefficients $a_{\alpha\beta}$ are also supposed to be smooth real-valued functions defined in Ω. The amount of smoothness required is what is needed for the application of the regularity theory of L^2 solutions, see for example the book by Friedman [3].

It is also supposed that L is symmetric, that is

$$(Lu, v) = (u, Lv)$$

for all u, v in $c_0(\Omega)$. [Notation: $(\ ,\ )$ is the L^2-inner product]. This implies the existence of real eigenvalues $\lambda_1 < \lambda_2 < \ldots$, with $\lambda_n \to +\infty$. Using the theory of symmetric compact operators in Hilbert spaces we have

$$\lambda_1||u||_0^2 \le (Lu, u) \tag{4}$$

for all $u \in H_0^m(\Omega) \cap H^{2m}(\Omega)$. $||\cdot||_0$ is the L^2-norm.

3. ON THE NONLINEAR PART OF (1)

We intend to obtain solutions in Sobolev spaces based in

L^2. So we shall make either one of the following assumptions on the function $f: \Omega \times R \to R$, besides continuity,

(5) There are constant $b \geq 0$ and $c(x) \in L^2(\Omega)$ such that

$$|f(x, s)| \leq b|s| + c(x), \; x \in \Omega, \; s \in R$$

or

(6) $m > N/2$, with no growth condition required on f.

The point of these hypotheses is to make the Niemytskii operator $F: u \to f(x, u)$ defined in L^2 or in H_0^m. As the reader should have already realized hypothesis (6) is made viewing the use of the Sobolev imbedding theorem. We recognize that assumption (6) is a little restrictive, for it precludes the applicability of our results to the Laplacian, say, even in the plane.* So for that matter our results do not include the results of [1] for second order operators. This is sort of annoying, and we hope to eliminate this deficiency. On the other hand, our results give simple proofs for some known results on the theory of ordinary differential equations, (compare with Tippett [4], Lees [5]), and also applies to operators arising in elasticity theory, like the biharmonic equation.

It is easy to see that

(i) Assumption (5) implies that the Niemytskii operator F defined by $(Fu)(x) = f(x, u(x))$ maps boundedly and continuously $L^2(\Omega)$ into itself.

(ii) Assumption (6) implies that F maps boundedly and continuously $H_0^m(\Omega)$ into $L^2(\Omega)$.

We shall show in the present paper that a very important role in the existence of solutions of the Dirichlet problem for (1) is played by the two limits below:

(7) $f_-(x) = \limsup_{s\to-\infty} \frac{f(x, s)}{s}$, $f_+(x) = \limsup_{s\to+\infty} \frac{f(x, s)}{s}$

In a previous paper [2] we have considered the non-resonant case

$$\lambda_n < \eta_n \leq f_-, f_+ \leq \eta_{n+1} < \lambda_{n+1}$$

where λ_n, λ_{n+1} are consecutive eigenvalues of L, and the resonant case

$$f_- = f_+ = \lambda_n.$$

In Section 4 we shall examine the non-resonant case

$$f_-, f_+ \leq \eta < \lambda_1$$

wher λ_1 is the first eigenvalue of L, and the resonant case where either f_- or f_+ could be equal to λ_1.

Of course there are natural questions to be formulated: what about $\lambda_n < f_- < \lambda_{n+1} < f_+ < \lambda_{n+2}$. This seems to be a hard problem. We would like to mention the paper by Ambrosetti and Prodi [6], which treats a problem of this sort.

4. THE NON-RESONANT PROBLEM

In this section the following basic assumption is made: there is constant η such that

(8) $$f_-(x), f_+(x) \leq \eta < \lambda_1$$

for all $x \in \Omega$. Then we have

1. Theorem. *Let* L *satisfy the hypotheses stated on Section 3. The nonlinearity* f *satisfies (5) (or (6)) and (8). Then there is a solution* $u \in H_0^m \cap H^{2m}$ *of (1).*

Remark. We first observe that without loss of generality we may suppose $\lambda_1 > 0$. Indeed instead of (1) consider the equation $Lu + ku = f(x, u) + ku$ where k is a large positive constant, i.e., $\lambda_1 + k > 0$. Then the first eigenvalue of the new linear operator is $\lambda_1 + k$, and the nonlinear part satisfies a hypothesis analogous to (8).

Proof on the case of Assumption (5). By the existence and regularity theory of solutions of the Dirichlet problem for $Lu = g(x)$ in Ω, it follows that the inverse operator of L (i.e., the solution operator, which we denote by A) is a map from L^2 to $H_0^m \cap H^{2m}$. So the solvability of Dirichlet problem for (1) is equivalent to the existence of fixed point for

(9) $$u = AFu,$$

where F is the Niemytskii map associated to f. Since AF is a compact operator in L^2, the idea now is to apply the Leray-Schauder theorem: "a solution u of (9) exists if there is an $R > 0$ such that $AFu \neq tu$ for all $||u|| = R$ and $t > 1$". Suppose by contradiction that there are sequences $t_n > 1$ and $||u_n|| \to \infty$ such that $t_n u_n = AFu_n$. This is equivalent to

(10) $$t_n Lu_n = f(x, u_n).$$

From (10) using (4) we have

(11) $$t_n \lambda_1 ||u_n||_0^2 \leq \int_\Omega f(x, u_n)u_n = \int_{|u_n| \leq R} f(x,u_n)u_n + \int_{|u_n|>R} f(x,u_n)u_n$$

where $R > 0$ is chosen in such a way that

$$\frac{f(x, s)}{s} \leq \eta + \varepsilon, \quad |s| > R$$

for a given $\varepsilon < \lambda_1 - \eta$. Denoting by $K = \sup\{|f(x, s)|: |s| \leq R, x \in \Omega\}$, it follows from (11) that

(12) $$\lambda_1 ||u_n||_0^2 \leq KR \text{ meas}(\Omega) + (\eta+\varepsilon)||u_n||_0^2,$$

which leads to a contradiction as $n \to \infty$.

Proof in the case of Assumption (6). In this case we look at the operator AF as compact operator on H_0^m. Thus, suppose by contradiction, that there are sequences $t_n > 1$ and $||u_n||_m \to \infty$ such that (10) holds. $||\cdot||_m$ is the Sobolev norm in H^m. It follows from (11) that

(13) $$t_n||u_n||_0^2 \le C_2 + C_3||u_n||_0^2.$$

From (10) using Garding's inequality we get

$$t_n c||u_n||_m^2 - t_n C_1||u_n||_0^2 \le K_R R \text{ meas}(\Omega) + (\eta+\varepsilon)||u_n||_0^2$$

which implies $||u_n||_m^2 \le C_4 + C_5||u_n||_0^2$, with $C_5 > 0$. So $||u_n||_0 \to \infty$, and a contradiction follows as in the proof of the first case.

Remark. One or both limits in (7) could be $-\infty$, and in this case the assumption (6) would have to be made. So this would include the type of equations studied by Hess [7], for second order operators L and monotone nonlinearities:

$$Lu + g(u) = f(x)$$

where $g\colon R \to R$ is a non-decreasing function and $|g(x)/s| \ge \phi(s)$, with $\phi(s) \to +\infty$ as $|s| \to \infty$. Observe that in this case both limits in (7) are $-\infty$. Again we emphasize that more hypotheses on f and L as in Hess yields stonger results; he is able to treat very general boundary problems, even variational inequalities.

5. THE RESONANT CASE

In this section we make the following assumption

(14) $$f_-(x),\ f_+(x) \le \lambda_1,$$

where equality to λ_1 should occur for all $x \in \Omega$, or only in a part of Ω, for at least one of f_+ or f_-. And it is precisely this fact that characterizes the resonance. If both f_- and f_+ are L^∞ functions we see that (5) follows. The interesting case also contained here is when either f_- or f_+ is $-\infty$. And in such a case we assume condition (6).

We start with the following auxiliary result.

2. Theorem. *Let L satisfy the conditions of Section 2. Let (5) (or (6)) and (14) be satisfied. Then the approximant*

equations

(15) $$Lu_n = f(x, u_n) - \frac{1}{n} u_n$$

have solutions $u_n \in H_0^m \cap H^{2m}$. *Moreover, either*

(*i*) $||u_n||_m$ *is bounded, and in this case, equation* (*1*) *has a solution in* $H_0^m \cap H^{2m}$

or

(*i*) *there is a subsequence, say* $||u_n||_m$, *going to* $+\infty$. *In this case, a subsequence of* $v_n = u_n/||u_n||_m$ *converges to an eigenfunction* v *of* L *corresponding to* λ_1, *and the two following inequalities hold*

(16) $$\int_\Omega [f(x, u_n) - \lambda_1 u_n]v_n > 0$$

and

(17) $$\int_\Omega [f(x, u_n) - \lambda_1 u_n]\ v > 0$$

for sufficiently large n.

Proof. The solvability of (15) is a direct consequence of Theorem 1.

If u_n is bounded in H_0^m, take a subsequence, call it u_n, such that $u_n \rightharpoonup u$ in H_0^m, $u_n \to u$ in L^2 and a.e. From (15) we have

$$a[u_n,\phi] = (f(x, u_n), \phi) - \frac{1}{n}(u_n, \phi)$$

and then

$$a[u,\phi] = (f(x, u), \phi),$$

which shows that u is an H_0^m solution of (1). The regularity theory then gives that $u \in H^{2m}$. Here $a[\cdot, \cdot]$ denotes the Dirichlet form associated with L.

Suppose now that $||u_n||_m \to +\infty$. Then there is a subsequence of $v_n = u_n/||u_n||_m$, call it v_n again, such that $v_n \rightharpoonup v$ in H_0^m, $v_n \to v$ in L^2 and a.e. So from (15) using (4) we have:

$$\lambda_1 ||v_n||_0^2 \leq a[v_n, v_n] = \int_\Omega \frac{f(x, u_n)}{||u_n||_m} v_n - \frac{1}{n}||v_n||_0^2. \tag{18}$$

Now given $\varepsilon > 0$ take $R > 0$ such that $\frac{f(x,s)}{s} < \lambda_1 + \varepsilon$ for $|s| > R$. So the integral in (18) can be estimated as follows

$$\begin{aligned}\int_\Omega \frac{f(x,u_n)}{||u_n||_m} v_n &= \int_{|u_n|\leq R} \frac{f(x,u_n)}{||u_n||_m} v_n + \int_{|u_n|>R} \frac{f(x,u_n)}{||u_n||_m} v_n \\ &\leq \frac{K}{||u_n||_m} ||v_n||_0 + (\lambda_1+\varepsilon)||v_n||_0^2.\end{aligned} \tag{19}$$

Using (19) and the fact that $\varepsilon > 0$ is arbitrary we get

$$\lim a[v_n,v_n] = \lambda_1 ||v||_0^2 \tag{20}$$

$$\lim \int_\Omega \frac{f(x,u_n)}{||u_n||_m} v_n = \lambda_1 ||v||_0^2. \tag{21}$$

Now we claim $v \neq 0$. In fact using (15) and Garding's inequality we get

$$c||v_n||_m^2 - c_1 ||v_n||_0^2 \leq \int_\Omega \frac{f(x,u_n)}{||u_n||_m} v_n - \frac{1}{n}||v_n||_0^2$$

or

$$c \leq (c_1 - \tfrac{1}{n})||v_n||_0^2 + \int_\Omega \frac{f(x,u_n)}{||u_n||_m} v_n$$

and passing to the limit using (21)

$$c \leq (c_1+\lambda_1)||v||_0^2$$

which implies $v \neq 0$, for otherwise $c = 0$.

Next we claim that v is an eigenfunction of L corresponding to λ_1. Indeed

$$\lambda_1 ||v_n - v||_0^2 \leq a[v_n-v, v_n-v] = a[v_n,v_n] - 2a[v,v_n] + a[v,v]$$

which implies, using (19)

$$0 \leq \lambda_1 ||v||_0^2 - a[v,v]$$

and this together with (4) gives

$$a[v,v] = \lambda_1 ||v||_0^2. \tag{22}$$

Now for an arbitrary $\phi \in H_0^m$ we have

$$\lambda_1 ||v+\varepsilon\phi||^2 \le a[v+\varepsilon\phi,\ v+\varepsilon\phi]$$

or, using (22)

$$2\lambda_1(v,\ \phi) + \lambda_1\varepsilon||\phi||^2 \le 2a[v,\ \phi] + \varepsilon a[\phi,\ \phi]$$

for arbitrary $\varepsilon > 0$. This implies $a[v,\ \phi] \ge \lambda_1(v,\ \phi)$, for all $\phi \in H_0^m$. Thus $a[v,\ \phi] = \lambda_1(v,\ \phi)$, i.e., v is an eigenfunction of L.

Now from (15) we obtain

$$a[v_n,\ v] = \int_\Omega \frac{f(x,u_n)}{||u_n||_m} v - \frac{1}{n}(v_n,\ v)$$

or using the fact that v is an eigenfunction

$$(v_n,\ v) = \frac{n}{||u_n||_m} \int_\Omega [f(x,\ u_n) - \lambda_1 u_n]v.$$

So for n sufficiently large, we have (17).

Analogously from (15) we have

$$\lambda_1||v_n||_0^2 \le a[v_n,\ v_n] = \int_\Omega \frac{f(x,u_n)}{||u_n||_m} v_n - \frac{1}{n}||v_n||_0$$

or

$$||v_n||_0^2 \le \frac{n}{||u_n||_m} \int_\Omega [f(x,\ u_n) - \lambda_1 u_n]v_n$$

which gives (16) for n sufficiently large. The proof of Theorem 2 is complete.

Next we use Theorem 2 to obatin results on the solvability of (1) under further restrictions on f.

3. Theorem. *Let the hypotheses of Theorem 2 be satisfied. Assume that* $f(x,\ u) = k(x,\ u)u + f(x)$, *where* $f(x) \in L^2(\Omega)$, $k(x,\ s) \in C^0(\bar{\Omega}\times R)$ *and* $f_+,\ f_- \in L^\infty$. *Then (1) has a solution* $u \in H_0^m \cap H^{2m}$ *if*

(23) $$\int_{v>0} f_+v^2 + \int_{v<0} f_-v^2 \ne \lambda_1||v||_0^2.$$

Proof. Using Theorem 2, it suffices to consider alternative (ii). From (15) we have

$$a[v_n, v] = \int_\Omega k(x, u_n)v_n v - \frac{1}{n}(v_n, v) + (\frac{f}{||u_n||_m}, v)$$

and passing to the limit

$$a[v, v] = \int_{v>0} f_+ v^2 + \int_{v<0} f_- v^2$$

which contradicts (23), in view of (22).

Now let us consider the equation

$$Lu - \lambda_1 u = g(u) + f(x) \tag{24}$$

where λ_1 is the first eigenvalue of L, $f \in L^2(\Omega)$ and g is a continuous real-valued function. Let us assume

(25) there is a constant c such that $g(s)s < c$

$$\frac{g(s)}{s} \to 0 \text{ as } s \to -\infty \tag{26}$$

$$\frac{g(s)}{s} \to -\infty \text{ as } s \to +\infty. \tag{27}$$

For example: $g(s) = -e^s$. See the Remark at the end for another example.

4. Theorem. *Let L satisfy the condition of Section 2. Assume (6), (25), (26) and (27). Then (24) has a solution* $u \in H_0^m \cap H^{2m}$ *if*

$$(f, v) < 0 \tag{28}$$

for all non-positive eigenfunctions of L corresponding to the first eigenvalue.

Proof. We use Theorem 2, since (26) and (27) imply (14). It suffices to analyse alternative (ii) of that theorem. From (16)

$$\int_\Omega [g(u_n) + f(x)]v_n > 0$$

which implies

$$(f, v_n) + \frac{c}{||u_n||_m} \text{meas } (\Omega) > \frac{1}{||u_n||_m}\int_\Omega [c - g(u_n)u_n]. \tag{29}$$

Suppose that $\Omega_+ = \{x: v(x) > 0\}$ has positive measure. Then $u_n(x) = v_n(x)||u_n||_m \to +\infty$ for $x \in \Omega_+$, and

$$\frac{c-g(u_n)u_n}{||u_n||m} = \frac{c}{||u_n||m} - g(u_n)v_n$$

goes pointwisely to $+\infty$, since (27) implies that $g(s) \to -\infty$, as $s \to +\infty$. Fatou's lemma gives a contradiction to (27). So $\text{meas}(\Omega_+) = 0$, i.e., $v \le 0$ a.e. From (27) we have

$$(f,v_n) + \frac{c}{||u_n||m}\text{meas}(\Omega) > 0$$

and passing to the limit we have a contradiction to (28).

Remark. One could have the assumptions

(26') $$\frac{g(s)}{s} \to -\infty \text{ as } s \to -\infty$$

(27') $$\frac{g(s)}{s} \to 0 \quad \text{as } s \to +\infty$$

instead of (26) and (27). For example, $g(s) = e^{-s}$. Then (28) should be replaced by: $(f,v) < 0$ for all non-negative eigenfunctions of L.

*Added in proof. The results presented here have been generalized by J. P. Gossez and the author so as to hold without hypothesis (6).

REFERENCES

1. Kazdan, U. L. and Warner, F. W., Remarks on some quasilinear elliptic equations, *Comm. Pure App. Math. XXXVIII* (1975), 567-597.

2. de Figueiredo, D. G., The Dirichlet problems for nonlinear elliptic equations: A Hilbert space approach, *Springer Lecture Notes 446*, 144-165.

3. Friedman, A., *Partial Differential Equations*, Holt-Rinehart-Winston, 1969.

4. Tippett, J., An existence-uniqueness theorem for two-point boundary value problems, *SIAM Math. Ann. 5* (1974), 153-157.

5. Lees, M., Discrete methods for nonlinear two-point boundary value problems, *Numerical Solutions for Partial Differential Equations*, Editor J. H. Bramble, Academic Press, 1966, 59-72.

6. Ambrosetti, A. and Prodi, G., On the inversion of some differentiable mappings with singularities between Banach spaces, *Annali di Mat.* *93*(1972), 231-246.

7. Hess, P., On semi-coercive nonlinear problems, *Indiana Math. J.* (1974), 645-654.

WEAK-INVARIANCE AND REST POINTS IN CONTROL SYSTEMS

M. L. J. Hautus*
Michael Heymann*
Center for Mathematical System Theory
University of Florida

and

Ronald J. Stern
Department of Mathematics
McGill University

1. INTRODUCTION

We consider an autonomous control system given by

<u>1.1</u> $\dot{x} = f(x,y)$

where $x \in \underline{\underline{R}}^n$ and $u \in \underline{\underline{R}}^m$ denote the state and control vectors, respectively. For a subset $\Omega \subset \underline{\underline{R}}^m$ we will denote by U_Ω the set of <u>admissible controls</u>, i.e., the space of Lebesgue measurable functions u: $[0,\infty) \to \Omega$. For an admissible control $u \in U_\Omega$, a solution of 1.1 (if it exists) will be denoted by $\phi_u(t)$, and if the solution satisfying the initial condition $x(0) = x_0$ is known to exist and to be unique on $[0,\infty)$ we shall denote it by $x(t) = \phi(t,0,x_0,u)$. In what follows it will be clear from the context whenever u represents a vector in Ω or a control in U_Ω.

A state $x \in \underline{\underline{R}}^n$ is called an Ω-<u>rest state</u> of system 1.1 if $0 \in f(x,\Omega) := \{f(x,u) \mid u \in \Omega\}$, that is, a state at which system 1.1 can be held indefinitely by a constant control. For the case where Ω consists of a single point, i.e., when

1.1 is independent of u and describes a classical dynamical system, the concept of Ω-rest state reduces to that of a critical point.

A subset $S \subset \underline{\underline{R}}^n$ is called <u>weakly</u> Ω-<u>invariant</u> if for each $x_o \in S$ there exists a control $u \in U_\Omega$ and a corresponding solution $\phi_u(t)$, such that $\phi_u(0) = x_o$ and $\phi_u(t) \in S$ for all $t \geq 0$. It is readily noted that weakly Ω-invariant sets are the control theoretic analogue of <u>positively weakly invariant</u> sets as studied in ROXIN [1965], which in turn generalize the concept of <u>positively invariant</u> sets of ordinary stability theory of dynamical systems (see e.g. BHATIA and SZEGÖ [1970]).

While positively invariant sets have been playing a major role in dynamical systems theory for quite some time, interest in weakly Ω-invariant sets has developed only quite recently (see FEUER and HEYMANN [1976a, 1976b]). Yet, there is some evidence that weakly Ω-invariant sets might play a similar role in control theory (FEUER and HEYMANN [1976a, 1976b], HEYMANN [1976], and see also STERN [1976] where some questions of optimal control in weakly Ω-invariant sets have been studied).

One important reason for the interest in weakly Ω-invariant sets is the fact that they play a significant role in various qualitative control theoretic questions such as, for example, constrained reachability and stabilizability. Of related interest is the question of existence of Ω-rest states in weakly Ω-invariant sets (see e.g. HEYMANN AND STERN [1975, 1976]).

In FEUER and HEYMANN [1976b] weakly Ω-invariant sets were characterized, and it was proved that a compact convex weakly Ω-invariant set contains an Ω-rest state. This result generalizes to a control theoretic setting the classical theorem of dynamical systems which asserts that a compact homeomorphically convex positively invariant set contains a critical point (see e.g. BHATIA and SZEGÖ [1970]). In HEYMANN and STERN [1976] various other results on the existence of Ω-rest states were obtained and, in particular, it was proved that an Ω-rest state also exists in a compact convex set with nonempty interior which is complementary weakly Ω-invariant, i.e., a set S such that $\overline{S^c}$ (the closure of its complement) is weakly Ω-invariant.

In the present paper the investigation of existence of Ω-rest states and related questions is further expanded. In Section 2 the main results of FEUER and HEYMANN [1976b] and of HEYMANN and STERN [1976] on weak invariance and existence of rest points are briefly reviewed. In Section 3 the concept of Ω-constrainedness is introduced and a general result (Theorem 3.2) on existence of Ω-rest states is proved. Significant generalizations of the results of Section 3 are obtained as corollaries to Theorem 3.2. Finally, in Section 4, attention is focused on control systems in the plane (i.e. in $\mathbb{R}^2$). Standard critical point theorems of dynamical systems theory are generalized to the control equation.

2. WEAKLY AND COMPLEMENTARY WEAKLY Ω-INVARIANT SETS.

We denote the Euclidean norm and inner product in $\mathbb{R}^n$ by $||.||$ and $\langle .,. \rangle$ respectively. Let $S \subset \mathbb{R}^n$ be a compact convex subset. For $x \in \partial S$, the boundary of S, we denote by V_x the set of unit outward normal vectors to S at x; that is

$$V_x = \{v \in \underline{R}^n | \langle v, y-x \rangle \leq 0 \text{ for all } y \in S, \|v\| = 1\}.$$

2.1 Definition. Consider system 1.1 and let $S \subset \underline{R}^n$ be compact and convex. For $x \in \partial S$ we say that f is weakly Ω-subtangential to S at x if there exists $u \in \Omega$ such that $\langle v, f(x, \mathbf{u}) \rangle \leq 0$ for all $v \in V_x$. f is said to be weakly Ω-subtangential to S provided the condition holds at each $x \in \partial S$.

The fundamental relation between the geometric property of weak Ω-substangentiality and the dynamic property of weak Ω-invariance is given in the following

2.2 Theorem. Consider system 1.1 and assume the following conditions hold:

   (a) Ω is a nonempty compact subset of $\underline{R}^m$.

   (b) $f(x, u)$ is continuous in both arguments and continuously differentiable in x.

   (c) $f(x, \Omega)$ is convex for each $x \in \underline{R}^n$.

Then a compact convex subset $S \subset \underline{R}^n$ is weakly Ω-invariant if and only if f is weakly Ω-subtangential to S.

A proof of Theorem 2.2 can be found in FEUER and HEYMANN [1976b]. (The reader is also referred to BEBERNES and SCHUUR [1970] where related results are obtained for contingent equations.) We also have the following

2.3 Theorem. Let the system data $\{f, \Omega\}$ satisfy conditions 2.2a - 2.2c and assume that $S \subset \underline{R}^n$ is nonempty, compact, convex and weakly Ω-invariant. Then S contains an Ω-rest state.

Theorem 2.3 was proved in FEUER and HEYMANN [1976b] using the Kakutani fixed point theorem and properties of attainable sets. A more direct proof using a fixed point theorem due to BROWDER [1968] was given in HEYMANN and STERN

[1976]. Below, in Section 3, a significant generalization of Theorem 2.3 will be derived using index theory and homotopy arguments.

We next have the following

2.4 Definition. *Consider system 1.1 and let* $S \subset \underline{\underline{R}}^n$ *be compact and convex. For* $x \in \partial S$ *we say that* f *is weakly* Ω-*supertangential to* S *at* x *if there exist vectors* $v \in V_x$ *and* $u \in \Omega$ *such that* $\langle v, f(x,u)\rangle \geq 0$. f *is said to be weakly* Ω-*supertangential to* S *if the condition holds at each* $x \in \partial S$.

The following lemma provides the relation between the properties of weak Ω-subtangentiality and weak Ω-supertangentiality.

2.5 Lemma. *Assume that the system data* $\{f,\Omega\}$ *satisfy* 2.2a - 2.2c. *Let* $S \subset \underline{\underline{R}}^n$ *be compact, convex, and have nonempty interior. Then* f *is weakly* Ω-*supertangential to* S *if and only if* -f *is weakly* Ω-*subtangential to* S.

While Lemma 2.5 is easily seen to hold true at *regular points* of ∂S (a point $x \in \partial S$ being regular if V_x consists of a single vector v_x), it is more difficult to verify in general. The general proof is based on the following considerations. (For details the reader referred to HEYMANN and STERN [1976].) Let T denote the set of regular points of ∂S. Assuming S is compact, convex, and has nonempty interior, define for each $x \in \partial S$ the sets

$$R_x: = \{w \in \underline{\underline{R}}^n | v_{x_i} \to w,\ x_i \in T,\ x_i \to x\},$$

$$\tilde{R}_x: = \{\tilde{w} \in \underline{\underline{R}}^n | \tilde{w} = \alpha\ w,\ \alpha \geq 0,\ w \in R_x\},$$

and the outward *normal cone* to S at x

$$\tilde{V}_x: = \{v \in \underline{\underline{R}}^n | \langle v, y-x\rangle \leq 0 \text{ for all } y \in S\}.$$

We then have the following lemma (due to ZARANTONELLO [1971], p. 280).

2.6 Lemma. *Let $S \subset \underline{\underline{R}}^n$ be compact, convex and have nonempty interior. Then for each $x \in \partial S$ we have $\tilde{V}_x = H(\tilde{R}_x)$ (where* $H(.)$ *denotes convex hull).*

With the aid of Lemma 2.6, Lemma 2.5 is proved using the (easy) fact that it holds at regular points and the fact that the set of regular points is dense in ∂S.

Next we have the following characterization of complementary weak Ω-invariance which follows readily using Lemma 2.5.

2.7 Theorem. *Assume that the system data* $\{f,\Omega\}$ *satisfy* 2.2a-2.2c, *and that in addition the following condition holds:*

(a) *The responses of* 1.1 *are uniformly bounded,* *i.e.,* *for each* $x \in \underline{\underline{R}}^n$ *and* $T > 0$ *there exists* $b < \infty$ *such that* $||\phi(t,0,x,u)|| < b$ *for all* $u \in U_\Omega$ *and all* $0 \leq t \leq T$.

Then a compact convex set $S \subset \underline{\underline{R}}^n$ *is complementary weakly* Ω-*invariant if and only if* f *is weakly* Ω-*supertangential to* S.

2.8 Remark. Recall (see e.g. LEE and MARKUS [1967]) that condition 2.7a together with conditions 2.2a - 2.2c insure that for each $x \in \underline{\underline{R}}^n$ and $u \in U_\Omega$ the solution $x(t) = \phi(t,0,x,u)$ of 1.1 exists and is unique on the semi-infinite interval $[0,\infty)$. In addition, the conditions insure that the *reachable set* (from x in time t) $F_t(x) := \{\phi(t,0,x,u) | u \in U_\Omega\}$ is compact and depends continuously on t and x for all $t \in [0,\infty)$ and all $x \in \underline{\underline{R}}^n$.

Upon combining Theorems 2.2 and 2.3 with Lemma 2.5 we obtain the following rest-point theorem for complementary weakly Ω-invariant sets.

2.9 Theorem. *Assume the system data* $\{f,\Omega\}$ *satisfy* 2.2a - 2.2c *and* 2.7a. *Assume* $S \subset \underline{\underline{R}}^n$ *has nonempty interior and is* compact convex, and complementary weakly Ω-invariant. Then S contains an Ω-rest state.

Just as in the case of Theorem 2.3, we shall see later that Theorem 2.9 can be generalized considerably using a homotopy argument and index theory.

3. Ω-CONSTRAINED SETS AND SOME GENERALIZATIONS.

We consider system 1.1, and throughout this section we will assume that conditions 2.2a - 2.2c and 2.7a hold. We shall need the following

3.1 Definition. A subset $S \subset \mathbb{R}^n$ is called Ω-constrained in time $t > 0$ provided that for all $x \in S$ we have

$$H(F_t(x)) \cap S \neq \emptyset \qquad \text{(nonempty intersection)}.$$

If S has nonempty interior it will be called strongly Ω-constrained provided there exists $t > 0$ such that for all $x \in \bar{S}$ (the closure of S) and all $\tau \geq t$

$$H(F_\tau(x)) \cap \text{int}(S) \neq \emptyset$$

where int(.) denotes interior. A point x is called t-hull periodic (resp. t-periodic) for $t > 0$ if $x \in H(F_t(x))$ (resp. $x \in F_t(x)$).

The main result of the present section is the following

3.2 Theorem. Consider system 1.1 and assume that conditions 2.2a - 2.2c and 2.7a hold. Assume further that

(a) S is a compact convex subset of $\mathbb{R}^n$ with nonempty interior.

(b) S is Ω-constrained in time $t^* > 0$.

If no point $x \in \partial S$ is t-hull periodic for any $t \in (0,t^*]$, then for each $\bar{u} \in \Omega$ there exists $\bar{x} \in$ int (S) such that $f(\bar{x},\bar{u}) = 0$.

To outline the proof of Theorem 3.2 we will first make use of some basic facts about the degree (or index) of mappings in $\mathbb{R}^n$ (see e.g. DUGUNDJI [1966]). Let Γ denote the metric space of nonempty compact convex subsets of $\mathbb{R}^n$ with Hausdorff topology. For topological spaces X and Y denote by $\underline{C}(X\to Y)$ the space of continuous mappings h: X → Y. For a subset $S \subset \mathbb{R}^n$ let V be a (multivalued) vector field on S, i.e., $V \in \underline{C}(S\to\Gamma)$. A point $x \in S$ is called a critical point of V if $0 \in V(x)$. The vector field is called regular on S if there are no critical points of V on S. If S is compact and convex and has nonempty interior and $h \in \underline{C}(\partial S\to\mathbb{R}^n)$ is a (single valued) vector field, defined and regular on ∂S (i.e., $h(x) \neq 0$ for all $x \in \partial S$) we denote by $\rho_{\partial S}(h)$ the degree of h on ∂S. Next we have the following

3.3 Lemma. Assume 3.2a holds and let $h \in \underline{C}(S\to\mathbb{R}^n)$ be a map whose degree $\rho_{\partial S}(h)$ on ∂S is nonzero. Then h is singular on S, i.e., there exists $x \in S$ such that $h(x) = 0$.

Outine of Proof. If h is regular on S then given any $x_0 \in \operatorname{int}(S)$, the map $F(t,x) := h[tx+(1-t)x_0]$, $0 \le t \le 1$, $x \in S$ is a homotopy and it is readily verified that $\rho_{\partial S}(h) = \rho_{\partial S}(F(1,.)) = \rho_{\partial S}(F(0,.)) = 0$ since $F(0,.)$ is constant on S. Hence a contradiction. □

3.4 Lemma. Let S satisfy 3.2a and assume $g \in \underline{C}(\partial S\to S)$ satisfies $g(x) \neq x$ for all $x \in \partial S$. Define the map $h \in \underline{C}(\partial S\to\mathbb{R}^n)$ by $h(x) = g(x) - x$. Then $\rho_{\partial S}(h) = (-1)^n$.

Outline of Proof. For $x_0 \in \operatorname{int}(S)$ let $h_0 \in \underline{C}(\partial S\to\mathbb{R}^n)$ be defined by $h_0(x) = x_0 - x$. Clearly $h(x)$ and $h_0(x)$ are never in opposite directions and hence $\rho_{\partial S}(h) = \rho_{\partial S}(h_0)$. From the definition of degree it follows immediately that $\rho_{\partial S}(h_0) = (-1)^n$ and the result follows. □

Let $D \subset \underline{\underline{R}}^n$ be open and let $V \in \underline{C}(D \to \Gamma)$ be a given vector field. Then it is well known that V admits continuous selections, i.e., there exist $v \in \underline{C}(D \to \underline{\underline{R}}^n)$ such that $v(x) \in V(x)$ for all $x \in D$. Next we have the following

3.5 Lemma. *Let v_1 and v_2 be any continuous selections of a vector field $V \in \underline{C}(D \to \Gamma)$ where $D \subset \underline{\underline{R}}^n$ is an open subset. If $S \subset D$ is any compact convex subset with nonempty interior such that V is regular on ∂S, then $\rho_{\partial S}(v_1) = \rho_{\partial S}(v_2)$.*

The proof of Lemma 3.5 is by a simple homotopy argument (see HAUTUS, HEYMANN and STERN [1976] for details). In view of the Lemma we can extend the definition of degree to multi-valued vector fields in $\underline{\underline{R}}^n$ as follows. We define the *degree* of a regular vector field $W \in \underline{C}(\partial S \to \Gamma)$, where $S \subset \underline{\underline{R}}^n$ is compact and convex with nonempty interior, as $\rho_{\partial S}(W) := \rho_{\partial S}(w)$ where w is any continuous selection of W. The following Lemma is an immediate consequence of Lemma 3.3 and Lemma 3.5 and of a well known fact for single valued vector fields (see e.g. CODDINGTON and LEVINSON [1955], Theorem 4.1, page 398).

3.6 Lemma. *Let S satisfy 3.2a and assume that $W \in \underline{C}(S \to \Gamma)$ is regular on S. Then $\rho_{\partial S}(W) = 0$.*

Next we shall need the following results:

3.7 Lemma. *Assume S satisfies 3.2a and let $G \in \underline{C}(\partial S \to \Gamma)$ be such that $G(x) \cap S \neq \emptyset$ for all $x \in \partial S$. Define $W \in \underline{C}(\partial S \to \Gamma)$ by $W(x) = G(x) - x$. If W is regular on ∂S then $\rho_{\partial S}(W) = (-1)^n$.*

3.8 Lemma. *Consider system 1.1 and assume that 2.2a - 2.2c as well as 2.7a and 3.2a hold. Let $Q(t,x) := H(F_t(x)) - x$ and let $V \in \underline{C}(S \to \Gamma)$ be defined by $V(x) := f(x,\Omega)$. Then*

(i) *V is regular on ∂S if and only if there exists*

$t_1 > 0$ such that $Q(t,.)$ is regular on ∂S for all $0 < t < t_1$.

(ii) If V is regular, then $\rho_{\partial S}(V) = \rho_{\partial S}(Q(t,.))$ for all $t > 0$ sufficiently small.

The proofs of Lemma 3.7 and 3.8 can be found in HAUTUS, HEYMANN and STERN [1976]. We can now prove Theorem 3.2.

3.9 Proof of Theorem 3.2. By the Ω-constrainedness at time t^*, $H(F_{t^*}(x)) \cap S \neq \emptyset$ for each $x \in \partial S$. Hence by Lemma 3.7 $\rho_{\partial S}(Q(t^*,.)) = (-1)^n$. The nonexistence of t-hull periodic points is equivalent to the regularity of $Q(t,.)$ for all $0 < t \leq t^*$. Hence $\rho_{\partial S}(Q(t,.)) = \rho_{\partial S}(Q(t^*,.))$ for all $0 < t \leq t^*$. Then by Lemma 3.8, V is regular on ∂S and $\rho_{\partial S}(V) = \rho_{\partial S}(Q(t^*,.)) = (-1)^n$. For every $\bar{u} \in \Omega$ the function v defined by $v(x) = f(x,\bar{u})$ is a continuous selection of V on S. Hence $\rho_{\partial S}(v) = (-1)^n$ and by Lemma 3.3 v has a critical point in S. □

The following Corollary of Theorem 3.2 significantly generalizes the statements of Theorems 2.3 and 2.9.

3.10 Corollary. Consider system 1.1, assume 2.2a - 2.2c, 2.7a and 3.2a hold, and assume further that S is either weakly Ω-invariant, or complementary weakly Ω-invariant. Then S contains an Ω-rest state. Moreover, if ∂S contains no Ω-rest states, then for each $\bar{u} \in \Omega$ there exists $\bar{x} \in \text{int}(S)$ such that $f(\bar{x},\bar{u}) = 0$.

Proof. Weak Ω-invariance implies that S is Ω-constrained in time t for each $t > 0$. If ∂S has no Ω-rest states, then the map V defined in Lemma 3.8 is regular on ∂S and hence, by the same lemma, $Q(t,.)$ is also regular for all t sufficiently small. Hence ∂S has no t-hull periodic points for small enough t. Theorem 3.2 then holds and the result follows. If S is complementary weakly Ω-invariant then by Theorem 2.7 f is weakly Ω-supertangential to S, and hence by Lemma 2.5 -f is weakly Ω-subtangential to S. By Theorem 2.2 it then follows that S is weakly Ω-invariant with respect to the system

(1.1a) $\dot{x} = -f(x,u)$.

Repeating the previous argument for the system (1.1a) completes the proof. □

3.11 Remark. In Corollary 3.10, when ∂S has Ω-rest states, it may be the case that Ω can be written as $\Omega = \Omega_I \cup \Omega_O$ with $\Omega_O \neq \emptyset$, such that for each $\bar{u} \in \Omega_I$ there exists $\bar{x} \in S$ for which $f(\bar{x},\bar{u}) = 0$, while for $\bar{u} \in \Omega_O$ there exist only $\bar{x} \in S^c$ (the complement of S) which satisfy $f(\bar{x},\bar{u}) = 0$. Yet this is not the general situation. In general, when ∂S has Ω-rest states, there are $\bar{u} \in \Omega$ for which $f(x,\bar{u}) = 0$ is unsolvable for x (in all of $\mathbb{R}^n$). This can easily be verified to be the case in the example $f(x,u) = x^2-u$, $\Omega = [-1,1]$, $S = [-1/2,1/2]$.

The Theorem 3.2 the existence of Ω-rest states was assured under the condition of Ω-constrainedness, coupled with the absence of t-hull periodic states in ∂S on $(0,t^*]$. In Corollary 3.10 the condition of weak Ω-invariance (or complementary weak Ω-invariance) served to replace (or rather to imply) the above conditions of the theorem. It is interesting to speculate about the possibility of removing the t-hull periodic point condition at the expense of strengthening the Ω-constrainedness condition (to require for example strong Ω-constrainedness instead). While this is in general impossible (as evidenced by an example in HAUTUS, HEYMANN and

STERN [1976]) it does however hold in the following special cases:

3.12 Theorem. *Assume conditions* 2.2a - 2.2c, 2.7a *and* 3.2a *all hold. Assume further that* f *is linear, i.e.,* $f(x,u) = Fx + Gu$ *for some real matrices* F *and* G. *If* S *is strongly* Ω-*constrained then* S *contains an* Ω-*rest state*.

3.13 Theorem. *Assume* 2.2a-2.2c, 2.7a *and* 3.2a *all hold and that* $n \leq 2$ (*where* n *is the dimension of the state space*). *If* S *is strongly* Ω-*constrained then* S *contains an* Ω-*rest state*.

3.14 Theorem. *Assume* 2.2a - 2.2c, 2.7a *and* 3.2a *all hold and that* Ω *consists of a single point*. *If* S *is strongly* Ω-*constrained then* S *contains an* Ω-*rest state*.

4. REST POINTS IN PLANAR CONTROL SYSTEMS.

A standard theorem in the theory of differential equations states that the trajectory of a periodic solution of an autonomous differential equation in the plane (i.e. $n = 2$) encloses a critical point. In the present section we shall deal with an extension of this theorem to control system 1.1 and some of its consequences.

Let K be a Jordan curve in the plane. For a real number $T > 0$, a mapping $x \in \underline{C}([0,T] \to \underline{\underline{R}}^2)$ is called a *proper parametrization* of K if the following conditions hold

4.1 $K = \{x(t) \mid 0 \leq t \leq T\}$

4.2 $x(t) \neq x(s)$ for all $0 \leq s < t < T$

4.3 $x(0) = x(T)$.

From the definition of a Jordan curve it follows that a proper parametrization always exists. If K is a Jordan curve and $v \in \underline{C}(K \to \underline{\underline{R}}^2)$ is a (single valued) vector field defined and regular on K, we denote (just as we did for convex sets in

Section 3) by $\rho_K(v)$ the index of v on K. If K is a Jordan curve we denote by enc(K) the bounded component of K^c and define $\overline{\text{enc}}(K) := K \cup \text{enc}(K)$. Finally, if $M \subset \underline{\underline{R}}^2$ is a compact connected subset, we define the <u>simply connected hull</u> of M, denoted SCH(M), as the complement of the unbounded component of M^c. It is then easily verified that SCH(M) is compact and simply connected (although it is not true in general that SCH(M) = M even if M is compact and simply connected).

Let $V \in \underline{C}(D \to \Gamma)$ be a multivalued vector field in $\underline{\underline{R}}^2$ where $D \subset \underline{\underline{R}}^2$ is open. Let K be a Jordan curve such that $\overline{\text{enc}}(K) \subset D$ and assume that V is regular on K. Then, just as in Section 3, the index $\rho_K(v)$ is the same for every continuous selection v of V. Hence, we define the index $\rho_K(V)$ of V by $\rho_K(V) = \rho_K(v)$ where v is an arbitrary continuous selection of V.

Next we have the following analog of Lemma 3.6:

<u>4.4</u> <u>Lemma</u>. <u>Let</u> $V \in \underline{C}(D \to \Gamma)$ <u>be a vector field and let</u> K <u>be a Jordan curve such that</u> $\overline{\text{enc}}(K) \subset D$. <u>If</u> V <u>is regular on</u> $\overline{\text{enc}}(K)$ <u>then</u> $\rho_K(V) = 0$.

We shall also make use of the following result due to LIFSHITZ [1946]:

<u>4.5</u> <u>Lemma</u>. <u>Let</u> $x \in \underline{C}([0,T] \to \underline{\underline{R}}^2)$ <u>be a proper parametrization of a Jordan curve</u> K, <u>and for</u> $0 < \delta < T$ <u>define</u> $v_\delta(t) := x(t^*+\delta) - x(t)$, $0 \le t \le T$, <u>where</u> $t^* = t$ <u>for</u> $0 \le t \le T - \delta$ <u>and</u> $t^* = t - T$ <u>for</u> $t > T - \delta$. <u>Then</u> $\rho_K(v_\delta) = 1$.

We can now state the following rest point theorem for control systems in the plane.

<u>4.6</u> <u>Theorem</u>. <u>Consider system</u> 1.1 <u>with</u> $n = 2$ <u>and assume the following hold</u>:

(a) $f(x,u)$ is continuous in both arguments for all $x \in \underline{\underline{R}}^2$ and $u \in \underline{\underline{R}}^m$.

(b) For all $x \in \underline{\underline{R}}^2$ the set $f(x,\Omega)$ is compact and convex.

Assume that for an admissible control $u \in U_\Omega$ there exists a solution $\phi_u(t)$ which for some $T > 0$ is a proper parametrization of a Jordan curve K on $[0,T]$. If K has no Ω-rest states of 1.1 then for each $\bar{u} \in \Omega$ there exists $\bar{x} \in \text{enc}(K)$ such that $f(\bar{x},\bar{u}) = 0$.

Proof. Assume that K has no Ω-rest states of 1.1, i.e., the vector field V defined by $V(x) := f(x,\Omega)$ is regular on K. For $\varepsilon > 0$ and all $x \in \underline{\underline{R}}^2$ define

$$V_\varepsilon(x) := \{y+\varepsilon z \mid y \in V(x);\ \|z\| \le 1\}.$$

Clearly $V_\varepsilon \in \underline{C}(\underline{\underline{R}}^2 \to \Gamma)$ and by the continuity of V and the compactness of K, there exists an $\varepsilon > 0$ such that V_ε is also regular on K. Since $V(x) \subset V_\varepsilon(x)$ for all $x \in \underline{\underline{R}}^2$ it is clear that $\rho_K(V_\varepsilon) = \rho_K(v)$ where v is any continuous selection of V. By Lemma 4.4 and the fact that every constant control provides a continuous selection of $f(.,\Omega)$, the proof will be complete upon showing that $\rho_K(V_\varepsilon) \ne 0$. By Lemma 4.5 this will be accomplished if we can show that for sufficiently small $\delta > 0$, $w_\delta(t) := \delta^{-1}[\phi_u(t+\delta)-\phi_u(t)] \in V_\varepsilon(\phi_u(t))$ for all $0 \le t \le T$ (with ϕ_u being extended periodically outside the interval $[0,T]$). In view of the uniform continuity of V (and of ϕ_u) on K there clearly exists a $\delta > 0$ such that $V(\phi_u(\tau)) \subset V_\varepsilon(\phi(t))$ for all $t \in [0,T]$ and all τ such that $|t-\tau| < \delta$. Now suppose that $w_\delta(t_0) \notin V_\varepsilon(\phi_u(t_0))$ for some $t_0 \in [0,T]$. Then there exists a vector $c \ne 0$ and a number α such that $\langle c, w_\delta(t_0)\rangle > \alpha$ and $\langle c,y\rangle \le \alpha$ for all $y \in V_\varepsilon(\phi_u(t_0))$.

Since $V(\phi_u(t)) \subset V_\varepsilon(\phi_u(t_0))$ for all $t_0 \le t \le t_0 + \delta$

and since $\dot{\phi}_u(t) \in V(\phi_u(t))$ a.e., it follows that

$$\langle c, w_\delta(t_o)\rangle = \delta^{-1} \int_{t_o}^{t_o+\delta} \langle c, \dot{\phi}_u(\tau)\rangle d\tau \le \alpha,$$

a contradiction. This completes the proof. □

As a consequence of Theorem 4.6 we also have

4.7 Theorem. *Consider system 1.1 with $n = 2$ and assume 4.6a and 4.6b hold. Assume that for some admissible control $u \in U_\Omega$ and some $T > 0$ there exists a solution $\phi_u(t)$ of 1.1 defined on $[0,T]$ such that $\phi_u(T) = \phi_u(0)$. Then SCH(L) contains an Ω-rest state of 1.1 where $L := \{\phi_u(t) \mid 0 \le t \le T\}$ is the trajectory of ϕ_u.*

The proof of Theorem 4.7 as well as that of the following Theorem can be found in HAUTUS, HEYMANN and STERN [1976].

4.8 Theorem. *Consider system 1.1 with $n = 2$, and assume that 4.6a and 4.6b hold. Let $S \subset \mathbb{R}^2$ be a compact, simply connected and locally connected subset. If for some admissible control $u \in U_\Omega$ there exists a solution $\phi_u(t)$ of 1.1 which is contained in S for all $t \in [0,\infty)$, then S contains an Ω-rest state.*

Theorem 4.5 implies the interesting fact that under conditions 4.6a and 4.6b, a planar control system (i.e., n =2) has Ω-rest states whenever there are bounded trajectories. If 4.6b fails to hold then 1.1 may not have any Ω-rest states in the plane even when there exist bounded trajectories as the following example illustrates

$$\dot{x}_1 = \sin u$$

$$\Omega = [0,\pi]$$

$$\dot{x}_2 = \cos u$$

When $n \ge 3$, the existence of bounded trajectories no longer insures the existence of Ω-rest states even when 4.6a and

4.6b hold. Indeed, this can be verified to be the case in the system

$$\dot{x} = ux - y$$
$$\dot{y} = x + uy \qquad \Omega = \{u \mid |u| \leq 1\}$$
$$\dot{z} = 1 - x^2 - y^2$$

which has bounded trajectories (e.g. starting at $x_o = y_o = \frac{\sqrt{2}}{2}$, $z_o = 0$ under the constant control $u(t) \equiv 0$) but has no Ω-rest states at all.

REFERENCES

1. Bebernes, J. W. and Schuur, J. D., The Wazewski topological method for contingent equations, *Ann. Di Mat. Pura ed Appl. T87* (1970), 271-279.

2. Bhatia, N. P., and Szego, G. P., *Stability Theory of Dynamical Systems* (1970), Springer Verlag (Band 161), Berlin.

3. Browder, F. E., The fixed point theory of multivalued mappings in topological vector spaces, *Math. Ann. 177* (1968), 283-301.

4. Coddington, E. A. and Levinson, N., *Theory of Ordinary Differential Equations* (1955), McGraw Hill, New York.

5. Dugundji, J., *Topology* (1966), Allyn and Bacon, Boston.

6. Feuer, A., and Heymann, M., Admissible sets in linear systems with bounded controls, *Intern. J. Contr. 23* (1976a), 381-392.

7. Feuer, A. and Heymann, M., Ω-Invariance in control systems with bounded controls, *J. Math. Anal. Appl. 53* (1976b), 266-276.

8. Hautus, M. L. J., Heymann, M., and Stern, R. J., Rest point theorems for autonomous control systems, *J. Math. Anal. Appl.* (1976), to appear.

9. Heymann, M., Control dominance, weak invariance and feedback, (1976), to appear.

10. Heymann, M. and Stern, R. J., Controllability of linear systems with positive controls: geometric considerations, *J. Math. Anal. Appl. 52* (1975), 36-41.

11. Heymann, M., and Stern, R. J., Ω-Rest points in autonomous control systems, *J. Diff. Equations 20* (1976), 389-398.

12. Lee, E. B., and Markus, L., *Foundations of Optimal Control Theory*, (1967), John Wiley and Sons, New York.

13. Lifshitz, J., Un teorema sobre transformaciones de curvas cerradas sobre si mismas, *Bol. Soc. Mat. Mex. 3*, (1946), 21-25.

14. Roxin, E. O., Stability in general control systems, *J. Diff. Equations 1* (1965), 115-150.

15. Stern, R. J., Linear time optimal control with state-dependent control restraints, (1976), to appear.

16. Zarantonello, E. H., Projections on convex sets, in *Contributions to nonlinear functional analysis* (1971) (E. H. Zarantonello, Ed.), Academic Press, New York.

M. L. J. Hautus on leave from Department of Mathematics, Technological University, Eindhoven, Netherlands.

Michael Heymann on leave from Department of Electrical Engineering, Technion-Isreal Institute of Technology, Haifa, Israel.

*Supported in part by US Army Research Grant DAAG29-76-G-0203 through the Center for Mathematical System Theory, University of Florida, Gainesville, FL 32611, USA.

HIGH ORDER CONTROLLED STABILITY AND CONTROLLABILITY

H. Hermes*
Department of Mathematics
University of Colorado

INTRODUCTION.

Let M be an analytic n-dimensional manifold and $\mathcal{D} = \{X^\alpha : \alpha \in U\}$ a collection of analytic tangent vector fields on M. We denote the tangent space of M at p by TM_p and let $\mathcal{D}(x) = \{X^\alpha(x) \in TM_x : \alpha \in U\}$. A solution of $\mathcal{D}$ is an absolutely continuous map φ taking a real interval I into M such that $\dot{\varphi}(t) \equiv d\varphi/dt \in \mathcal{D}(\varphi(t))$ p.p. in I. For any $t \geq 0$ the set of points in M attainable at time t by solutions of $\mathcal{D}$ which initiate from p at time 0 is denoted by $A(t,p,\mathcal{D})$.

If $q \in M$ and there is a solution of $\mathcal{D}$ such that $\varphi(0) = p$ and $\varphi(t) = q$ we say p can be controlled to q in time t by $\mathcal{D}$. Let $X \in \mathcal{D}$ and $T^X(\cdot)p$ denote the solution of $\dot{x} = X(x)$ $x(0) = p$. If $T^X(t)p \in \operatorname{int} A(t,p,\mathcal{D})$ for all $t > 0$ then given any $t_1 > 0$ there exists some nbd. N of $T^X(t_1)p = p^1$ such that every point in N can be controlled to $p = T^{-X}(t_1)p^1$ in time t_1 by $-\mathcal{D} = \{-X^\alpha : \alpha \in U\}$. This means the system $-\mathcal{D}$ is "locally controllable" along the reference solution $T^{-X}(\cdot)p$ at p. If the reference vector field X satisfies $X(p) = 0$ then $T^X(t)p \equiv p$ and $p \in \operatorname{int} A(t,p,\mathcal{D})$ for all $t > 0$ means there is "controlled stability" for the rest solution p of the system $-\mathcal{D}$.

The usual description of a control system is $\dot{x} = f(x,u)$ with u an admissible control if measurable with values $u(t) \in U \subset R^m$. For notational convenience assume $0 \in U$ and $f(x,0) \equiv X(x)$ is the reference vector field. If we write $Y^\alpha(x) = f(x,\alpha) - f(x,0)$ for $\alpha \in U$, then the study of the above control system is equivalent to the study of the system of vector fields

(1) $$\mathcal{D} = \{X+Y^\alpha : \alpha \in U\}.$$

We consider the set of all analytic vector fields, V(M), on M as a real Lie algebra with product the Lie product [X,Y]. Notationally, $(\mathrm{ad}\, X,Y) = [X,Y]$ and inductively $(\mathrm{ad}^{k+1} X,Y) = [X,(\mathrm{ad}^k\, X,Y)]$. For any subset $\mathcal{C} \subset V(M)$, $L(\mathcal{C})$ denotes the Lie algebra generated by $\mathcal{C}$. Define

(2) $$S^1 = \{(\mathrm{ad}^j\, X,Y^\alpha) : \alpha \in U,\ j = 0,1,\ldots\}.$$

For a system $\mathcal{D}$ in the form given in (1), an easy modification of theorem 3.2 in [1] yields

1. Proposition. Let $\mathcal{D}$ be as given in (1). A necessary and sufficient condition that int $A(t,p,\mathcal{D}) \neq \phi$ for $t > 0$ is that $\dim L(S^1)(p) = n$.

From relatively standard linear, or first order, theory we have

2. Proposition. Let co. denote the convex hull of a set. For $\mathcal{D}$ as in (1) assume that $0 \in$ int. co.$\{Y^\alpha(x):\alpha\in U\}$ for x near p, where int. denotes interior relative to span$\{Y^\alpha(x):\alpha\in U\}$. Then a sufficient condition that $T^X(t)p \in$ int.cl.$A(t,p,\mathcal{D})$ for all $t > 0$ is that dim. span $S^1(p) = n$.

In order to simplify the notation, statement of hypotheses and results, we restrict attention to less general systems than given in (1). Specifically let

(3) $\mathcal{E}^K = \{X+\sum_{i=2}^m \alpha_i Y^i: |\alpha_i| \le K, i = 2,\dots,m \le n\}$

where X, Y^i are analytic vector fields in M and $K > 0$. If the control values α_i are merely in $\mathbb{R}^1$ (i.e., as a control system, controls have components in the closed unit ball in the space of regular countably additive measures) the system will be designated $\mathcal{E}^\infty$. For a system $\mathcal{E}^K$, or $\mathcal{E}^\infty$, the set analagous to S^1 as given in (2) will again be designated S^1 and

(4) $S^1 = \{(\mathrm{ad}^j X, Y^i): j \ge 0, i = 2,\dots,m\}$.

Propositions 1 and 2 show that a necessary and sufficient condition that $\mathrm{int}.A(t,p,\mathcal{E}^K) \ne \phi$ for all $t > 0$ is $\dim L(S^1)(p) = n$ while a sufficient condition that $T^X(t)p \in \mathrm{int}.A(t,p,\mathcal{E}^K)$ for all $t > 0$ is that $\dim \mathrm{span}\, S^1(p) = n$.

For the system $\mathcal{E}^\infty$ the influence of the vector fields $Y^2, \dots, Y^m$ can "override" the influence of X. An argument similar to that given in [2, lemma 6.4] or (3, lemma 3.4] shows

3. <u>Proposition</u>. $\dim L\{Y^2,\dots,Y^m\}(p) = n$ is a sufficient condition that $T^X(t)p \in \mathrm{int}\, A(t,p,\mathcal{E}^\infty)$ for all $t > 0$.

For the C^k topology of vector fields on M, $\dim L\{Y^2,\dots,Y^m\}(p) = n$ is a generic condition when $m \ge 3$. Example 1.1 of [4] shows that proposition 3 is not valid for $\mathcal{E}^K$ with $K \ne \infty$.

Two approaches have been developed to study the problem of determining if $T^X(t)p \in \mathrm{int}\, A(t,p,\mathcal{E}^K)$ for all $t > 0$. The first is "geometric" and depends on $L\{Y^2,\dots Y^m\}$ defining an involutive distribution of dimension less than n (or in the case $X(p) = 0$ that $\dim L\{Y^2,\dots,Y^m\}(p) < n$). This approach will be discussed in section 1; further results on the geometric approach may be found in [4], [6], [7] and [8].

The second approach, which is basically a high order Pontriagin maximum principle, is to analytically try to compute the tangent space to $A(t_1,p,E^K)$ at $T^X(t_1)p = p^1$. Basically, if $\varepsilon > 0$ and $\gamma: [0,\varepsilon) \to A(t,p,E^K)$ with $\gamma(0) = p^1$ then $\dot{\gamma}(0)$ is a tangent vector to $A(t_1,p,E^K)$ at p^1. If the set of all such tangent vectors fills $\mathbb{R}^n$, then $p^1 = T^X(t_1)p \in \text{int } A(t_1,p,E^K)$. We shall briefly illustrate the results of some computations of this type in section 2. Related work on this approach may be found in [9], [10], [11].

1. THE GEOMETRIC APPROACH.

Suppose, now, that $X(p) = 0$ and $\dim L\{Y^2,\ldots,Y^m\}(p) = n - 1$. By theorem of Nagano, [5], $L\{Y^2,\ldots,Y^m\}$ has an $(n-1)$ dimensional integral manifold through p which we denote $N^{n-1}(p)$. If X is tangent to, or always "points to one side" of, $N^{n-1}(p)$ intuitively it is expected that $A(t,p,E^K)$ would, respectively, have empty interior or lie on one side of $N^{n-1}(p)$. Thus $T^X(t)p \equiv p$ would be on the boundary of $A(t,p,E^K)$. Analytically this leads to the following. Let $V^2,\ldots,V^n \in L\{Y^2,\ldots,Y^m\}$ be linearly independent at p and Z a non-zero one-form such that $\langle Z(p),V^i(p)\rangle = 0$, $i = 2,\ldots,n$. For $\nu = (\nu_2,\ldots,\nu_n)$, with ν_i a nonnegative integer, $\nu! = \nu_2!\ldots\nu_n!$ and $|\nu| = \sum_2^n \nu_i$ define

$$a(\nu) = \langle Z(p),(\text{ad}^{\nu_n} V^n,(\ldots,(\text{ad}^{\nu_2} V^2,X)..)(p)\rangle$$

$$\varphi_r(s) = \sum_{|\nu|=r} (1/\nu!)(-s_2)^{\nu_2}\ldots(-s_n)^{\nu_n} a(\nu).$$

4. Proposition. Assume that $X(p) = 0$ and $\dim L\{Y^2,\ldots,Y^m\}(p) = n - 1$. A necessary and sufficient condition that $A(t,p,E^K)$ or $A(t,p,E^\infty)$ have nonempty interior for all $t > 0$ is that there exists an integer $r \geq 1$ such that $\varphi_r(s) \not\equiv 0$ for s in some nbd. of $0 \in \mathbb{R}^{n-1}$. Let r* be the

smallest such integer (assuming such an r exists).

(i) If $\varphi_{r*}(s)$ changes sign in every nbd. of $0 \in \mathbb{R}^{n-1}$ then $T^X(t)p \in \text{int } A(t,p,E^\infty)$ for all $t > 0$. Since for odd r, $\varphi_r(-s) = -\varphi_r(s)$, if r* is odd this will be the case.)

(ii) If $\varphi_{r*}(s)$ is definite in some nbd. of $0 \in \mathbb{R}^{n-1}$, $T^X(t)p \in \partial A(t,p,E^K)$, $T^X(t)p \in \partial A(t,p,E^\infty)$, for small $t > 0$.

(iii) If $\varphi_{r*}(s)$ is semi-definite, then $\sum_{r=1}^{\infty} \varphi_r(s)$ changing sign in every nbd. of $0 \in \mathbb{R}^{n-1}$ is a necessary and sufficient condition that $T^X(t)p \in \text{int } A(t,p,E^\infty)\ \forall\ t > 0$.

(iv) If $m = n$ and $Y^2,\ldots,Y^n$ are involutive and linearly independent at p statements (i) and (iii) hold with E^∞ replaced by E^k.

1. Remark. If $r^* = 1$ this is a first order test; $r^* > 1$ gives a higher order test.

2. Remark. In dimension $n = 2$ the hypotheses are minimal and the result conclusive. Indeed, if $Y^2(p) = 0$ the initial point is a rest solution for any control. If $Y^2(p) \neq 0$, (iv) holds and if r* exists either $\varphi_{r*}(s)$ changes sign in every nbd. of $0 \in \mathbb{R}^1$ (implying $T^X(t)p \in \text{int } A(t,p,E^K)$) or $\varphi_{r*}(s)$ is definite is some nbd. (implying $T^X(t)p \in \partial A(t,p,E^K)$ for small $t > 0$). Examples, worked in detail, for the case $n = 2$ and a case $n = 3$ with (iv) holding can be found in [6].

1. Example. Let $M = \mathbb{R}^3$, $X(x) = (0,0,x_3)$, $Y^2(x) = (x_3^2,0,1)$, $Y^3(x) = (0,x_3,0)$, $p = 0$.

Computing shows $[Y^2,Y^3](x) = (0,-1,0)$ while other elements of $L\{Y^2,Y^3\}$ vanish. Thus $\dim L\{Y^2,Y^3\}(p) = 2$ and $V^2 = Y^2 \in L\{Y^2,Y^3\}$, $V^3 = [Y^2,Y^3] \in L\{Y^2,Y^3\}$ are linearly independent at p. Next, $(\mathrm{ad}^j X, V^2)(x) = ((-2)^j x_3^2, 0, 1)$, $j \geq 0$; and $(\mathrm{ad}^j X, Y^3)(p) = 0$ for $j \geq 0$ hence dim span $S^1(p) = 1$ and the linear test fails. Choose $Z(p) = (1,0,0)$ so $\langle Z(p), V^i(p)\rangle = 0$, $i = 2,3$. Also, $(\mathrm{ad}\, X, V^3)(x) = 0$, hence $a(1,0) = \langle Z(p), (\mathrm{ad}\, V^2, X)(p)\rangle = 0$, $a(0,1) = \langle Z(p), (\mathrm{ad}\, V^3, X)(p)\rangle = 0$ and $\varphi_1(s) \equiv 0$. Next, $(\mathrm{ad}^2 V^2 X) = (4x_3, 0, 0) \Rightarrow a(2,0) = 0$; $(\mathrm{ad}\, V^3, (\mathrm{ad}\, V^2, X)) = 0 \Rightarrow a(1,1) = 0$ and $(\mathrm{ad}^2 V^3 X) = 0 \Rightarrow a(0,2) = 0$ so $\varphi_2(s) \equiv 0$. Finally, $(\mathrm{ad}^3 V^2, X)(x) = (-4,0,0) \Rightarrow a(3,0) = -4$, $(\mathrm{ad}\, V^3, (\mathrm{ad}^2\, V^2, X)) = (\mathrm{ad}^2 V^3, (\mathrm{ad}\, V^2, X)) = (\mathrm{ad}^3 V^3, X) = 0$ showing $a(2,1) = a(1,2) = a(0,3) = 0$ and $r^* = 3$ with $\varphi_3(s) = 4s_2^3$. From proposition 4, (i), we conclude $T^X(t)p \equiv p \in \mathrm{int}\, A(t,p,E^\infty)$ for all $t > 0$.

2. Example. (This example is to show that one cannot, in general, replace E^∞ by E^K in proposition 4 (i).)

Let $M = \mathbb{R}^4$, $p = 0$, $X(x) = (x_2, 0, 0, x_3^2)$, $Y^2(x) = (0,1,0,x_3^2)$ and $Y^3(x) = (0,0,1,0)$. If we consider the system $\dot{x} = X(x) + u_2(t)Y^2(x) + u_3(t)Y^3(x)$, $x(0) = p$ with $|u_i(t)| \leq 1$, $i = 2,3$ we see $\dot{x}_4(t) = x_3^2(1+u_2(t)) \geq 0$ therefore $T^X(t)p \equiv p \in \partial A(t,p,E^1)$ for all $t \geq 0$.

Computing, $[Y^2,Y^3](x) = (0,0,0,2x_3)$; $[[Y^2,Y^3],Y^3](x) = (0,0,0,2)$; higher order products of Y^2 and Y^3 vanish; $\dim L\{Y^2,Y^3\}(p) = 3$. We choose $V^2 = Y^2$, $V^3 = Y^3$. $V^4 = [[Y^2,Y^3],Y^3]$.

Next, $(\mathrm{ad}\, X, Y^2)(x) = (1,0,0,0)$, $(\mathrm{ad}^j\, X, Y^2) = 0$ if $j \geq 2$; $(\mathrm{ad}\, X, Y^3)(x) = (0,0,0,2x_3)$, $(\mathrm{ad}^j\, X, Y^3) = 0$ if $j \geq 2$. Thus dim span $S^1(p) = 3$ and the linear test fails to give information.

By inspection of V^2, V^3, V^4 we see to choose $Z(p) = (1,0,0,0)$. Then $a(1,0,0) = \langle Z(p), [V^2,X](p)\rangle = -1$; $a(0,1,0) = \langle Z(p), [V^3,X](p)\rangle = 0$; $a(0,0,1) = \langle Z(p), [V^4,X](p)\rangle = 0$ so

$\varphi_1(s) = s_2$; $r^* = 1$ which is odd so φ_1 changes sign in every nbd. of $0 \in \mathbb{R}^3$ yet $T^X(t)p \equiv p \in \partial A(t,p,E^1)\ \forall\ t \geq 0$. On the other hand, proposition 4 (i) shows that $p \in \text{int}\ A(t,p,E^\infty)$. Here it is the bound, K, on the control values which determines whether p is on the boundary, or interior, of $A(t,p,E^K)$.

2. CALCULATION OF TANGENT CONES TO ATTAINABLE SETS.

We shall again consider the system E^K as given by (3) and assume $X(p) = 0$.

Let $t_1 > 0$ be given and $\tau, s \geq 0$, $0 \leq \tau + s \leq t_1$. Then $T^X(\tau) \circ T^{X+\alpha_i Y^i}(s) \circ T^X(t_1-\tau-s)p \in A(t_1,p,E^K)$ if $|\alpha_i| \leq K$. But $T^X(t_1-\tau-s)p = T^X(-\tau)p = p$ thus if we define $q(s) = T^X(\tau) \circ T^{X+\alpha_i Y^i}(s) \circ T^X(-\tau)p$ then, for some $\varepsilon > 0$, $q\colon [0,\varepsilon) \to A(t_1,p,E^K)$ and $q(0) = p$. The one sided derivative $\lim_{s^+ \to 0} \frac{d}{ds} q(s) = \alpha_i \sum_{\nu=0}^{\infty} \tau^\nu/\nu!\ (\text{ad}^\nu X, Y^i)(p)$ is a tangent vector to $A(t_1,p,E^K)$ at p. One may show, [7, prop. 1.2], by extending this technique, that the vectors $(\text{ad}^\nu X,Y^i)(p)$ and their negatives, are tangent to $A(t_1,p,E^K)$ at p. Also any vector in the span of $\{(\text{ad}^\nu X,Y^i)(p)\colon i = 2,\ldots,m;\ \nu \geq 0\}$ is a tangent vector. This yields the previously mentioned first order condition: a sufficient condition that $p \in \text{int}\ A(t,p,E^K)$ for all $t > 0$ is $\dim \text{span}\ S^1(p) = n$, with S^1 as in (4).

If $q'(0) = \lim_{s^+\to 0} d/ds\ q(s) = 0$ $q''(0) = \lim_{s^+\to 0} d^2/ds^2\ q(s)$ is an admissible tangent vector to $A(t,p,E^K)$ at p. In this way (and other ways) one can attempt to compute "higher order" tangent vectors and thereby derive higher order tests. Second order tests can be found in [9], [10] and [11]. It is customary, for a system such as E^K, to call a test second order (or quadratic) if it includes examining elements of $L(S^1)$ which are products of a pair of elements of S^1, i.e., elements of the form $[(ad^j X,Y^i), (ad^k X,Y^\nu)]$. We shall next outline the derivation of a special high order test and illustrate the type of results and difficulties which can occur.

We consider the system E^K as given in (3) with the assumption that $X(p) = 0$. Let $\mu_2,\ldots,\mu_m$ be reals with $|\mu_i| \le K$, $i = 2,\ldots,m$, and such that $\sum_{i=2}^m \mu_i Y^i(p) = 0$. For example, if some $Y^j(p) = 0$ we could choose $\mu_j = K$ and all remaining $\mu_i = 0$ In any case, one could choose all $\mu_i = 0$ which will make the test, to be derived, the linear test. With this assumption, for any $t_1 > 0$ and $s,\tau \ge 0$, $0 \le s + \tau \le t_1$, $j = 2,\ldots,m$.

(5) $T^{X+\Sigma\mu_i Y^i}(\tau) \circ T^{X+KY^j}(s) \circ T^{X+\Sigma\mu_i Y^i}(-\tau)p \in A(t_1,p,E^K)$.

The one sided derivative (i.e. for $s \ge 0$) of the left side of (5) with respect to s, at $s = 0$, yields the tangent vector Z to $A(t_1,p,E^K)$ at p with

(6) $$Z = \sum_{\nu=0}^{\infty} \tau^\nu/\nu! \quad (ad^\nu(X+\sum\mu_i Y^i),\ X + KY^j)(p).$$

Lemma. Let X, W, V be analytic vector fields on M, and for each integer j, $\nu = (\nu_1,\ldots,\nu_{2^j-1})$ $\gamma = (\gamma_1,\ldots,\gamma_{2^j-1})$ be multi-indices with each ν_i,γ_i a nonnegative integer; $|\nu| = \Sigma\nu_i$, $|\gamma| = \Sigma\gamma_i$, $\gamma_1 \ge 0$, γ_{i+1}, ν_i nonzero for $i \ge 1$ if the sum of the preceding indices is less than j and zero if this sum equals j. Then $(ad^j(X+W),V) = \sum_{|\nu|+|\gamma|=j} (ad^{\nu_{2^j-1}} X,(ad^{\gamma_{2^j-1}} W, (ad^{\nu_{2^j-1}-1}X, (\ldots(ad^{\nu_1}X,(ad^{\gamma_1}W,V)\ldots)$.

Proof. Easy induction on j.

Notation. For a given integer $j \ge 0$ and ν,γ as above, let

(7) $L(j,\nu,\gamma,X,W)V = (ad^{\nu_{2^j-1}} X,(\ldots(ad^{\nu_1}X,(ad^{\gamma_1}W,V)\ldots)$.

For ease of presentation, consider the case $m = 3$ with $Y^2(p) = 0$ so that we may replace $X + \Sigma\mu_i Y^i$ by $X + \mu Y^2$. For this case, using the above notation, lemma, and the observation that $X(p) = Y^2(p) = 0$ implies $(L(j,\nu,\gamma,X,Y^2)X)(p) = 0$

(8) $$Z = K\sum_{j=0}^{\infty} \tau^j/j! \left(\sum_{k=0}^{j} \mu^k \sum_{|\gamma|=k,|\nu|=j-k} (L(j,\nu,\gamma,X,Y^2)Y^3)(p)\right)$$
$$= K\sum_{k=0}^{\infty} \mu^k \sum_{j=k}^{\infty} \tau^j/j! \sum_{|\gamma|=k,|\nu|=j-k} (L(j,\nu,\gamma,X,Y^2)Y^3)(p).$$

1. Remark. The coefficient of $(\mu)^0$ is $K \sum_{j=0}^{\infty} \tau^j/j!$ $(ad^j X,Y^3)(p)$. The coefficient of $(\mu)^1$ is
$$K \sum_{j=1}^{\infty}\tau^j/j! \left\{ \sum_{\gamma=0,\dots,j-1} (ad^{j-\gamma-1}X,[Y^2,(ad^{\gamma}X,Y^3)](p)\right\},$$
etc.

2. Remark. Since only one factor Y^3 appears in each term defining the tangent vector Z, replacing Y^3 by $-Y^3$ yields the conclusion that -Z is also a tangent vector to $A(t_1,p,E^K)$ at p.

Define

(9) $$V^{j,k} = K \sum_{\substack{|\gamma|=k \\ |\nu|=j-k}} L(j,\nu,\gamma,X,Y^2)Y^3$$

$$Q^0 = \{V^{j,0}: j = 0,1,\dots\} = \{(ad^j X,Y^3): j \geq 0\}$$
$$Q^1 = \{V^{j,1}: j = 1,2,\dots\}, \text{ etc.}$$

In a fashion somewhat similar to the proof of proposition 1.2 in [7], one may show any element in span $\cup_{i=0}^{m} Q^i(p)$ can be obtained as a tangent vector to $A(t_1,p,E^K)$. This is a tedious and difficult verification, but yields

5. Proposition. Consider the n-dimensional, analytic, system

(E^K) $\dot{x} = X(x) + u_2(t)Y^2(x) + u_3(t)Y^3(x)$, $x(0) = p$

with $|u_i(t)| \leq K$, $i = 1,2$ and $X(p) = Y^2(p) = 0$. With $V^{j,k}$ as in (9) and Q^i as above, a sufficient condition that $T^X(t)p = p \in \text{int } A(t,p,E^K)$ for all $t > 0$ is that for some integer m, dim span $(\cup_{i=0}^{m} Q^i(p)) = n$.

3. Example. Let $M = \mathbb{R}^3$, $X(x) = (x_2,0,0)$, $Y^2(x) = (0,x_3,x_1^2)$, $Y^3(x) = (0,0,1)$, $p = 0$, $K = 1$. Since both $X(p)$ and $Y^2(p) = 0$, $(ad^j X,Y^2)(p) = 0$ for $j \geq 0$. Also, $[X,Y^3] = 0$ hence rank

$S^1(p) = \text{rank } Q^0(p) = 1$. Next, $V^{1,1}(p) = (L(1,0,1,X,Y^2)Y^3)(p) = [Y^2,Y^3](p) = (0,1,0)$; $V^{2,1}(p) = [Y^2,[X,Y^3]](p)+[X,[Y^2,Y^3]](p) = (1,0,0)$. Thus dim span $(Q^0(p) \cup Q^1(p)) = 3$ and $p \in \text{int } A(t,p,E^1)$ for all $t > 0$. Note that in this example, dim $L\{Y^2,Y^3\}(p) = 2$; the integral manifold of $L\{Y^2,Y^3\}$ thru p is the plane $x_1 = 0$ while X(x) "points to both sides" of this integral manifold for x in every nbd. of zero on this manifold. Proposition 4 (i) will show, for this example, that $p \in \text{int } A(t,p,E^\infty)$ for all $t > 0$. Our result here is sharper, i.e., $p \in \text{int } A(t,p,E^1)$ for all $t > 0$.

Concluding remarks. The test given in proposition 5 can easily be extended to the case $\sum_{i=2}^m \mu_i Y^i(p) = 0$, rather than for the simplifying assumption $Y^2(p) = 0$. The case $Y^2(p) = 0$, with $m > 3$, is a trivial extension of proposition 5. In any case, the result is far from necessary since, for example, terms such as $[Y^3,[Y^2,Y^3]](p)$ do not occur in the test given in proposition 5. A general necessary and sufficient condition that $T^X(t)p \in \text{int } A(t,p,E^K)$ with E^K as in (3) remains an intriguing and difficult problem.

Proofs and extensions of the results stated here will appear elsewhere.

REFERENCES

1. Sussman, H. J. and Jurdjevic, V., Controllability of nonlinear systems, *J. Diff. Eqs.* 12(1972), 95-116.

2. Sussman, H. J. and Jurdjevic, V., Control systems on lie groups, *J. Diff. Eqs.* 12(1972), 313-329.

3. Hirschorn, R. M., Global controllability of nonlinear systems, (to appear *SIAM J. Control*).

4. Hermes, H., Local controllability and sufficient conditions in singular problems, II, (to appear *SIAM J. Control*).

5. Nagano, T., Linear differential systems with singularities and an application of transitive lie algebras, *J. Math. Soc. Japan 18*(1966), 398-404.

6. Hermes, H., High order algebraic conditions for controllability, *Proceedings Algebraic Methods in Systems Theory*, (Udine Italy), 1975.

7. Hermes, H., Local controllability and sufficient conditions in singular problems (I), *J. Diff. Eqs. 20*(1976), 213-232.

8. Hermes, H., *On Local Controllability*, Banach Center Publications, Warsaw, Poland, Vol. 1 (1976), 103-106.

9. Krener, A., The high order maximal principle and its application to singular extremals, (to appear *SIAM J. Control*).

10. Krener, A., The high order maximal principle, *Geometric Methods in Systems Theory*, Mayne-Brockett Eds., D. Reildell Pub. Co., Dordrecht, Holland (1973), 174-184.

11. Knobloch, H., Higher order approximations of attainable sets, (preprint).

---

*This research was supported by NSF grant MSP 71-02649.

ON THE BEHAVIOR OF LINEAR UNDAMPED ELASTIC SYSTEMS PERTURBED BY FOLLOWER FORCES

E. F. Infante*
Lefschetz Center for Dynamical Systems
Division of Applied Mathematics
Brown University

and

J. A. Walker**
Department of Mechanical Engineering
and Astronautical Sciences
Northwestern University

I. INTRODUCTION

A number of problems that arise in the theory of linear elastic structures [1] can be exemplified by the following partial differential equation and boundary conditions:

$$\frac{\partial^2 u}{\partial t^2} + \frac{\partial^4 u}{\partial \xi^4} + p\,\frac{\partial^2 u}{\partial \xi^2} = 0, \; t > 0, \; 0 < \xi < 1, \tag{1}$$
$$u(0,t) = u_\xi(0,t) = u_{\xi\xi}(1,t) = u_{\xi\xi\xi}(1,t) = 0, \; t \geq 0,$$

with arbitrary intitial conditions preassigned for $u(\xi,0)$ and $u_t(\xi,0)$, $\xi \in (0,1)$. After appropriate normalizations, $u(\xi,t)$ represents the displacement at a point ξ and time t of a slender elastic rod imbedded in a rigid medium at $\xi = 0$ and subjected, at $\xi = 1$, to a compressive force of magnitude p always tangential to the rod.

This particular problem known as Beck's Problem [1], is one of many nonconservative problems encountered in the theory of elastic structures; the problem is called nonconservative in that the total energy is not necessarily constant given the follower nature of the load.

In the more familiar Euler problem [1], in which the load p is always in the vertical direction (and is represented by the above partial differential equation but with boundary conditions $u(0,t) = u_{\xi\xi}(0,t) = u(1,t) = u_{\xi\xi}(1,t) = 0$) energy is conserved; as is well known, a bifurcation of static solutions occurs at a critical value of p, the so-called Euler static buckling load. No such phenomenon arises in the case of Beck's problem; bifurcation of static solutions does not occur. Rather, as the load p is increased, a critical value is surpassed and certain oscillatory solutions no longer remain bounded; i.e., dynamic instability occurs.

This phenomenon has been investigated in the engineering literature [1] through the use of separation of variables and subsequent analysis of the eigenvalues corresponding to the first few eigenvectors. In this manner, a critical value of the load p is determined below which the motions described by these eigenvectors are bounded, whereas for p above this value at least one such motion is unbounded. For Beck's problem, the critical value of p thus obtained is approximately 20.05. Since, in the stable case, all eigenvalues are purely imaginary, such an analysis is not necessarily conclusive [2,4,8].

The purpose of this note is to outline another approach to the study of the stability of these problems, namely a Liapunov approach. For this purpose, motivated by (1), we wish to consider a generalization of this problem.

Let H be a separable, complex Hilbert space with inner produce $\langle , \rangle$, and the linear operator $K: (\mathcal{D}(K) \subset H) \to H$; consider the class of linear evolution equations of the form

$$\ddot{y}(t) + Ky(t) = 0, \quad t \geq 0,$$
$$y(0) = \phi \tag{2}$$
$$\dot{y}(0) = \psi.$$

The relationship between this equation and (1) is clear; abusing notation, we may take $Ky(t) = \dfrac{\partial^4 y(\xi,t)}{\partial\xi^4} + p\,\dfrac{\partial^2 y(\xi,t)}{\partial\xi^2}$, $\mathcal{H} = L_2(0,1)$ and $\mathcal{D}(K) = \{y \in W_2^4(0,1) \mid y(0) = \dfrac{\partial y(0)}{\partial\xi} = \dfrac{\partial^2 y(1)}{\partial\xi^2} = \dfrac{\partial^3 u(1)}{\partial\xi^3} = 0\}$. Recalling certain characteristics of more general elastic systems, we shall henceforth make the following assumptions about the operator K.

Assumptions:

i) the point spectrum $\Delta(K)$ of K is symmetric about the real axis;

ii) the operator K can be written as $K = A + B$ with $\mathcal{D}(K) = \mathcal{D}(A) \subset \mathcal{D}(B) \subset \mathcal{H}$, [in the case of (1), let $Ay = \dfrac{\partial^4 y}{\partial\xi^4}$, $By = p\,\dfrac{\partial^2 y}{\partial\xi^2}$];

iii) A is a positive selfadjoint operator with compact resolvent $R(\zeta,A)$ for some complex number ζ, with $0 < \lambda_1 < \lambda_2 < \ldots$, each eigenvalue λ_n being of (algebraic and geometric) multiplicity one, $n = 1,2,\ldots$;

iv) there exist nonnegative constants c_1, c_2 such that $||By||^2 \leq c_1 ||y||^2 + c_2 \langle y, ay \rangle$ for every $y \in \mathcal{D}(A)$ [in the case of (1), we may take $c_1 = 0$, $c_2 = p^2$].

The object of this investigation is to determine, for (2) and under Assumptions i)-iv), conditions that insure the well-posedness of the problem and that are sufficient for stability, in the sense of Liapunov, of the equilibrium. More precisely, we wish to view the operator B as perturbing the selfadjoint operator A and we wish to determine conditions, dependent only on a knowledge of the eigenvalues of A and on the measure of the perturbation B given by the constants c_1 and c_2 of Assumption iv), that insure stability of the equilibrium in the sense of Liapunov.

In this brief summary, the proofs of the stated propositions and theorems are omitted because of their length; they will appear in a future paper.

II. WELL-POSEDNESS

Given the operator $K = A + B$, with A selfadjoint and positive, let $A^{\frac{1}{2}}$ be the unique positive selfadjoint square root of A [5]. Define a second Hilbert space $\chi = \mathcal{D}(A^{\frac{1}{2}}) \times H$ with inner product

$$\langle x_1, x_2 \rangle_\chi = \langle A^{\frac{1}{2}} y_1, A^{\frac{1}{2}} y_2 \rangle + \langle z_1, z_2 \rangle, \quad x_i = (y_i, z_i) \in \chi, \quad i = 1,2,$$

and note that, for $x_i = (y_i, z_i) \in \mathcal{D}(A) \times H \subset \chi$, $i = 1,2$, we have

$$\langle x_1, x_2 \rangle_\chi = \langle y_1, Ay_2 \rangle + \langle z_1, z_2 \rangle.$$

We may now view (2) in this new Hilbert space as

$$\begin{aligned} \dot{x}(t) &= Fx(t), \quad t \geq 0, \\ x(0) &= \varphi \in \mathcal{D}(F), \end{aligned} \tag{3}$$

where $F:(\mathcal{D}(F) \subset \chi) \to \chi$ is the linear operator defined on $\mathcal{D}(F) = \mathcal{D}(A) \times \mathcal{D}(A^{\frac{1}{2}})$ by $Fx = (z, -Ky)$ for $x = (y,z) \in \mathcal{D}(F)$.

In this particular format, it is possible to apply the ideas of [6] to show that, under our assumptions, the following result holds.

1. Proposition. F is the infinitesimal generator of a linear C_0-group $\{S(t)\}_{-\infty < t < \infty}$ on χ.

This proposition disposes of the question of well-posedness, guaranteeing that a unique strong solution of (3), hence of (2), exists and is defined for all time. We now turn to the more difficult question of stability.

III. STABILITY

If the perturbing operator B were to vanish, K = A would be selfadjoint; in this case, $\langle x,Fx\rangle_{\chi} = 0$ for every $x \in \mathcal{D}(F)$, from which it follows that $||S(t)||_{(\chi,\chi)} = 1$ for every t; then, the equilibrium x = 0 of (3) is stable. If $B \neq 0$ and not symmetric, the question of stability is much more difficlut to resolve.

The Hille-Phillips-Yoshida Theorem implies that a necessary condition for the stability of the equilibrium of (3) is that there be no point with positive real part in $\sigma(F)$, the spectrum of F. Since $(\zeta I-F)^{-1}$ exists if and only if $(-\zeta^2 I-K)^{-1}$ exists, and the point spectrum $\Delta(K)$ is assumed symmetric about the real axis, it follows that $\Delta(F)$ is polar symmetric. But then, for stability, all elements of $\Delta(F)$ must be purely imaginary, which implies that $\Delta(K)$ must consist solely of nonegative real numbers. We wish to determine conditions sufficient for $\Delta(K)$ to consist solely of nonnegative real numbers, conditions dependent only on the knowledge of $\{\lambda_n\}$, n = 1,2,..., and of the constants c_1, c_2 of Assumption iv).

For this purpose, define the positive quantities

$$r_1 = \frac{\lambda_2-\lambda_1}{2}, \quad r_n = \min\{\frac{\lambda_{n+1}-\lambda_n}{2}, \frac{\lambda_n-\lambda_{n-1}}{2}\}, \quad n = 2,3,...,$$
$$\alpha_n = \frac{r_n}{\sqrt{c_1+c_2(\lambda_n+2r_n)}}, \quad n = 1,2,... \tag{4}$$

Through the use of these quantities, it is possible to prove the following simple result.

2. Proposition. If $\alpha_n > 1$, $n = 1,2,\ldots$, then $K = A + B$ is a closed operator with compact resolvent $R(\zeta,K)$ for $\zeta \notin \sigma(K)$; furthermore, the spectrum $\sigma(K)$ consists solely of the set of eigenvalues $\Delta(K)$, each $\gamma \in \Delta(K)$ is real and of (geometric and algebraic) multiplicity one, and $\Delta(K)$ may be so ordered that $\Delta(K) = \{\gamma_n\}$ with $|\gamma_n - \lambda_n| \le \frac{r_n}{\alpha_n} < r_n$, $n = 1,2,\ldots$.

This proposition is proven following the methods employed in [3] in the study of perturbations of spectral operators with discrete spectra, in particular Theorem XIX. 2.7. It is noted that this result gives a condition sufficient for all eigenvalues γ_n of K to be real. The following transparent but useful proposition gives a condition that guarantees that these eigenvalues are positive.

3. Proposition. If there exists a linear operator $H: (\mathcal{D}(A) \subset \mathcal{H}) \to \mathcal{H}$ such that for every nonzero $y \in \mathcal{D}(A)$

$$0 < \mathrm{Re}\langle Hy,y\rangle, \; 0 < \mathrm{Re}\langle Hy,Ky\rangle,$$

then every real eigenvalue of K is positive.

This result, with $H = A$, yields the useful sufficient condition $\|BA^{-1}\| < 1$ for the positivity of the real eigenvalues; moreover, through Assumption iv), it can be easily seen that this last condition is satisfied if $c_1 + c_2\lambda_1 < \lambda_1^2$.

The last two propositions state conditions sufficient for $\Delta(F)$ to be purely imaginary. However, even purely imaginary $\sigma(F)$ may not be sufficient to guarantee stability under our assumptions [2,4,8]. The following result gives an approach to this difficulty that seems fruitful.

4. Proposition. Suppose there exists a bounded linear selfadjoint operator $G: \mathcal{H} \to \mathcal{H}$ such that $GK: (\mathcal{D}(A) \subset \mathcal{H}) \to \mathcal{H}$ is symmetric and, for some $\mu > 0$

$$\mu\|y\|^2 \le \langle y,Gy\rangle, \quad \forall \quad y \in \mathcal{H},$$

$$\mu\langle y,Ay\rangle \le \langle y,GKy\rangle, \quad \forall \quad y \in \mathcal{D}(A);$$

then the equilibrium $x = 0$ of (3) is stable and the group $\{S(t)\}_{-\infty < t < \infty}$ generated by F is bounded.

This proposition is an application of an idea proposed in [9]. The relationship of this proposition to Liapunov theory is brought to the fore upon noting that the functional $V:\chi \to \mathcal{R}$, defined as the continuous extension to χ of the functional

$$\langle y,GKy\rangle + \langle z,Gz\rangle, \quad x = (y,z) \in \mathcal{D}(F) \subset \chi,$$

is a Liapunov functional for (3) [9].

This result motivates the search for such an operator G, or for the determination of conditions sufficient for the existence of such an operator.

In the finite dimensional case it has been shown [10] that the existence of such a G is both necessary and sufficient for the stability of the equilibrium $x = 0$; moreover, it has been shown that no such G exists unless the eigenvectors of K are complete. But in the finite dimensional case, the eigenvectors of K are complete if and only if each eigenvalue γ_n of K has equal algebraic and geometric multiplicities.

No equally simple completeness result is known in the infinite dimensional case of interest here, although Proposition 2 states conditions sufficient for every γ_n to have (algebraic and geometric) multiplicity one. However, a result along similar lines is possible. For this purpose, given a closed linear operator $L:(\mathcal{D}(L)\subset\mathcal{H}) \to \mathcal{H}$, with spectrum $\sigma(L)$ consisting solely of denumberable isolated eigenvalues

$\{\beta_n\}$, n = 1,2,..., let

$$P_n(L) = \frac{1}{2\pi i}\oint_{\Gamma_n} R(\zeta,L)d\zeta,$$

where Γ_n is a rectifiable simple closed curve which encloses β_n in its interior and has $\sigma(L) - \beta_n$ in its exterior. $P_n(L)$ is the eigenprojection associated with β_n, and $P_n(L)P_m(L) = 0$ for $m \neq n$. It is known that if β_n has equal algebraic and geometric multiplicities, then the range of $P_n(L)$ is spanned by the eigenvectors of L associated with β_n. If L is self-adjoint with compact resolvent, then [5] these eigenprojections are complete (i.e., s-lim $\sum_n P_n(L) = I$) and selfadjoint (i.e., $P_n(L) = P_n^*(L)$) and the family $\{P_n(L)\}$ is said to be orthogonal; in particular, $||I-2\sum_{n\in J} P_n(L)|| = 1$ for every finite index set J of positive intergers.

The operator K = A + B may not be selfadjoint; yet, it is possible to obtain the following result guaranteeing the completeness of the eigenprojections.

5. Proposition. If $\sum_n \frac{1}{\alpha_n^2} < \infty$ and $\alpha_n > 1$, n = 1,2,..., then the set of eigenprojections $\{P_n(K)\}$ is complete. Moreover, there exists a $\nu \geq 1$ such that

$$||I-2\sum_{n\in J} P_n(K)|| \leq \nu \leq 1 + 2\sum_n \frac{1}{\alpha_n(\alpha_n-1)} + 2\left(\sum_n \frac{1}{\alpha_n^2}\right)^{\frac{1}{2}}$$

for every finite index set J of positive integers.

This proposition, whose proof is rather lengthy, gives conditions for the completeness of the $\{P_n(K)\}$ which require only knowledge of the eigenvalues of A and of the constants c_1 and c_2 of Assumption iv); moreover, the quantity $\nu - 1$ is recognized to be a measure of the nonorthogonality of the $\{P_n(K)\}$. These eigenprojections and their adjoints $\{P_n^*(K)\}$

leads to the following characterization of the possible operators G of Proposition 4.

6. Proposition. Let $\sigma(K) = \{\gamma_n\}$, each eigenvalue γ_n real, isolated and of (algebraic and geometric) multiplicity one; assume that the $\{P_n(K)\}$ are complete. If there exists a bounded linear selfadjoint operator $G:H \to H$ such that $GK:(\mathcal{D}(A) \subset H) \to H$ is symmetric, then $G = \text{s-lim} \sum_n \eta_n P_n^*(K) P_n(K) y$, where the $\{\eta_n\}$ are real and bounded.

This result characterizes the operators G to be considered, but does not prove the existence of such an operator with the properties required by Proposition 4. However, it is possible to obtain the following result.

1. Theorem. Let $\sigma(K) = \{\gamma_n\}$, each eigenvalue γ_n real, isolated and of equal algebraic and geometric multiplicities; assume that the eigenprojections $\{P_n(K)\}$ are complete and that there exists a $\nu \geq 1$ such that $||I-2 \sum_{n \in J} P_n(K)|| \leq \nu$ for every finite index set J of positive intergers. Then there exists a bounded linear selfadjoint operator $G:H \to H$ such $GK:(\mathcal{D}(A) \subset H) \to H$ is selfadjoint and for every $y \in H$,

$$\frac{1}{\nu^2}||y||^2 \leq \langle y, Gy \rangle \leq \nu^2 ||y||^2.$$

Furthermore, if $||BA^{-1}|| < 1$, then for every $y \in \mathcal{D}(A)$,

$$\frac{1}{\nu^2}(1-||BA^{-1}||)\langle y, Ay \rangle \leq \langle y, GKy \rangle \leq \nu^2(1+||BA^{-1}||)\langle y, Ay \rangle;$$

moreover, the equilibrium $x = 0$ of (3) is stable and the group $\{S(t)\}_{-\infty < t < \infty}$ generated by F is bounded.

The conditions of this theorem can be made to strictly depend on the eigenvalues of A and on the constants c_1 and c_2 of Assumption iv), by assuming the conditions of Proposition 5 and by imposing that $c_1 + c_2\lambda_1 < \lambda_1^2$, which guarantees that $||BA^{-1}|| < 1$. Furthermore, this theorem and Proposition

5 provide computable estimates for a Liapunov functional $V:\chi \to R$, estimates that may be very useful in the study of further (perhaps nonlinear) perturbations of (2).

Although the proof of this theorem is nonconstructive and involves an application of the Markov-Kakutani fixed point theorem, Proposition 6 implies the existence of an explicit representation for G in terms of the $\{P_n(K)\}$. Upon combining the assumptions of Proposition 5 with those of Theorem 1, it is possible to obtain further information based upon a particular choice of G in terms of the $\{P_n(K)\}$.

2. Theorem. If $\sum_n \frac{1}{\alpha_n^2} < \infty$ and $\alpha_n > 1$ for $n = 1,2,\ldots$, then there exists a bounded linear selfadjoint operator $G:H \to H$ such that $GK:(\mathcal{D}(A)\subset H) \to H$ is selfadjoint, $G - I$ is compact and, defining $\mu = \sum_n \frac{3\alpha_n - 2}{\alpha_n(\alpha_n - 1)^2} + \left(\sum_n \frac{1}{\alpha_n^2}\right)^{\frac{1}{2}}$, G satisfies

$$||G-I|| \le \mu,$$

$$\frac{1}{1+\mu}||y||^2 \le \langle y,Gy\rangle \le (1+\mu)||y||^2, \ \forall \ y \in H.$$

Furthermore, if $||BA^{-1}|| < 1$, then for every $y \in \mathcal{D}(A)$,

$$\frac{1}{1+\mu}(1-||BA^{-1}||)\langle y,Ay\rangle \le \langle y,GKy\rangle \le (1+\mu)(1+||BA^{-1}||)\langle y,Ay\rangle;$$

moreover, the equilibrium of (3) is stable and the group $\{S(t)\}_{-\infty < t < \infty}$ generated by F is bounded.

The assumptions of this theorem imply those of Theorem 1, and therefore its conclusions as well. The additional information is obtained by specifying that the desired operator be of the form $G = \text{s-lim} \sum_n P_n^*(K)P_n(K)$.

IV. BECK'S PROBLEM

For illustrative purposes, let us briefly return to Beck's problem, which we used to motivate our investigation.

With $H = L_2(0,1)$, $Ay = \frac{\partial^4 y(\xi)}{\partial \xi^4}$ and $\mathcal{D}(A) = \{y \epsilon W_2^4(0,1) \mid y(0) = \frac{\partial y(0)}{\partial \xi} = \frac{\partial^2 y(1)}{\partial \xi^2} = \frac{\partial^3 y(1)}{\partial \xi^3} = 0\}$, the eigenvalues of the self-adjoint operator A are computed to be $\lambda_n = \beta_n^4$, where $\cos \beta_n \cosh \beta_n = -1$. Simple computations [7] lead to values of λ_i and r_i given by

$$\lambda_1 \approx (1.875)^4 \quad , \quad r_1 \approx 236 \quad ,$$

$$\lambda_2 \approx (4.694)^4 \quad , \quad r_2 \approx 236 \quad ,$$

$$\lambda_3 \approx (7,855)^4 \quad , \quad r_3 \approx 1,658 \ ,$$

$$\vdots \qquad\qquad\qquad \vdots$$

$$\lambda_n \approx \left(n-\frac{1}{2}\right)^4 \pi^4 \quad , \quad r_n \approx \frac{1}{2}\left[\left(n-\frac{1}{2}\right)^4 - \left(n-\frac{3}{2}\right)^4\right], \text{ for n large,}$$

Since $By = p\frac{\partial^2 y(\xi)}{\partial \xi^2}$, Assumption iv) is satisfied with $c_1 = 0$ and $c_2 = p^2$; from this we have that $\alpha_n = \frac{r_n}{p\sqrt{\lambda_n + 2r_n}}$, from which we obtain

$$\alpha_1 \approx \frac{236}{p\sqrt{485}} \quad ,$$

$$\alpha_2 \approx \frac{236}{p\sqrt{957}} \quad ,$$

$$\alpha_3 \approx \frac{1,657}{p\sqrt{7,115}} \quad ,$$

$$\vdots$$

$$\alpha_n \approx 0(n), \text{ for n large.}$$

It is then clear that $\sum_n \frac{1}{\alpha_n^2} < \infty$; the condition $\alpha_n > 1$ for $n = 1,2,\ldots$ is most restrictive for $n = 2$, which imposes the condition $p < 7.6$. Finally, to insure that $||BA^{-1}|| < 1$, we may insist that $c_1 + c_2\lambda_1 < \lambda_1^2$, which immediately leads to the condition $p < 3.5$.

We therefore conclude, from Theorem 1, that the equilibrium of Beck's problem will be stable if $p < 3.5$. It should be noted that this value is much more conservative than the one obtained by the method of separation of variables.

REFERENCES

1. Bolotin, V. V., *Nonconservative Problems of the Theory of Elastic Stability*, Pergamon Press, New York, 1963.

2. Delfour, M. C., Generalization de resultats de R. Datko sur les fonctions de Lyapunov quadratiques definies sur un espace de Hilbert, (Report Centre de Recherches Mathematiques, CRM-457), Universite de Montreal, January, 1974.

3. Dunford, N. and Schwartz, J. T., *Linear Operators*, Wiley-Interscience, New York, 1971.

4. Hille, E., and Phillips, R. S., Functional Analysis and Semigroups, *American Mathematical Society Colloquium Publication*, *31* (1957), Providence, R.I.

5. Kato, T., *Perturbation Theory for Linear Operators*, Springer-Verlag, New York, 1966.

6. Lumer, G., and Phillips, R. S., Dissipative Operators in a Banach Space, *Pacific J. Math.* *11* (1961), 679-698.

7. Meirovitch, L., *Analytical Methods in Vibrations*, MacMillan Co., 1967.

8. Slemrod, M., Asymptotic behavior of C_0 semigroups as determined by the spectrum of the generator, *Indiana Math. J.*, (Sept. 1976), to appear.

9. Walker, J. A., On the application of Liapunov's direct method to linear dynamical systems, *J. Math. Anal. Applic.* *53* (1976), 187-220.

10. Walker, J. A., On the application of Liapunov's direct method to linear lumped-parameter elastic systems, *ASME J. Appl. Mech.* *41* (1974), 278-284.

---

*This work was supported in part by the Office of Naval Research under Grant NONR N0014-67-A-0009, and in part by the U.S. Army Research Office under AROD DAHC04-75-G-0077.

**This work was supported by the National Science Foundation under GK 40009.

RANDOM OPERATOR EQUATIONS

R. Kannan *
Department of Mathematical Sciences
University of Missouri-St. Louis

1. INTRODUCTION

In this paper we outline some of the recent developments in selected topics in the theory of random operator equations. A random operator equation is an equation of the type $T(\omega,x) = y(\omega)$ where $y(\omega)$ is a known function and $T(\omega)$ is an operator from $\Omega \times X$ into Y where X and Y are Banach spaces and Ω is a measurable space. The chief feature in the theory of random operator equations is that besides the questions of existence, uniqueness etc. there are also the important questions of measurability of the solution and the statistical properties of the random solution.

In this paper we will concern ourselves with only two aspects of the theory of random equations: existence of random solutions and applications to questions of stochastic approximation. There are several other important questions e.g., limit theorems, approximation methods, estimates on the moments of the solution, probability distribution of the solution etc., which we do not discuss here.

The randomness of the solution of a random equation is usually proved as follows: for each fixed ω in Ω we obtain the existence of a solution from the deterministic case and then prove the randomness of the map which associates with each $\omega \in \Omega$, the set of all solutions of the deterministic equation. If however there are iterative techniques to obtain the solution in the deterministic case, then the randomness follows by treating the solution as a limit of a sequence of random variables. In this paper we give two examples of this technique in Section 3. However when there are no iterative techniques available or when the equation itself involves multi-valued random operators one has to extract further information from the deterministic case on the structure of the set of solutions. Situations explaining this technique are given in the following sections where we also outline some basic results from the theory of multi-valued random maps.

A problem of fundamental importance in the theory of random equations concerns the relationship between the expected solution of a random equation and the solution of the deterministic equation obtained by replacing the random quantities in the equation by their expected values. Consider the random operator equation

$$T(\omega)x(t) = y(t,\omega).$$

Let $\tilde{T} = E\{T(\omega)\}$ and $\tilde{y}(t) = E\{y(t,\omega)\}$; and consider the deterministic operator equation

$$\tilde{T}\,\tilde{x}(t) = \tilde{y}(t).$$

The question can now be stated as follows: Is

$$E\{x(t,\omega)\} = \tilde{x}(t)?$$

An interesting application of random fixed point theorems involving contraction mappings is the work of Hanš [17,18,19] to the methods of stochastic approximation. In view of our methods to establish existence of random solutions for random operator equations without resorting to iterative techniques, it would be of interest to generalize the work of Hanš and the Czech school to more general class of nonlinear operators and apply them to nonlinear random differential and integral equations.

Random operators and questions involving random operators have been studied recently by several authors. An excellent introduction to the recent contributions is the book of Bharucha-Reid [5]. Thus, for example, an important class of random operators are random differential operators. There are three different ways in which a random differential equation may arise: i) the coefficients may be random ii) random boundary conditions and iii) the differential operator is deterministic but the nonlinearities are random. An example of i) is a nonlinear differential equation of the form

$$x'' + a(\omega)x' + b(\omega)x = f(x,\omega)$$

where the coefficients $a(\omega)$ and $b(\omega)$ are random variables with a given joint distribution. Stochastic differential equations with random coefficients arise in engineering applications where they describe the influence of fluctuations in system parameters on overall performance. As in the deterministic case the question of random Green's function and related topics arise (Bharucha-Reid [5]).

As an example of ii) we consider

$$x'' + \lambda x = g(x)$$

$$x(0) = 0, \; x'(1) + \xi(\omega)x(1) = 0.$$

Such problems arise, for example, in the study of transverse vibrations of an elastic string. The string is fixed at

$x = 0$. And the support coefficient $\xi(\omega)$ is a nonnegative real valued random variable. For a discussion of the associated eigenvalue problem and related question see Bharucha-Reid [5].

Finally an equation of the type

$$x'' + \text{grad } G(x) = p(t,\omega)$$
$$x(0) = x(2x),\ x'(0) = x'(2x)$$

is an example of iii). This represents the equation of motion of a conservative system subject to random periodic disturbances. Another important class of random equations stem from random initial value problems (possibly involving multivalued nonlinear operators) and random nonlinear Volterra integral equations.

2. PREREQUISITES

Let $(\Omega,\mathcal{B},\mu)$ be a probability space with a probability measure μ; that is, Ω is a nonempty set, $\mathcal{B}$ is the σ-algebra of subsets of Ω and μ is a probability measure. We say that the probability space is complete if $B \in \mathcal{B}$, $\mu(B) = 0$ and $B_0 \subseteq B$ implies that $B_0 \in \mathcal{B}$.

A function g from Ω into a metric space Y is called a Y-valued generalized random variable if the inverse image, under the function g, of each Borel set $B \in \mathcal{B}_Y$ belongs to $\mathcal{B}$ where $\mathcal{B}_Y$ is the σ-algebra generated by closed subsets of Y.

The mapping T from $\Omega \times \Gamma$ into Y, where Γ is an arbitrary set is called a random operator if for each fixed $\gamma \in \Gamma$, the function $T(\cdot,\gamma)$ is a generalized random variable. With these definitions we can obtain the following [17].

2.1. Theorem. If $g_1,g_2,\ldots$ is a sequence of generalized random variables with values in a metric space X converging a.s. to the mapping $g : \Omega \to X$, then g is a generalized random variable.

2.2. Theorem. Let g be a generalized ν.v. with values in a separable Banach space X and let T be an a.s. continuous random operator of the space $\Omega \times X$ into a metric space Z. Then the mapping W of Ω into Z defined by, for every $\omega \in \Omega$, $W(\omega) = T[\omega, g(\omega)]$ is a generalized ν.v. with values in Z.

A random operator $T(\omega) : X \to Y$ is said to be continuous at $x_0 \in X$ if $\lim_{n\to\infty} ||x_n - x_0|| = 0$ implies $\lim_{n\to\infty} ||T(\omega,x_n) - T(\omega,x_0)|| = 0$ almost surely. A random operator T from $\Omega \times X$ into Y, where Ω is a complete probability space, X a separable metric space and Y a metric space, is said to be separable if there exists a countable set $S \subset X$ and a negligible set $N \in \mathcal{B}$, $\mu(N) = 0$, such that

$$\{\omega : T(\omega,x) \in K;\ x \in F \cap S\}\ \Delta\ \{\omega : T(\omega,x) \in K,\ x \in F\} \subset N$$

for every closed set K in $\mathcal{B}_Y$ and every F in $\mathcal{B}_X$.

For a further study of separable random operators, one is referred to ([3], [14], [28]). It is easy to see that the above definition of separability is equivalent to the following: there exists a negligible set $N \in \mathcal{B}$ and a countable set $S \subset X$ such that for $\omega \notin N$ and each $x \in X$ there exists a sequence $\{x_i\} \in S$ such that $x_i \to x$ and $T(\omega,x_i) \to T(\omega,x)$. We can now state the following [3]:

2.3. Theorem. Let X be a separable Banach space and $T(\omega) : \Omega \times X \to X$ be a continuous random operator. Then $T(\omega)$ is separable.

2.4. Theorem. Let $T(\omega) : \Omega \times X \to E$ be a separable random operator, where E is a compact subset of a separable Banach space X. Then $||T(\omega)||$ is a nonnegative real valued random variable.

Let T be a random operator from $\Omega \times X$ into Y. An equation of the type $T[\cdot,x(\cdot)] = y(\cdot)$ where y is a given random variable with values in Y is called a random operator equation. Any mapping $x(\omega) : \Omega \to X$ which satisfies $T(\omega,x(\omega)) = y(\omega)$ a.s. is said to be a wide-sense solution of the above random operator equation. Any X-valued v.v. which satisfies $\mu\{\omega : T(\omega,x(\omega)) = y(\omega)\} = 1$ is said to be a random solution of the above equation. Thus if the wide-sense solution is also measurable it will be a random solution. Clearly there may exist wide-sense solutions that are not random. For a more detailed study of these concepts, see [17].

3. ITERATIVE TECHNIQUES AND RANDOM OPERATOR EQUATIONS

In this section we will discuss two classes of random nonlinear problems where the solution of the corresponding deterministic cases may be obtained by iterative techniques. The availability of these techniques together with Theorem 2.1 would ensure the existence of a random solution. We first state an important tool in the proof of the random contraction mapping theorem [17].

3.1. Theorem. Let T be a continuous random operator on a separable Banach space X to itself such that

$$\mu\left(\bigcup_{m=1}^{\infty}\bigcup_{n=1}^{\infty}\bigcap_{x\in X}\bigcap_{y\in X}\{\omega : ||T^n(\omega,x)-T^n(\omega,y)|| \le (1-\tfrac{1}{m})||x-y||\}\right) = 1$$

where for every $\omega \in \Omega$, $x \in X$ and $n = 1,2,\ldots$ we set $T^1(\omega,x) = T(\omega,x)$ and $T^{n+1}(\omega,x) = T[\omega,T^n(\omega,x)]$. Then there exists a generalized random variable ϕ with values in X satisfying

$$\mu\{\omega : T[\omega,\phi(\omega)] = \phi(\omega)\} = 1.$$

If there exists another generalized random variable ψ with the property $T[\omega,\psi(\omega)] = \psi(\omega)$ with probability one, then $\psi(\omega) = \phi(\omega)$ with probability one.

The above theorem can be proved by using Theorems 2.1 and 2.2. Using Theorem 3.1 Hanš [17] proves the following random version of Banach's contraction mapping theorem.

3.2. Theorem. Let T be a continuous random operator from $\Omega \times X$ into X, X being a separable Banach space and let $k(\omega)$ be a nonnegative real valued random variable such that

$$\mu\{\omega : k(\omega) < 1\} = 1$$

and

$$\mu\{\omega : ||T(\omega,x)-T(\omega,y)|| \le k(\omega)||x-y||\} = 1$$

for every pair of elements $x, y \in X$. Then there exists a generalized random variable ϕ with values in X such that

$$\mu\{\omega : T(\omega,\phi(\omega)) = \phi(\omega)\} = 1.$$

As a direct application of Theorem 3.2 we have the following theorem on the randomness of the inverse of a random operator.

3.3. Theorem. Let T be a random operator on a separable Banach space X such that

$$\mu\{\omega : ||T(\omega,x)-T(\omega,y)|| \le k(\omega)||x-y||\} = 1$$

where $k(\omega)$ is as in Theorem 3.2. Then for every real $\lambda \ne 0$ such that $k(\omega) < |\lambda|$ a.s. there exists a random operator $S(\omega)$ which is the inverse of $T(\omega) - \lambda I$.

Random contraction mapping theorems have been used to establish existence of random solutions of nonlinear integral and differential equations ([5]). Another interesting version of the random contraction mapping theorem may be seen in [16]. An application of Theorem 3.2 may be seen in [20] where they obtain an algorithm for the identification of a random linear discrete-time system described by a random difference equation. In a later section in this paper we will discuss an interesting application of Theorem 3.2 to a stochastic approximation situation.

We now consider another class of random nonlinear problems where the existence of a random solution can be obatined by iterative methods. Thus we now consider random nonlinear integral equations, involving nonlinearities which give rise to monotone operators, of the type

$$u(x,\omega) + \int_{\Sigma} k(x,y,\omega) f(y,u(y),\omega)\,dy = a(x,\omega)$$

where i) Σ is a σ-finite measure space

ii) $k(x,y,\omega)$ is a function from $\Sigma \times \Sigma \times \Omega$ into the space of all linear transformations on R^n

iii) f is a given nonlinear mapping from $\Sigma \times R^n \times \Omega$ into R^n

iv) a is a known function and u is an unknown function in the class of all n-vector functions on $\Sigma \times \Omega$.

Such an equation is referred to as a nonlinear random Hammerstein equation. In order to consider this equation for the existence of random solutions we transform it into an operator equation. Let

i) N be the random Nemytskii operator defined by

$$N(\omega,u)(y) = f(y,u(y),\omega)$$

ii) K be the random linear operator corresponding to the kernel $k(x,y,\omega)$ defined by

$$K(\omega,u)(x) = \int_{\Sigma} k(x,y,\omega)u(y,\omega)dy.$$

The random Hammerstein equation may now be written as

$$u + KNu = a$$

which is a random nonlinear operator equation. In order that the above operator equation makes sense we have to introduce hypotheses on $k(x,y,\omega)$ and the nonlinear function $f(y,u,\omega)$ so that K and N are well-defined random operators and the composition KN is well-defined over a suitable space and maps the space into itself.

Random Hammerstein equations where the nonlinearities are of Lipschitz type have been studied by Bharucha-Reid [5]. Essentially the method consists in treating the operator form of the nonlinear Hammerstein equation as a fixed point of the random operator defined by

$$T_1 u = a - KNu$$

over a suitable Banach space. One then imposes suitable growth hypotheses on K and N so that KN is a contraction map over a suitable closed convex subset into itself.

We now state a result [22] on the existence of a unique random solution of the nonlinear Hammerstein equation where we obtain the measurability by means of an iterative technique.

<u>3.4</u>. <u>Theorem</u>. Let a separable reflexive Banach space X and a real Hilbert space S be such that (X,S,X^*) is in normal position. Let $N\colon X^* \to X$ be a random nonlinear hemicontinuous, bounded, monotone operator and let $K\colon X \to X^*$ be a random, linear symmetric continuous monotone operator. Then

a) K can be written as J^*J where J^* is a random linear operator from S to X^* and J is the dual of J^*.

b) The sequence $\{J^* v_n(\omega) + a(\omega)\}$ where $v_n(\omega)$ is defined by

$$v_{n+1} = \frac{n}{n+1} v_n - \frac{1}{n+1} JN [J^* v_n + a],$$

for any arbitrary measurable initial choice $v_1(\omega)$ in S, converges to the unique random solution $u(\omega)$ of the Hammerstein equation

$$u(\omega) + KNu(\omega) = a(\omega).$$

<u>Outline of proof</u>: We recall that a real Banach space X is said to be in normal position (Amann [1]) with respect to a real Hilbert space S if

i) $X^* \subset S \subset X$, algebraically and topologically

ii) $\langle u,v \rangle_S = (u,v)$, $u \in X^*$, $v \in S$

iii) S is dense in X.

Since the restriction of K to S is a positive continuous operator it has a square root J_1. We extend J_1 continuously to an operator $J \in B(X,S)$ and then show that $K = J^*J$. It can then be shown that J and J^* are random if K is so. We then reduce the nonlinear Hammerstein equation to an equation of the type $(I+T)v = 0$ where $T : S \to S$ is maximal monotone. We now utilize an iterative scheme due to Bruck [8] and, by virtue of Theorem 2.1, conclude the existence of a unique random solution.

4. INVERSE OF A RANDOM OPERATOR AND ITS MEASURABILITY

Let $T(\omega)$ be a random operator with values in $L(X,S)$ where X and S are separable. Then the inverse $T^{-1}(\omega)$, if it exists, is an operator in $L(S,X)$ which maps $T(\omega)X$ into X almost surely. The following theorem on the randomness of the inverse of a linear operator is due to Hanš [7] (see also Nashed and Salehi [27]).

4.1. Theorem. Let $(\Omega,\mathcal{B},\mu)$ be a complete probability space, X a separable Banach space and Y a normed linear space. Let T be a separable random linear operator from $\Omega \times X$ into Y such that a.s. $T(\omega,\cdot)$ is invertible and its inverse $T^{-1}(\omega,\cdot)$ is bounded. Then T^{-1} is a linear random operator from $\Omega \times Y$ into X.

This theorem follows from a general theorem due to Nashed and Salehi [27]. In this paper they study the measurability of the generalized inverse of a random linear operator (not necessarily bounded).

4.2. Theorem [27]. Let $(\Omega,\mathcal{B},\mu)$ be a complete probability space, X a separable metric space and Y a metric space. Let T be a separable random operator from $\Omega \times X$ onto Y such that a.s. $T(\omega,\cdot)$ is invertible and its inverse $T^{-1}(\omega,\cdot)$ is continuous. Then T^{-1} is also a random operator from $\Omega \times Y$ into X.

Proof. Let the superscript c denote the complementation, $S(\cdot,r)$ and $\bar{S}(\cdot,r)$ denote the open and closed balls of radius r around $\cdot$ respectively. To prove the measurability of T^{-1} it suffices to show that for an arbitrary $y \in Y$ and a closed ball $\bar{S}(x',r)$ the event $\{\omega : T^{-1}(\omega,y) \in \bar{S}(x',r\}$ is in B. We note that

$$\{\omega : T^{-1}(\omega,y) \in \bar{S}(x',r)\} = \bigcup_{x\in\bar{S}(x',r)} \{\omega : T(\omega,x) = y\}.$$

We assert that

$$\text{(a)} \quad \bigcup_{x\in\bar{S}(x',r)} \{\omega : T(\omega,x)=y\} = \bigcap_{n=1}^{\infty} \bigcup_{x\in\bar{S}(x',r+1/n)} \{\omega : T(\omega,x') \in S(y,1/n)\}$$

It suffices to show that the right-hand side is contained in the left-hand side. Let $\omega_0 \in \bigcap_{n=1}^{\infty} \bigcup_{x\in\bar{S}(x',r+1/n)} \{\omega : T(\omega,x') \in S(y,1/n)\}$ Then for each n, there exists $x_n \in S(x',r+1/n)$ such that $T(\omega_0,x_n) \in S(y,1/n)$. It follows that $\lim_{n\to\infty} T(\omega_0,x_n) = y$.

Because $T^{-1}(\omega_0,\cdot)$ is continuous it follows that x_n converges to $T^{-1}(\omega_0,y)$, which we denote by x. Hence $T(\omega_0,x) = y$. Moreover $x \in \bar{S}(x',r)$. This implies that ω_0 belongs to $\bigcup_{x\in\bar{S}(x',r)} \{\omega: T(\omega,x)=y\}$. But

(b) $$\Big[\bigcap_{n=1}^{\infty} \bigcup_{x\in S(x',r+1/n)} \{\omega: T(\omega,x) \in S(y,1/n)\}\Big]^c$$

$$= \bigcup_{n=1}^{\infty} \bigcap_{x\in S(x',r+1/n)} \{\omega: T(\omega,x) \in S^c(y,1/n)\}.$$

Because of separability of T, $\bigcap_{x\in S(x',r+1/n)} \{\omega: T(\omega,x) \in S^c(y,1/n)\}$ is measurable. Hence by (a) and (b) the result follows.

We now apply Theorem 4.2 to a random nonlinear Hammerstein equation [27].

4.3. Theorem. Let X be a separable reflexive Banach space and let

i) $N: X^* \to X$ be a hemicontinuous, monotone, bounded and random operator

ii) $K: X \to X^*$ is a random linear monotone operator such that

$$(Ku,u) \geq \mu \, ||Ku||^2_{X^*}, \quad \mu > 0.$$

Then the random Hammerstein equation

$$u + KNu = a$$

has a unique random solution for each $a(\omega) \in X^*$.

Outline of proof: The Hammerstein equation may be rewritten as

$$Tu = K^{-1}(u-a) + Nu \ni 0.$$

The existence of a unique solution for each fixed ω follows from the fact that T is a maximal monotone, coercive operator. In order to prove the randomness of the solution, it suffices to show $(I + KN)^{-1}$ satisfies the hypotheses of Theorem 4.2. The details of these arguments may be seen in [27].

Thus the crucial step in proving the randomness of the inverse of random operators lies in extracting additional information on the structure of $(I + KN)^{-1}$ for each fixed ω. However in many applications to random differential and integral equations the operator $(I + KN)^{-1}$ is not necessarily single-valued. This necessitates the study of multivalued random operators.

5. RANDOM CORRESPONDENCES

A random correspondence is a multivalued function that assigns to each element ω of a measurable space Ω a subset of a topological space X. Random correspondences have been recently studied by several authors ([2], [10], [13], [26], [30]). In most of these papers however an assumption is made of either the measurable structure on T being a Radon measure or the space S being Euclidean. We will consider a correspondence $T : \Omega \to X$ as a subset of $\Omega \times X$. A correspondence $T : \Omega \to X$ is said to be random if and only if $T^{-1}(B)$ is measurable for each closed subset B of X. Further if $T : \Omega \times X \to Y$ where Y is a topological space, then the randomness of T is defined in terms of the product σ-algebra $\mathcal{B} \times B_X$ generated by the sets of the type $A \times B$, $A \in \mathcal{B}$, $B \in \mathcal{B}_X$. The randomness of $T : \Omega \to X$ clearly implies that $\mathrm{dom}\,T$ is measurable.

5.1. Theorem. Let X be a separable metric space and $T : \Omega \to X$ have closed values i.e., $T(\omega)$ is a closed subset of X for each $\omega \in \Omega$. Then the following are equivalent:

i) T is random

ii) $T^{-1}(C)$ is measurable for all open sets C in X

iii) $T^{-1}(C)$ is measurable for all compact sets C in X

iv) $d(x,T(\omega))$ is a measurable function of $\omega \in \Omega$ for each $x \in X$.

5.2. Theorem [10]. For a closed valued correspondence $T : \Omega \to X$, a separable metric space, the following are equivalent:

i) T is random

ii) domT is measurable and there is a countable (or finite) family $\{x_i\}$ of measurable functions $x_i : \Omega \to X$ such that $T(\omega) = c\ell\{x_i(\omega)\}$ for all $\omega \in$ domT.

This important characterization leads to the following basic theorem on single-valued random selections.

5.3. Theorem ([10,[26]). If $T : \Omega \to X$ (a separable metric space) be a closed-valued random correspondence, there exists at least one random selection i.e., a random function x : domT $\to X$ such that $x(\omega) \in T(\omega)$ for all $\omega \in$ domT.

5.4. Theorem [22]. Let X be a separable metric space and Y a metric space. Further let $T : \Omega \times X \to Y$ be measurable in Ω and continuous in X. Then T is measurable.

However if in addition the measure space $(\Omega,\mathcal{B},\mu)$ is complete then one has the following very important theorem due to Debreu [13] (see also Sainte-Beuve [31]).

5.5. Theorem. Let $T : \Omega \to X$, a separable metric space, be closed-valued. Then the following are equivalent:

i) T is a random correspondence

ii) gph T is $\mathcal{B} \times \mathcal{B}_X$-measurable.

Application of Theorem 5.5: We now provide an application of Debreu's theorem to establish the existence of a single-valued random solution of a random nonlinear boundary value problem. Thus we consider the existence of random periodic solutions for the Liénard problem:

$$x'' + \frac{d}{dt}\left(\frac{dG}{dx}\right) + g(x,\omega) = p(t,\omega)$$

$$x(\omega,0) = x(\omega,2\pi),\ x'(\omega,0) = x'(\omega,2\pi)$$

where i) $\int_0^{2\pi} p(t,\omega)dt = 0$

ii) $\frac{g(x,\omega)}{|x|} \to 0$ as $|x| \to \infty$

iii) $x(\omega)g(\omega,x) \geq 0$ for $|x| > b$

iv) $G(\omega,x) \in C^2(R,R)$ in x.

For each ω the existence of a periodic solution under the above hypotheses may be seen in [12]. We now outline the method of proof of the randomness of the solution ([23]).

If L denotes the linear differential operator defined by $Lx = -x''$, where $x \in H^2[0,2\pi]$ and $x(0) = x(2\pi)$, $x'(0) = x'(2\pi)$ and $N : \mathcal{D}(N) = \Omega \times H^1[0,2\pi] \to L^2[0,2\pi]$ be defined by $Nx = \frac{-d}{dt}\left(\frac{dG}{dx}\right) - g(x,\omega) + p(t,\omega)$ then the random Liénard problem may be written in the form

$$Lx + Nx = 0.$$

If we denote by P the operator which assigns to each $x(t,\omega) \in L^2[0,2\pi]$ the element $\int_0^{2\pi} x(t,\omega)dt \in L^2[0,2\pi]$ then clearly $P : \Omega \times L^2[0,2\pi] \to \mathcal{N}(L)$. Thus $H = [L|_{\mathcal{R}(L)}]^{-1}$ is defined. Proceeding as in [12] we can rewrite the Liénard problem into an equivalent system of operator equations

$$x + H(I-P)Nx = x^*$$

$$PNx = 0.$$

The operator L can be written in the form TT^* where

$Tx = -x'$, $x \in \mathcal{D}(T) = H^1[0,2\pi]$, $x(0) = x(2\pi)$.

$T^*x = x'$, $x \in \mathcal{D}(T^*) = H^1[0,2\pi]$, $x(0) = x(2\pi)$

This induces a natural decomposition of H(I-P) in the form J^*J where $J^* = (I-P)[T^*|\mathcal{D}(T^*) \cap N(T^*)^{\perp}]^{-1}$ and $J = [T|\mathcal{D}(T) \cap N(T)^{\perp}]^{-1}(I-P)$. Proceeding as in [21] we can obtain that $x = J^*y + x^*$ is a solution of the Liénard problem where y and x^* satisfy

$$y + JN(J^*y+x^*) = 0$$

$$PN(J^*y+x^*) = 0.$$

But this system of equations can be written as a single operator equation of the type

$$(y,x^*) + T(y,x^*) = 0$$

over $R(L) \times N(L)$, where T is defined by

$$T(y,x^*) = (JN(J^*y+x^*),\ PN(J^*y+x^*)).$$

Clearly $T : R(L) \times N(L) \to R(L) \times N(L)$ and is compact for each ω. Writing I + T as T_1 we obtain

$$T_1(\omega,z) = 0,\ z = (y,x^*).$$

Clearly T_1 is a random operator because P, J, N and J^* are so. Also T_1 is continuous in z for each fixed ω, because of the continuity properties of P, J, N and J^*. But this implies by Theorem 5.4 that T_1 is $\Omega \times X$ measurable, where $X = R(L) \times N(L)$. It follows that gph Γ is measurable where $\Gamma : \Omega \to X$ is the multivalued function which assigns to each $\omega \in \Omega$ the set of solutions of $T_1(\omega,z) = 0$. By Theorem 5.5, it follows that Γ is measurable since $(\Omega,\mathcal{B},\mu)$ is complete. And since the continuity properties of T_1 for each ω imply that Γ is closed valued, it follows by Theorem 5.3 that there exists a random solution of the Liénard problem.

6. MULTIVALUED VERSION OF THE NASHED AND SALEHI THEOREM

In [4] Bharucha-Reid raises the question of random analogues of fixed point theorems for multivalued contraction and nonexpansive mappings. Deterministic analogues of initial value problems of the type $x'(t) \in R(t,x(t),\omega)$, $x(0) = x_0(\omega)$ have been studies recently by Castaing [11], naturally raising questions regarding the existence of random solutions. Nonlinear Volterra equations of the type $u(t) + \int_0^t a(t-x)g(u(s),\omega)ds = f(t,\omega)$, $0 \le t < \infty$ in a Hilbert space H, a(t) a real function on $[0,\infty)$ and g is a multivalued nonlinear operator from H into itself occur in the study of mechanical systems with memory. Brézis [7] has studied evolution equations involving multivalued monotone operators. It is evident that in order to apply the ideas of the earlier sections to random equations involving multivalued random operators, we have to obtain a multivalued analogue of the Nashed-Salehi theorem. If we follow the proof of their theorem we realize that the continuity of T^{-1} (for fixed ω) has to be replaced by a similar hypothesis for multivalued operators. In [25] we have studied the case of random nonlinear boundary value problems at resonance involving maximal monotone operators. We have shown that if we replace continuity of T^{-1} for fixed ω in the single valued case by demicontinuity of T^{-1} for fixed ω in the multivalued case then one can obtain the existence of random solutions for such boundary value problems. Thus we prove that the random nonlinear boundary value problem $Lu + Nu = f$ where L is a positive self-adjoint random differential operator over a real Hilbert space such that dim $\mathcal{N}(L)$ is finite (need not be zero) and N is a random nonlinear maximal monotone operator is solvable for random solutions under appropriate hypotheses on f and N. The demicontinuity of the appropriate operator T^{-1} in the proof of the theorem follows from the maximality of N.

The interesting point here is that such a concept of demicontinuity is true for wider classes of nonlinear problems e.g., multivalued nonexpansive mappings, situations involving Schauder fixed point theorem etc. In fact, we have proved in [24] that if we consider an analogous boundary value problem to the one considered above where N need not be maximal monotone but satisfies some compactness hypotheses, we can still obtain the demicontinuity of the appropriate operator T^{-1} that arises in the proof. It will be interesting to study random analogues of the Schauder, Krasnoselskii, Kakutani fixed point theorems using the above concept of demicontinuity. These, together with the random Volterra equations, will be studied elsewhere. It must be pointed out here that these situations immediately raise the question of separability for multivalued operators and hence it would be useful to obtain characterizations of separability as in Section 2.

7. STOCHASTIC APPROXIMATION

In this section we will describe some applications of the random contraction mapping theorem to questions in stochastic approximation. The approach of Hanš and the Czech school is different from the direction taken by Robbins-Monroe [29] which has been extensively developed by Burkholder [9], Blum [6] and others. A comparative study of the results obtained by these techniques may be seen in the papers of Hanš. We will give three different applications of random fixed point theorems and related techniques. An application of the Robbins-Monroe technique and related works in stochastic approximation to problems of prediction and control synthesis may be seen in Gardner [15]. Our earlier discussions on the existence of random solutions in various other situations where the stringent condition of contraction mappings is not required lead us to believe that we are now in a position to generalize the results of Hanš et. al. [17] to general nonlinear problems, thereby answering a question raised by Hanš [17, p. 195]. We first state and prove the following theorem due to Hanš [17]:

7.1. Theorem. Let T be a generalized stochastic process mapping the Cartesian product space $\Omega \times [0,\infty) \times X$ into the space X and almost surely continuous with respect to both the arguments $t \geq 0$ and $x \in X$ simultaneously. Let there exist an element $\hat{x} \in X$, a real-valued random variable c, and let the following assumptions be satisfied:

(a) $\mu\{\omega: c(\omega) < 1\} = 1$

(b) $\mu\{\omega: \lim_{t\to\infty} ||t^{-1}\int_0 T(\omega,s,\hat{x})ds - \hat{x}|| = 0\} = 1;$

(c) $\mu\{\omega: ||T(\omega,t,x) - T(\omega,t,y)|| \leqq c(\omega)||x-y||\} = 1$

for every $t \geqq 0$ and every two elements x and y from X. Further, let $x_t(\cdot)$ be the solution of the random operator equation

$$\xi_0(\cdot) = T[\cdot,0,\xi_0(\cdot)];$$

$$\xi_t(\cdot) = t^{-1}\int_0^t T[\cdot,x,\xi_s(\cdot)]ds \quad \text{for } t > 0.$$

Then x_t is for every $t \geqq 0$ a generalized random variable and we have

$$\mu\{\omega : x_t(\omega) \text{ is continuous in } t\} = 1$$

and

$$\mu\{\omega: \lim ||x_t(\omega)-\hat{x}|| = 0\} = 1.$$

Proof. Let us denote by C_∞ the space of all continuous mappings f of the space $[0,\infty)$ into the space X such that the relation $\lim_{t\to\infty}||f(t)-\hat{x}|| = 0$ or, as we shall write hereafter, such that $f(t) \to \hat{x}$ holds. Introducing the distance function ρ for every pair of elements f and g from C_∞ by $\rho(f,g) = \sup_{t\geq 0}||f(t)-g(t)||$ the space C_∞ becomes a separable metric space whose σ-algebra of all Borel subsets is the σ-algebra generated by the class of sets $\{f:f(t) \in B\}$, where t runs over $[0,\infty)$ and B runs over $\mathcal{B}_X$. Further, let us denote by S the operator on the Cartesian product space $\Omega \times C_\infty$ defined by $[S(\omega,f)](0) = T[\omega,0,f(0)]$ and by $[S(\omega,f)](t) = t^{-1}\int_0^t T[\omega,s,f(s)]ds$ for every $f \in C_\infty$, $t > 0$, and every $\omega \in E$, where E equals the intersection over all $t \geq 0$, $x \in X$, and $y \in X$ of the sets occurring in (a), (b), (c), and the set of those $\omega \in \Omega$ for which $T(\omega,t,x)$ is continuous in both t and x simultaneously. For every $\omega \in \Omega - E$, $f \in C_\infty$, and $t \in [0,\infty)$ let us put $[S(\omega,f)](t) = \hat{x}$. First of all let us prove that the mapping S maps the Cartesian product space $\Omega \times C_\infty$ into the space C_∞. Choose arbitrarily $\omega \in C$, $f \in C_\infty$, and $\delta > 0$. Then there exists a real number t_0 such that for every $t \geq t_0$ we have $||f(t)-\hat{x}|| \leq \delta/3$ and simultaneously $||t^{-1}\int_0^t T(\omega,s,\hat{x})ds-\hat{x}|| \leq \delta/3$. Then for every $t \geq t_0\rho(f,h)3/\delta$, where he denotes the function for which $h(t) = \hat{x}$ for every $t \in [0,\infty)$, we can write

$$||[S(\omega,f)](t)-\hat{x}|| \leq t^{-1}\int_0^t ||T(\omega,s,f(s) - T(\omega,s,\hat{x}))||ds$$
$$+ ||[S(\omega,h)](t)-\hat{x}|| \leq c(\omega)(t_0/t)\rho(f,h)+c(\omega)(1-t_0/t)(\delta/3)+\delta/3 \leq \delta.$$

Since the $\Omega - E$ part is trivial we have proved that S maps the Cartesian product space $\Omega \times C_\infty$ into the space C_∞. Now, let us prove that S is a contraction mapping. Thus, let us take arbitrarily $f \in C_\infty$, $g \in C_\infty$, and $\omega \in E$. (For $\omega \in \Omega - E$ we get a singular case.) Then using (c) we have $\rho[S(\omega,f),S(\omega,g)] \leqq c(\omega)\rho(f,g)$. The mapping S being a random contraction transformation of the Cartesian product space $\Omega \times C_\infty$ into the complete separable metric space C_∞, we can apply theorem 3.3, which asserts that there exists a generalized random variable ϕ with values in the space C_∞ so that $S[\omega,\phi(\omega)] = \phi(\omega)$ holds with probability one. However, we have $E \in \mathcal{B}$, $\mu(E) = 1$; hence if we define x_t for every $\omega \in \Omega$ and every $t \in [0,\infty)$ by $x_t(\omega) = [\phi(\omega)](t)$ we immediately obtain all the assertions of theorem 7.1.

Another interesting application of the random contraction mapping theorem using strong laws of large numbers for Banach space valued random variables is the following one due to Hanš [18].

<u>7.2</u>. <u>Theorem</u>. Let $(\Omega,\mathcal{B},\mu)$ be a probability space, X a separable metric space with the σ-algebra $\mathcal{B}_X$ of all Borel subsets of the space X, c a function on Ω, ϕ, f_1, f_2, ... a sequence of generalized random variables and T_1, T_2, ... a sequence of continuous random transforms of the Cartesian product $\Omega \times X$ into the space X so that the following conditions are satisfied:

$$0 \leqq c(\omega) < 1 \text{ for every } \omega \in \Omega;$$

$$\mu\{\omega: \lim_{n\to\infty} \rho(f_n(\omega), \phi(\omega)) = 0\} = 1;$$

$$\mu\{\omega: \lim_{n\to\infty} \rho(T_n(\omega, \phi(\omega)), \phi(\omega)) = 0\} = 1;$$

$$\mu\{\omega:\rho(T_n(\omega,x),\ T_n(\omega,f_n(\omega))) \leqq c(\omega)\ \ (x,f_n(\omega))\} = 1$$

for every $n = 1, 2, \ldots$ and $x \in X$.

Let us choose the generalized random variable V_1 arbitrarily and define, for every $n = 1, 2, \ldots$, the mapping V_{n+1} of the space Ω into the space X, for every $\omega \in \Omega$, by the formula

$$V_{n+1}(\omega) = T_n(\omega, V_n(\omega)).$$

Then $V_1, V_2, \ldots$ is a sequence of generalized random variables which converges almost surely to the generalized random variable ϕ.

A third application of random contraction mapping theorem is the following problem from experience theory [19].

7.3. Theorem. Let $(\Omega, \mathcal{B}, \mu)$ be a probability space, X a separable Banach space, B_X the σ-algebra of all Borel subsets of X and T a generalized stochastic process of the Cartesian product space $\Omega \times [0,\infty) \times X$ into X such that it is for every $\omega \in \Omega$ continuous with respect to the arguments t and x separately. Further let the following conditions hold:

there exists a real-valued random variable c with the property

$\mu\{\omega: c(\omega) < 1\} = 1$ so that

$\mu\{\omega: \|T(\omega,t,x_1) - T(\omega,t,x_2)\| \leqq c(\omega)\ \|x_1 - x_2\| = 1$

for every $t \geqq 0$ and every pair $x_1, x_2 \in X$;

there exists a generalized random variable ϕ so that

$\mu\{\omega: \lim_{t\to\infty} \|T(\omega,t,\phi(\omega)) - \phi(\omega)\| = 0\} = 1.$

Put

$$x_t(\omega) = \int_{\max(0,\ t-2)}^{\max(0,\ t-1)} T(\omega,s,x_s(\omega))\,ds$$

for every $t \geqq 0$, where the integral in (5) is meant in the sense of Bochner.

Then $x_t(\omega)$ is for every fixed $t \geqq 0$ a generalized random variable with values in the space X and the conditions (3) and (4) are fulfilled.

As remarked at the beginning of this section it will be of importance to obtain these results for more general classes of nonlinearities by using our earlier results. A particularly significant problem is to establish the existence of a random solution to the initial value problem of Volterra integral equation mentioned at the beginning of Section 6 with monotone operators as the class of nonlinearities and obtain results analogous to those of this section.

REFERENCES

1. Amann, H., Existence theorems for equations of Hammerstein type, *Applicable Anal.* *1*(1972), 385-397.

2. Aumann, R. J., Integrals of set-salued functions, *J. Math. Anal. Appl.* *12*(1965), 1-12.

3. Bharucha-Reid, A. T. and Mukherjea A., Separable random operators, I., *Rev. Roumaine Math. Pures Appl.* *14*(1969), 1553-1561.

4. Bharucha-Reid, A. T., Fixed point theorems in probabilistic analysis, *Bull. Amer. Math. Soc.* *82*(1976), 641-657.

5. Bharucha-Reid, A. T., *Random Integral Equations*, Academic Press, 1972.

6. Blum, J. R., Multidimensional stochastic approximation method, *Ann. Math. Stat.* *25*(1954), 737-744.

7. Brézis, H., Montonicity methods in Hilbert spaces and some applications to nonlinear partial diff. equations, *Contributions to Nonlinear Functional Analysis*, (E. Zarantonello, ed.) Academic Press, 1971.

8. Bruck, R. E., A strongly convergent iterative solution of $0 \in U(x)$ for a maximal monotone operator U in Hilbert space, *Jour. Math. Anal. Appl.* *48*(1974), 114-126.

9. Burkholder, D. L., On a class of stochastic approximation procedures, *Ann. Math. Stat. 27*(1956), 1044-1059.

10. Castaing, C., Sur les multi-applications measurables, *Rev. Francaise d'Informatique et de Recherche Opérationelle 1*(1967), 91-126.

11. Castaing, C., Sur les équations differentielles, *C. R. Acad. Sci. Paris* (1966), 63-66.

12. Cesari, L. and Kannan, R., Periodic solutions in the large of nonlinear ordinary differential equations, *Rendiconti di Math. 8*(1975), 633-654.

13. Debreu, G., Integration of correspondences, *Proc. Fifth, Berkeley Symp. on Math. Stat. and Probl, Vol. II. Part I*, 351-372.

14. Doob, J. L., *Stochastic Processes*, John Wiley, New York, 1953.

15. Gardner, Jr., L. A., Stochastic approximation and its application to problems of prediction and control synthesis, *Internat. Sympos. Nonlinear Differential Equations and Nonlinear Mechanics* (J. P. LaSalle and S. Lefschetz, eds.) Academic Press, New York, 1963, 241-258. MR 26 #5708.

16. Grenander, U., *Probabilities on Algebraic Structures*, Wiley, New York: Almqvist and Wiksell, Stockholm, 1963, MR 34 #6810.

17. Hanš, O., Random operator equations, *Proc. 4th Berkeley Sympos. on Math. Statist. and Probability (1960), Vol. II*, Univ. California Press, Berkeley, Calif., 1961, 185-202, MR 26 #4185.

18. Hanš, O., Random fixed point theorems, *Trans. 1st Prague Conf. on Information Theory, Statist. Decision Functions, and Random Processes*, (Liblice, 1956), Czechoslovak Acad. Sci., Prague, 1957, 105-125, MR 20 #7332.

19. Hanš, O. and Driml, M., Continuous stochastic approximations, *Trans. 2nd Prague conf. on Information*, 113-122.

20. Jury, E. I. and Oza, K. G., System identification and the principle of random contraction mapping, *SIAM J. Control 6*(1968), 244-257, MR 38 #5486.

21. Kannan, R. and Locker, John, Nonlinear boundary value problems and operators TT^*, *Jour. Diff. Eqns.*, (to appear).

22. Kannan, R. and Salehi, H., Random nonlinear equations and monotonic nonlinearities, *Jour. Math. Anal. Appl.* (to appear).

23. Kannan, R. and Salehi, H., Measurability of solutions of nonlinear equations, *Proc. Conf. Nonlinear Func. Anal.*, Michigan State University, Marcel Dekker, 1976.

24. Kannan, R., Alternative Problems and Nonlinear Operator Equations, Nonlinear Analysis Volume in honor of Professor Rothe, Academic Press

25. Kannan, R., Monotonicity and measurability, *Proc. International Conf. Nonlinear Analysis,* (Arlington) Academic Press, 1976.

26. Kuratowski, K. and Ryll-Nardzewski, C., A general theorem on selectors, *Bull. Acad. Polon. Sci. Ser. Math. Sci. Astronom. Phys.* *13*(1965), 397-403.

27. Kampé De Feriet, J., Random integrals of differential equations, *Lectures on Modern Mathematics,* T. L. Saaty, ed., John Wiley, New York, 1965.

28. Neveu, J., *Mathematical Foundations of the Calculus of Probability,* Holden-Day, San Francisco, 1965.

29. Robbins, H. and Monroe, S., A stochastic approximation method, *Ann. Math. Stat.* *22*(1951), 400-407.

30. Rockafellar, R. T., Measurable dependence of convex sets and functions on parameters, *Jour. Math. Anal. Appl.* *28*(1969), 4-25.

31. Sainte-Beuve, M. F., Sur la generalization d'um theoreme de section measurable de von Neumann-Aumann, *Jour. Functional Analy.* *17*(1974), 112-129.

*Partially supported by a NSF grant.

SOME ASYMPTOTIC PROPERTIES OF SOLUTIONS OF THE NAVIER-STOKES EQUATIONS

George H. Knightly*
Department of Mathematics
University of Massachusetts

1. INTRODUCTION

We consider here some questions concerning rates of decay, in space and time, of nonstationary disturbances in an otherwise steady motion of a viscous incompressible fluid past a finite body bounded by rigid walls. We suppose the steady motion has at infinity the velocity $\beta^* V$ in a fixed direction V. Under suitable hypotheses we prove that the distrubances are carried downstream with speed $\beta^* > 0$ and damped in time as $t^{-1/2}$. The steady motions considered have paraboloidal wake regions in the direction V. We show, under additional assumptions, that the disturbance also possesses a wake region, which is again being swept down stream at the rate β^*.

One situation covered by the above description is the starting problem, in which the body and surrounding fluid are both at rest until some initial instant, $t = 0$, when the body begins accelerating smoothly in the direction -V. Acceleration continues until time $t = 1$ when the body has achieved the velocity $-\beta^* V$, which it maintains thereafter. If $x = (x_1, x_2, x_3)$ denotes the spatial coordinate in a frame translating in the direction -V with the body then, in the flow region Ω corresponding to the exterior of the body, the motion is governed by the Navier-Stokes equations

(1a) $$\frac{\partial w}{\partial t} - \nu\Delta w + w\cdot\nabla w + \nabla p = 0,$$

(1b) $$\nabla\cdot w = 0,$$

relating the velocity vector $w = w(x,t)$ and scalar pressure $p = p(x,t)$, where ν is the constant kinematic viscosity of the fluid and the pressure includes a term compensating for the acceleration of the coordinates. In addition we assume boundary conditions

(1c) $$w = \gamma \text{ on } \partial\Omega,$$

(1d) $$\lim_{|x|\to\infty} w(x,t) = \beta(t)V,$$

and the initial condition

(1e) $$w(x,0) = \alpha(x),$$

where $\alpha(x) = 0$ for the starting problem, the speed of translation of the body is specified by $\beta = \beta(t)$, $\beta(0) = 0$, $\beta(t) \equiv \beta^*$ if $t \geq 1$, while $\gamma = \gamma(x,t)$ accounts for other components of velocity at the boundary. For example, if the fluid adheres to the walls and the body rotates about an axis of symmetry then γ is not zero. Furthermore, we suppose that $\gamma(x,0) = 0$ and $\gamma(x,t) = \gamma^*(x)$ if $t \geq 1$. In this paper we take $\nu = 1$.

The stationary problem associated with (1),

(2a) $$-\Delta w + w\cdot\nabla w + \nabla p = 0,$$

(2b) $$\nabla\cdot w = 0,$$

(2c) $$w = \gamma^* \text{ on } \partial\Omega,$$

(2d) $$\lim_{|x|\to\infty} w(x) = \beta^* V,$$

was shown by Finn [1,2,3] to have, for $\beta^* - \gamma^*$ sufficiently small, a Physically Reasonable solution $w(x) = \beta^* V + b(x)$ having a wake region in the direction V. If one also assumes, as we shall here, that $|\gamma^*(x)| \leq \beta^*$ on $\partial\Omega$, then this wake property may be characterized by the estimate, valid for $x \in \Omega$,

(3) $$|b(x)| \le \beta^* C Q_0(x,0),$$

where C is a constant independent of β^* and $Q_\varepsilon(x,t)$, $0 \le \varepsilon \le \frac{1}{2}$, is defined for all $x \in \mathbb{R}^3$ and $t \ge 0$ by (Here $\theta \in [0,\pi]$ is defined by $|x| \cos\theta = x\cdot V$, $\xi = |x|\cos\theta$ and $\eta = |x|\sin\theta$.)

(4) $$Q_\varepsilon(x,t) = \begin{cases} (1+t+|x|^2)^{-1/2} & \text{in the paraboloid } \xi \ge 0,\ \eta^2 \le \xi, \\ (1+t+|x|^2)^{-1/2-\sigma(1-2\varepsilon)} & \text{along the surface } \xi \ge 0,\ \xi^{1+2\sigma} = \eta^2 \text{ for } 0 \le \sigma \le \frac{1}{2}, \\ (1+t+|x|^2)^{-1+\varepsilon} & \text{otherwise, i.e., outside the cone } \xi \ge 0,\ 0 \le \eta \le \xi. \end{cases}$$

If we seek a solution of problem (1) in the form

(5) $$w(x,t) = \beta(t)V + b(x) + u(x,t),$$

we shall see in the next section that the disturbance velocity, u, may be represented in nonlinar integral form

(6) $$u = u_0 + F(u)$$

in terms of a fundamental solution tensor for the Oseen equations. Here u_0 is a sum of integrals depending on $b, \alpha, \beta, \gamma, \beta^*$ and γ^* and on the stress Tu at the boundary, while F(u) contains terms linear and quadratic in u.

Our assumptions on the quantities determining u_0 will imply the inequality

(7) $$|u_0(x,t)| \le \beta^* C Q_\varepsilon(x-\beta^* tV, t)$$

for some $\varepsilon \in (0,\frac{1}{2}]$ and constant C independent of β^* and ε. Then we prove for β^* sufficiently small, say $0 < \beta^* < \beta_0(\varepsilon)$, that u satisfies

(8) $$|u(x,t)| \le \beta^* \varepsilon^{-1/2} C Q_\varepsilon(x-\beta^* tV, t).$$

with constant C independent of ε and β^*. In general when

$\alpha \neq 0$ we obtain (7) and (8) with $\varepsilon = \frac{1}{2}$. For the starting problem $\alpha = 0$, under stronger assumptions on the boundary stress we get (7) and (8) for some $\varepsilon \in (0,\frac{1}{2})$. The paraboloidal nature of the wake region for u is a consequence of (8) and the decay properties of $Q_\varepsilon(x,t)$ for $0 \leq \varepsilon < \frac{1}{2}$, while the convection of disturbances downstream is evident from the dependence on $x-\beta^* tV$ in (8) for $0 \leq \varepsilon \leq \frac{1}{2}$.

The derivation of the estimate (8) is obtained in §3 by an iterative procedure utilizing (6) and (7) and based upon the fact that for each $\varepsilon \in (0,\frac{1}{2}]$ the operator F preserves the class of vector fields dominated by $Q_\varepsilon(x-\beta^* tV,t)$. This latter property apparently fails at $\varepsilon = 0$ due to the appearance of logarithmic factors in the estimates.

2. REPRESENTATION OF THE DISTURBANCE

We suppose that $w = \beta^* V + b(x)$, $p = \pi(x)$ is a solution of problem (2) and we look for a solution of problem (1) in the form $w = \beta(t)V + b(x) + u(x,t)$, $p = \pi(x) - \beta'(t)V\cdot x + q(x,t)$. If we set $U(t) = (\beta(t) - \beta^*)V$, then we find that the disturbance u must be a solution of the problem

(9a) $\frac{\partial u}{\partial t} - \Delta u + \beta^* V\cdot\nabla u+\nabla q = -\{U\cdot\nabla b+U\cdot\nabla u+b\cdot\nabla u+u\cdot\nabla b+u\cdot\nabla u\}$,

(9b) $\nabla\cdot u = 0$,

(9c) $u = -U + \gamma - \gamma^*$ on $\partial\Omega$,

(9d) $\lim_{|x|\to\infty} u(x,t) = 0$ uniformly in t,

(9e) $u(x,0) = \alpha(x) - b(x)$.

An existence theory for a related generalized problem has been developed by Heywood [4,5] (see also [6] where the linearized starting problem is discussed). Here we assume that (9) has a classical solution u and we study some of its properties under additional hypotheses.

In order to obtain the representation (6) for a solution u of problem (9) we introduce the fundamental solution E of the Oseen equations

$$\frac{\partial u}{\partial t} - \Delta u + \beta^* V\cdot\nabla u + \nabla q = 0,\ \nabla\cdot u = 0. \tag{10}$$

The components of the tensor E (see Oseen [10], also [8]) are defined by the following relations

$$\Phi_0(r,t) = (16\pi^3 t)^{-1/2}\int_0^1 \exp[-a^2r^2/(4t)]da \tag{11}$$

$$\Phi(x,t) = \Phi_0(|x-\beta^* tV|,t)$$

$$E_{ij} = -\Delta\Phi\delta_{ij} + \frac{\partial^2\Phi}{\partial x_i\partial x_j},\ i,j = 1,2,3,$$

$$E = (E_{ij}(x,y,t-\tau)),\ i,j = 1,2,3,$$

where δ_{ij} is the Kronecker delta symbol. If E_j denotes the $j^{\underline{th}}$ column of $E(x-y,t-\tau)$, then when $t > \tau$ the pair $(u,q) = (E_j,0)$ satisfies (10) in the (x,t)-variables and satisfies the adjoint system

$$\frac{\partial u}{\partial \tau} + \Delta u + \beta^* V\cdot\nabla u + \nabla q = 0,\ \nabla\cdot u = 0 \tag{12}$$

in the (y,τ)-variables. E becomes singular at $(y,\tau) = (x,t)$ in such a way that for any smooth, divergence-free vector field $u(x,t)$ we have

$$\lim_{\tau\to t^-}\int_G u(y,\tau)\cdot E(x-y,t-\tau)dy = u(x,t) + \int_S \frac{y-x}{4\pi|y-x|^3}u(y,t)\cdot n d\sigma_y \tag{13}$$

where G is any bounded region in $\mathbb{R}^3$, S its boundary, $x \in G$ and n is the unit exterior normal on S.

Let $G = \Omega(R) = \{y \in \Omega\colon |x-y| < R\}$ for large R, let u,q be a smooth solution of problem (9) in the class

$$u(x,t) = o(1),\ \nabla u(x,t), q(x,t) = o(|x|) \tag{14}$$

as $|x| \to \infty$, locally uniformly in t, and let b(x) be bounded. Then the identity

$$u\cdot\left[\frac{\partial E}{\partial \tau} + \Delta E + \beta^* V\cdot\nabla_y E\right] + E\cdot\left[\frac{\partial u}{\partial \tau} - \Delta u + \beta^* V\cdot\nabla u + \nabla q\right]$$

$$= -E\cdot[U\cdot\nabla b + U\cdot\nabla u + b\cdot\nabla u + u\cdot\nabla b + u\cdot\nabla u]$$

may be integrated by parts over $\Omega(R) \times (0,t)$, using (13) and (14) and letting $R \to \infty$ to obtain equation (6) with

$$(15)\qquad u_0(x,t) = \sum_{i=1}^{7} F_i(x,t),\quad F(u)(x,t) = \sum_{i=8}^{10} F_i(x,t),$$

where

$$(16)\quad F_1(x,t) = \int_\Omega [b(y) - \alpha(y)]\Delta\Phi(x-y,t)dy,$$

$$F_2(x,t) = \int_{\partial\Omega} [\beta^* V - \gamma^*(y)]\cdot n\nabla\Phi(x-y,t)d\sigma_y,$$

$$F_3(x,t) = \int_{\partial\Omega} [U(t)-\gamma(y,t)+\gamma^*(y)]\cdot n(y-x)[4\pi|y-x|^3]^{-1}d\sigma_y,$$

$$F_4(x,t) = \int_0^t\int_{\partial\Omega} [\gamma(y,\tau)\cdot nE(x-y,t-\tau)\cdot(U(\tau)-\gamma(y,\tau)+\gamma^*(y))$$
$$+ (\gamma(y,\tau)-\gamma^*(y))\cdot nE(x-y,t-\tau)\cdot(\beta^* V-\gamma^*(y))]d\sigma_y d\tau,$$

$$F_5(x,t) = \int_0^t\int_{\partial\Omega} n\cdot TE(x-y,t-\tau)\cdot[U(\tau)-\gamma(y,\tau)+\gamma^*(y)]d\sigma_y d\tau,$$

$$F_6(x,t) = \int_0^t\int_{\partial\Omega} n\cdot Tu(y,\tau)\cdot E(x-y,t-\tau)d\sigma_y d\tau,$$

$$F_7(x,t) = \int_0^t\int_\Omega U(\tau)\cdot\nabla E(x-y,t-\tau)\cdot b(y)dyd\tau,$$

$$F_8(x,t) = \int_0^t\int_\Omega U(\tau)\cdot\nabla E(x-y,t-\tau)\cdot u(y,\tau)dyd\tau,$$

$$F_9(x,t) = \int_0^t\int_\Omega [b(y)\cdot\nabla E(x-y,t-\tau)\cdot U(y,\tau) +$$
$$u(y,\tau)\cdot\nabla E(x-y,t-\tau)\cdot b(y)]dyd\tau,$$

$$F_{10}(x,t) = \int_0^t\int_\Omega u(y,\tau)\cdot\nabla E(x-y,t-\tau)\cdot u(y,\tau)dyd\tau.$$

Here the components of the stress tensor, Tu, are given by

$$(17)\qquad (Tu)_{ij} = -q\delta_{ij} + \left(\frac{\partial u_i}{\partial x_j} + \frac{\partial u_j}{\partial x_i}\right),\quad i,j = 1,2,3,$$

and TE is formed using (17) with $q = 0$.

The following properties of the kernels involved in (16) may be proved using the developments in [8, section 3], where additional properties of Φ_0 are also found:

(18)

(i) $\Delta\Phi(x-y,t-\tau)=-[4\pi(t-\tau)]^{-3/2}\exp[-|x-y-\beta^*(t-\tau)V|^2/(4(t-\tau))]$,

(ii) $\left|\frac{\partial^m\Phi(x-y,t-\tau)}{\partial y^m}\right| \le C[(t-\tau)^{1/2}+|x-y-\beta^*(t-\tau)V|]^{-1-m}, m=0,1,2,\ldots,$

with constant C depending only on m.

A disturbance u satisfying (9) and (14) has the representation (6), which we wish to use to derive certain properties of u. To ensure that we are studying the given disturbance and not some other solution of (6), we require the next result, which is easily proved by an argument similar to one in the proof of Theorem 3 in [9].

1. Lemma. Suppose that U(t) and b(x) are bounded. Given $u_0(x,t)$ the equation (6) has at most one solution u(x,t) bounded and continuous on $\Omega \times [0,T]$.

3. STATEMENT OF THE RESULTS

Here we shall formulate the main results and outline their proof. The boundary, $\partial\Omega$, is assumed to be sufficiently smooth (three times continuously differentiable is enough). The letter C denotes constants that are independent of β^* and ε, at least on bounded intervals of those parameters; the value of C in successive usages may differ.

First we specify conditions under which u_0 possesses the bound (7). Here $|D^m\gamma(x,t)|$ and $|D^m\gamma^*(x)|$ denote bounds for tangential $m^{\underline{th}}$ derivatives on $\partial\Omega$.

2. Lemma. Suppose that $\gamma(x,t)$ and $\gamma^*(x)$ are three times continuously differentiable on $\partial\Omega$, b(x) satisfies (3) and the following estimates hold:

(19) $|\alpha(x)| \le \beta^* C(1+|x|^2)^{-1/2}$, for all $x \in \Omega$;

(20) $0 < \beta^* \leq 1, |\beta(t)| \leq \beta^*, |D^m\gamma(x,t)| \leq \beta^*, |D^m\gamma^*(x)| \leq \beta^*,$

for $m = 0,1$, $t > 0$ and $x \in \partial\Omega$;

u is a disturbance which

(21) $|Tu(x,t)| \leq \beta^* C(1+t)^{-1}$, for $x \in \partial\Omega$, $t > 0$.

Then

(22) $|u_0(x,t)| \leq \beta^* CQ_{\frac{1}{2}}(x-\beta^* tV,t)$ for $x \in \Omega$, $t > 0$.

If, in addition, $\alpha = 0$ and for some $\varepsilon \in [0,\frac{1}{2})$

(23) $|Tu(x,t)| \leq \beta^* C(1+t)^{2\varepsilon-2}$ for $x \in \partial\Omega$, $t > 0$

then

(24) $|u_0(x,t)| \leq \beta^* CQ_\varepsilon(x-\beta^* tV,t)$ for $x \in \Omega$, $t > 0$.

Some of the details of the proof of Lemma 2 are given in section 4.

1. Remark. While the hypotheses (19), (20) restrict the data of problem (1), assumptions such as (21) or (23) require knowledge of the solution. The decay rate (21) is possessed by solutions of a Cauchy problem for the Navier-Stokes equations studied in [7] (see also [9]). What is actually needed below is the estimate (7) for u_0. It may happen that (7) can be verified under hypotheses different from those stated in Lemma 2.

1. Theorem. Let $u_0(x,t)$ be a continuous vector field for $x \in \Omega$, $t > 0$, satisfying (7) for some $\varepsilon \in (0,\frac{1}{2}]$. Let $b(x)$ satisfy (3) and suppose that $u(x,t)$ is a bounded continuous solution of the corresponding equation (6). Then there exist positive numbers $\beta_0(\varepsilon)$ and C such that for $0 < \beta^* < \beta_0(\varepsilon)$,

(25) $|u(x,t)| \leq \beta^* C\varepsilon^{-1/2} Q_\varepsilon(x-\beta^* tV,t)$, $x \in \Omega$, $t > 0$.

Since disturbances u in the class (14) admit the representation (6), we have the following corollary to Theorem 1.

1. Corollary. Let u be a solution of problem (9) in the class (14). If u_0 and b and ε are as in Theorem 1 and if β^* is sufficiently small, then u satisfies (25).

If we introduce the linear operator $Lu = F_8(u) + F_9(u)$ and the bilinear form

$$N(u,v) = \int_0^t \int_\Omega u\cdot\nabla E\cdot v\,dy\,d\tau,$$

then $F_{10}(u) = N(u,u)$ and equation (6) may be written as

$$u = u_0 + Lu + N(u,u). \tag{26}$$

Our proof of Theorem 1 is based on the next lemma, which will be proved in section 4.

3. Lemma. Suppose that b satisfies (3) and $|\beta(t)| \le \beta^*$. If v^1 and v^2 are vector fields satisfying

$$|v^i(x,t)| \le k_i Q_\varepsilon(x-\beta^* tV,t) \quad i = 1,2 \tag{27}$$

for $x \in \Omega$, $t > 0$, some constants k_1, k_2 and some $\varepsilon \in (0,\frac{1}{2}]$, then there are constants A, B and H, independent of ε, β^*, k_1 and k_2, such that for $x \in \Omega$, $t > 0$,

$$|Lv^i(x,t)| \le k_i(\varepsilon^{-1}\beta^* A + \varepsilon^{-1}\beta^{*1/2}B)Q_\varepsilon(x-\beta^* tV,t), \tag{28}$$

$$|N(v^1,v^2)| \le k_1 k_2 H\varepsilon^{-1} Q_\varepsilon(x-\beta^* tV,t). \tag{29}$$

Proof of Theorem 1. We insert a parameter λ in equation (26) to obtain

$$u = u_0 + \lambda Lu + \lambda N(u,u). \tag{30}$$

We shall determine for u a series expansion of the form

$$u = \sum_{n=0}^{\infty} \lambda^n v_n \tag{31}$$

and show, for β^* sufficiently small, that the series converges for λ = 1 to a solution of (26), which is the original solution because of Lemma 1. Inserting (29) into (28) yields the recursion relations

$$v_0 = u_0, \quad v_{n+1} = Lv_n + \sum_{j=0}^{n} N(v_j, v_{n-j}), \quad n = 0,1,2,\ldots. \tag{32}$$

According to (7) there holds for suitable constant K,

$$|v_0(x,t)| \le \beta^* K Q_\varepsilon(x-\beta^* tV,t).$$

We set $W_0 = K$. If $|v_j(x,t)| \le \beta^* W_j Q_\varepsilon(x-\beta^* tV,t)$ for $j = 0,1,\ldots,n$, then from Lemma 3 we have

$$|v_{n+1}(x,t)| \le \beta^* W_{n+1} Q_\varepsilon(x-\beta^* tV,t),$$

where

$$(33) \qquad W_{n+1} = \varepsilon^{-1}(\beta^* A+\beta^{*1/2}B)W_n + \varepsilon^{-1}\beta^* H \sum_{j=0}^{n} W_j W_{n-j}.$$

Now the equation

$$(34) \qquad W = K + \lambda\varepsilon^{-1}(\beta^* A+\beta^{*1/2}B)W + \lambda\varepsilon^{-1}\beta^* HW^2$$

has a solution $W(\lambda)$ analytic in a neighborhood of $\lambda = 0$. This solution has the expansion

$$(35) \qquad W(\lambda) = \sum_{n=0}^{\infty} \lambda^n W_n,$$

with $W_0 = K$ and W_n given by (33) for $n > 0$. A calculation based on equation (34) shows that the circle of convergence of the series in (35) includes $\lambda = 1$ whenever

$$(36) \qquad 0 \le \beta^* \le \varepsilon^2[B + (2A\varepsilon+4\varepsilon HK)^{1/2}]^{-2}.$$

Thus when (36) holds we have

$$|u(x,t)| \le \sum_{n=0}^{\infty} |v_n(x,t)| \le \beta^* W(1) Q_\varepsilon(x-\beta^* tV,t),$$

where $W(1) = \sum_{n=0}^{\infty} W_n \le C\varepsilon^{-1/2}$ and the constant C may be chosen independently of $\varepsilon \in (0,\frac{1}{2}]$.

4. PROOF OF THE ESTIMATES

In this section we provide the ideas and some of the details of the proofs of Lemmas 2 and 3. We shall be estimating the terms F_i, $i = 1,\ldots,10$ defined in (16) at an arbitrary evaluation point (x,t), $x \in \Omega$, $t > 0$. The integration variables will be (y,τ) and we shall use the notation $\zeta = x - \beta^* tV$, $z = y - \beta^* \tau V$, $R = |\zeta|$, $P = |z|$, $r = |\zeta - z|$, $\rho = |y|$. The following observations will be useful; as in

(4) we take $\xi = \xi(x) = |x|\cos\theta$, $\eta = \eta(x) = |x|\sin\theta$ for $\theta = \theta(x)$ defined by $\cos\theta = |x|^{-1}x\cdot V$.

4. Lemma. Let $k \geq 1$, $\sigma, \varepsilon \in [0,\frac{1}{2}]$. There are constants C depending only on k such that if $0 < \xi$ and $\xi^{1+2\sigma} = k\eta^2$ then

(37) $$Q_\varepsilon(x,t) \leq Q_\varepsilon(x,0) \leq C(1+|x|^2)^{-1/2-\sigma(1-2\varepsilon)},$$

or, alternatively,

(38) $$Q_\varepsilon(x,t) \leq C(1+t+|x|^2)^{-1/2}\xi^{1-2\varepsilon}\eta^{4\varepsilon-2}.$$

Proof of Lemma 3. We shall discuss only the bound (28) for the term F_9; the estimates (28) for F_8 and (29) for F_{10} are similar in nature and in some respects simpler. For $\varepsilon \in (0,\frac{1}{2}]$, define

$$J = J(x,t;\beta^*,\varepsilon) = \int_0^t\int_{\mathbb{R}^3} |\nabla E(x-y,t-\tau)|Q_0(y,0)Q_\varepsilon(z,\tau)\,dy\,d\tau.$$

If we establish the inequality

(39) $$J \leq \varepsilon^{-1}\beta^{*-1/2}BQ_\varepsilon(\zeta,t),\quad x \in \mathbb{R}^3,\ t > 0,$$

with constant B independent of ε and β^*, then the estimate (28) for F_9 is easily obtained using (3) and (27).

In proving (39) we consider separately the three cases: (i) $R \leq 2$, $t \leq 4$; (ii) $t \leq R^2$, $R \geq 2$; (iii) $R^2 \leq t$, $t \geq 4$, and we shall divide the region of integration into several subregions in cases (ii) and (iii). We indicate the integral over the $j^{\underline{th}}$ subregion by

$$J_j = [A_{1j};A_{2j};A_{3j};A_{4j}],$$

where A_{1j} denotes the inequalities specifying the subregion while A_{2j}, A_{3j} and A_{4j} denote the bounds used for $|\nabla E(x-y,t-\tau)|$, $Q_0(y,0)$ and $Q_\varepsilon(z,\tau)$, respectively (multiplicative constants C are omitted). Once the bounds A_{2j}, A_{3j} and A_{4j} are substituted into the integral the estimate for J_j is computed by integrating over the (possibly larger) subregion determined by the first two or three inequalities listed in A_{1j}.

Case (i). When $R \le 2$, $t \le 4$ we have

$$J \le J_1 = [0<\tau<t,\ 0<r<\infty;\ (t-\tau+r^2)^{-2};1;1] \le t^{1/2} \le Q_\varepsilon(\zeta,t).$$

Case (ii). When $t \le R^2$. $R \ge 2$ we have $J \le \sum_{j=1}^{9} J_j$, with J_j given below. In $J_1,\ldots,J_8$ we always require $r > R/2$ and we set $\theta = \theta(z)$, $\xi = \xi(z)$, $\eta = \eta(z)$.

$$J_1 = [t<t,\ P<2R,\ \theta>\pi/4,\ \rho>R;\ R^{-4};\ R^{-1};\ P^{2\varepsilon-2}],$$

$$J_2 = [\tau<t,\ r>R/2,\ P>2R,\ \theta>\pi/4,\ \rho>R;\ r^{-4};\ R^{-1};\ R^{2\varepsilon-2}],$$

$$J_3 = [\tau<t,\ P<R,\ P<\rho<R,\ \theta>\pi/4;\ R^{-4};\ p^{-1};\ \tau^{-1/2}P^{2\varepsilon-1}],$$

$$J_4 = [\tau<t,\ \rho<R,\ P>\rho,\ \theta>\pi/4;\ R^{-4};\ \rho^{-1};\ \tau^{-1/2}\rho^{2\varepsilon-1}].$$

$$J_5 = [\tau<t,\ \eta<\xi^{1/2},\ \xi<2R;\ R^{-4};\ \xi^{-1};\ \tau^{-1/2}],$$

$$J_6 = [\tau<t,\ \eta<\xi^{1/2},\ \xi>2R;\ \xi^{-4};\ \xi^{-1};\ \tau^{-1/2}],$$

$$J_7 = [\tau<t,\ \xi^{1/2}<\eta<\xi,\ \xi<2R;\ R^{-4};\ \xi^{-1};\ \xi^{1-2\varepsilon}\eta^{4\varepsilon-2}\tau^{-1/2}],$$

$$J_8 = [\tau<t,\ \xi^{1/2}<\eta<\xi,\ \xi>2R;\ \xi^{-4};\ \xi^{-1};\ \xi^{-2\varepsilon}\eta^{4\varepsilon-2}].$$

Each of the terms $J_1,\ldots,J_8$ is dominated by $\varepsilon^{-1}R^{2\varepsilon-2}$, with the ε^{-1} appearing only in J_7 and J_8. The contribution, J_9, of integration over $r < R/2$ depends on the position of ζ. If $\theta(\zeta) \ge \pi/4$, then $r < R/2$ implies $\theta(z) \ge \pi/12$. From (37) with $\sigma = 1/2$ we find that

$$J_9 = [\tau<t,\ r<R/2,\ \theta(\zeta)>\pi/4;\ (t-\tau+r^2)^{-2};R^{-1};R^{2\varepsilon-2}] \le R^{2\varepsilon-2}.$$

If ζ lies in the paraboloid $\eta^2(\zeta) \le \xi(\zeta)$, then

$$J_9 = [\tau<t,\ r<R/2;\ (t-\tau+r^2)^{-2};\ R^{-1};\ R^{-1}] \le R^{-1}.$$

If ζ lies on one of the surfaces $\xi(\zeta)^{1+2\sigma} = \eta^2(\zeta)$ for $\sigma \in (0,\frac{1}{2})$, we divide the ball $r < R/2$ into three parts and set $J_9 = J_{91} + J_{92} + J_{93}$, where

$$J_{91} = [\tau<t,|\xi(z)-\xi(\zeta)|<R/2,\ 12\eta^2(z)<\xi(z);[t-\tau+R^{1+2\sigma} + (\xi(z)-\xi(\zeta))^2]^{-2};\ R^{-1};\ R^{-1}],$$

$$J_{92} = [\tau<t,|\xi(z)-\xi(\zeta)|<R/2,\ \xi(z)<12\eta^2<\xi^{1+2\sigma}(z);[t-\tau+R^{1+2\sigma} + (\xi(z)-\xi(\zeta))^2]^{-2};\ R^{-1};\ \eta^{4\varepsilon-2}(z)\xi^{-2\varepsilon}(z)],$$

$$J_{93} = [\tau<t, r<R/2, 12\eta^2(z)>\xi^{1+2\sigma}(z);\ (t-\tau+r^2)^{-2};\ R^{-1};\ R^{-1-2\sigma(1-2\varepsilon)}].$$

We find that for $j = 1,2,3$ $J_{9j} \le \varepsilon^{-1}R^{-1-2\sigma(1-2\varepsilon)}$. Then in case (ii), since $R^2 \ge C(1+t+R^2)$ we get $J \le C\varepsilon^{-1}Q_\varepsilon(\zeta,t)$.

Case (iii). When $R^2 \le t$, $t \le 4$, we have $J \le \sum_{j=1}^{15} J_j$ with J_j defined below. In $J_1,\dots,J_4$ we always require that $\theta(z) \ge \pi/4$.

$$J_1 = [\tau<t/2,\ P<3t^{1/2},\ \rho>t^{1/2};\ t^{-2};\ t^{-1/2};\ P^{2\varepsilon-2}],$$

$$J_2 = [\tau<t/2,\ P>3t^{1/2},\ \rho>t^{1/2};\ p^{-4};\ t^{-1/2};\ P^{2\varepsilon-2}],$$

$$J_3 = [\tau<t/2,\ \rho<t^{1/2},\ P>\rho;\ t^{-2};\ \rho^{-1};\ \rho^{2\varepsilon-2}],$$

$$J_4 = [\tau<t/2,\ P<\rho<t^{1/2};\ t^{-2};\ P^{-1};\ P^{2\varepsilon-2}],$$

$$J_5 = [t/2<\tau<t,\ r<\infty,\ \rho>\beta^*\tau/2;\ (t-\tau+r^2)^{-2};(\beta^*t)^{-1/2};\ t^{\varepsilon-1}].$$

In J_6 and J_7, note that when $\rho < \beta^*\tau/2$ then $\rho>\beta^*\tau/2\ge\beta^*t/4$ and $Q_\varepsilon(z,\tau) \le (t+\beta^{*2}t^2)^{\varepsilon-1} \le \beta^{*-1/2}t^{\varepsilon-5/4}$

$$J_6 = [t/2<\tau<t,\ r<\infty,\ r<\rho<\beta^*\tau/2;\ (t-\tau+r^2)^{-2};\ (1+r^2)^{-1/2};\ \beta^{*-1/2}t^{\varepsilon-5/4}],$$

$$J_7 = [t/2<\tau<t,\ \rho<\beta^*\tau/2,\ \rho<r;\ (t-\tau+\rho^2)^{-2};\ (1+\rho^2)^{-1/2};\ \beta^{*-1/2}t^{\varepsilon-5/4}].$$

In $J_8,\dots,J_{14}$ we set $\eta = \eta(z)$, $\xi = \xi(z)$. Then

$$J_8 = [\tau<t/2,\ \xi<2t^{1/2},\ \eta^2<\xi;\ t^{-2};\ \xi^{-1};\ \tau^{-1/2}],$$

$$J_9 = [\tau<t/2,\ 2t^{1/2}<\xi<\infty,\ \eta^2<\xi;\ \xi^{-4};\ \xi^{-1};\ \xi^{-1}],$$

$$J_{10} = [\tau<t/2,\xi<2t^{1/2},\xi^{1/2}<\eta<\xi;t^{-2};\xi^{-1};\tau^{-1/2}\eta^{4\varepsilon-2}\xi^{1-2\varepsilon}],$$

$$J_{11} = [\tau<t/2,2t^{1/2}<\xi<\infty,\xi^{1/2}<\eta<\xi;\xi^{-4};\xi^{-1};\tau^{-1/2}\eta^{4\varepsilon-2}\xi^{1-2\varepsilon}].$$

In $J_{12},\dots,J_{15}$, $\theta(z) \le \pi/4$ and $\tau \ge t/2$. Therefore $\rho \ge \beta^*t/2$. J_{12} corresponds to integration over the paraboloid $\eta^2 < \xi$ when $r > R/2$. It is evaluated differently when $R \ge 1$ than when $R < 1$. When $R < 1$ we set $J_{12} = J_{121} + J_{122}$, where

$$J_{121} = [t/2<\tau<t,\ r<3,\ \eta^2<\xi;\ (t-\tau+r^2)^{-2};\ (\beta^*t)^{-1/2};t^{1/2}],$$

$$J_{122} = [t/2<\tau<t, 1<\xi<\infty, \eta^2<\xi; (t-\tau+\xi^2)^{-2}; (\beta^* t)^{-1/2}; t^{-1/2}].$$

When $R \geq 1$ we have $J_{12} = J_{121} + J_{122}$, where now

$$J_{121} = [t/2<\tau<t,\ \xi<2R,\ \eta^2<\xi,\ r>R/2;\ (t-\tau+R^2)^{-2}; \\ (\beta^* t)^{-1/2};\ t^{-1/2}].$$

$$J_{122} = [t/2<\tau<t,\ 2R<\xi<\infty,\ \eta^2<\xi,\ r>R/2;\ (t-\tau+\xi^2)^{-2}; \\ (\beta^* t)^{-1/2};\ t^{-1/2}].$$

In either event, we obtain $J_{12} \leq \beta^{*-1/2} t^{-1} \log t \leq \varepsilon^{-1} \beta^{*-1/2} t^{\varepsilon-1}$.

$$J_{13} = [t/2<\tau<t,\ \xi<2R,\ \xi^{1/2}<\eta<\xi,\ r>R/2;\ (t-\tau+R^2)^{-2}; \\ (\beta^* t)^{-1/2};\ t^{-1/2}\xi^{1-2\varepsilon}\eta^{4\varepsilon-2}],$$

$$J_{14} = [t/2<\tau<t,\ 2R<\xi,\ \xi^{1/2}<\eta<\xi,\ r>R/2;\ (t-\tau+\xi^2)^{-2}; \\ (\beta^* t)^{-1/2};\ t^{-1/2}\xi^{1-2\varepsilon}\eta^{4\varepsilon-2}].$$

Each term $J_1, \ldots, J_{14}$ is dominated by $\varepsilon^{-1}\beta^{*-1/2}t^{\varepsilon-1}$. The remaining contribution, J_{15}, depends upon the position of ζ and corresponds to integration over $t/2<\tau<t$, $r<R/2$, $\theta(z)<\pi/4$. If $\theta(\zeta) > \pi/4$, then

$$J_{15} = [t/2<\tau<t,\ r<R/2,\ \theta(z)<\pi/4;\ (t-\tau+r^2)^{-2};\ (\beta^* t)^{-1/2}; \\ t^{-1/2}R^{2\varepsilon-1}] \leq \beta^* t^{\varepsilon-1}.$$

If ζ lies inside the paraboloid $\eta^2(\zeta) \leq \xi(\zeta)$, then

$$J_{15} = [t/2<\tau<t,\ r<R/2;\ (t-\tau+r^2)^{-2};\ (\beta^* t)^{-1/2};\ t^{-1/2}] \leq \\ (\beta^* t)^{-1/2}.$$

Finally, if $\xi^{1+2\sigma}(\zeta) = \eta^2(\zeta)$ for some $\sigma \in (0,\frac{1}{2})$, then we write $J_{15} = J_{151} + J_{152} + J_{153}$, where

$$J_{151} = [t/2<\tau<t,\ |\xi(z)-\xi(\zeta)|<R/2,\ 12\eta^2(z)<\xi(z); \\ (t-\tau+R^{1+2\sigma}+|\xi(z)-\xi(\zeta)|^2)^{-2};\ (\beta^* t)^{-1/2};\ t^{-1/2}],$$

$$J_{152} = [t/2<\tau<t,\ |\xi(z)-\xi(\zeta)|<R/2,\ \xi(z)<12\eta^2(z)<\xi^{1+2\sigma}(z); \\ (t-\tau+R^{1+2\sigma}+|\xi(z)-\xi(\zeta)|^2)^{-2};\ (\beta^* t)^{-1/2}; \\ t^{-1/2}\xi^{1-2\varepsilon}\eta^{4\varepsilon-2}],$$

$$J_{153} = [t/2<\tau<t,\ r<R/2;\ (t-\tau+r^2)^{-2};\ (\beta^* t)^{-1/2}; \\ t^{-1/2}R^{2\sigma(2\varepsilon-1)}].$$

We find that $J_{15} \le \beta^* \varepsilon^{-1} t^{-1/2-\sigma(1-2\varepsilon)}$. Since $t \ge C(1+t+R^2)$ in case (iii), the above estimates complete the proof of (39).

Now we give a brief discussion of proof of Lemma 2. It suffices to prove (22) and (24) for F_i, $i = 1,\ldots,7$, under the stated hypotheses. We first make some simplifying observations.

When (3) and (19) hold, then $|F_1(x,t)| \le C\beta^* G(\zeta,t)$, where

$$G(\zeta,t) = \int_{\mathbf{R}^3} (1+\rho^2)^{-1/2} |\Delta\Phi_0(|\zeta-y|,t)| dy$$

From Lemma 1 of [7] we have $G(\zeta,t) \le CQ_{1/2}(\zeta,t)$. Thus (22) holds for F_1.

It turns out that for $i = 2,\ldots,7$ F_i satisfies (24) independently of the assumptions on α and the boundary stress. Therefore, since (24) implies (22), once (22) is proved for F_6 it then holds for u_0.

The terms F_1 and F_7 involve volume integrals. A proof of (24) for these terms (when $\alpha = 0$ in F_1) proceeds along the lines of the proof of Lemma 3 and will not be given here.

In the terms $F_2,\ldots,F_5$ involving integration over $\partial\Omega$ we shall show that

(40) $$|F_i(x,t)| \le \beta^* C(t+R^2)^{-1}$$

when $t > 2$ or when $R > m = \max(4,4d)$, where d is the radius of the smallest ball containing $\partial\Omega$. For this range of R and t (40) implies (24). It then remains to show that F_i is bounded when $t \le 2$ and $R \le m$. A proof of this fact follows the usual methods of potential theoretic estimates near the boundary and utilizes the smoothness assumptions on $\partial\Omega$ and the boundary data. We omit the details of this argument.

Next we verify (40) for $i = 2,\ldots,5$. From (18)(ii) and (20) we have

$$F_2(x,t) \le \beta^* C\int_{\partial\Omega}(t+|\zeta-y|^2)^{-1}d\sigma_y \le \beta^* C(t+\delta^2(\zeta))^{-1},$$

where $\delta(\zeta)$ denotes the distance from ζ to $\partial\Omega$. If $R > 2d$, then $\delta(\zeta) \ge R-d \ge R/2$ and (40) follows for $i = 2$. When $i = 3$ the data terms in the integrand are zero for $t > 1$. Then $F_3(x,t) \equiv 0$ for $t > 1$ and,

$$|F_3(x,t)| \le \beta^* C\ \delta(x)^{-2}.$$

When $R \ge m$ one sees using $\beta^* \le 1$, $t \le 1$ that $\delta(x) \ge R/4$ and (40) follows for $i = 3$.

In F_4 and F_5 the integrands are zero for $\tau > 1$. If $\tau > 2$, then from (18)(ii) and (20) we have $|E(x-y),t-\tau)| \le Ct^{-3/2}$, $|TE(x-y,t-\tau)| \le Ct^{-2}$ and

(41) $$|F_4(x,t)| \le \beta^* Ct^{-3/2},\quad |F_5(x,t)| \le \beta^* Ct^{-2}.$$

So (40) follows from (41) when $t > 2$ and $R^2 \le t$, $i = 4,5$. If $R > m$, $\tau \le 1$ and $\beta^* \le 1$, then $r \ge R/2$. In this case (18)(ii) and (20) give $|E(x-y,t-\tau)| \le CR^{-3}$, $|TE(x-y,t-\tau)| \le CR^{-4}$ and

(42) $$|F_4(x,t)| \le \beta^* CR^{-3},\quad |F_5(x,t)| \le \beta^* CR^{-4}.$$

Then (40) follows from (42) when $R > m$ and $t \le R^2$, completing the proof of (40) for $i = 4,5$.

Finally, we suppose that (23) holds for some $\varepsilon \in [0,\frac{1}{2}]$ and we consider F_6. If $R^2 \le t$, then

(43) $$|F_6(x,t)| \le \beta^* C[t^{-3/2}\int_0^{t/2}(1+\tau)^{2\varepsilon-2}d\tau + (1+t)^{2\varepsilon-2}I] \le \beta^* Ct^{2\varepsilon-2},$$

where, for $\partial\Omega$ smooth, we have

(44) $$I \le \int_0^t \int_{\partial\Omega}(t-\tau+r^2)^{-3/2}d\sigma \le C.$$

If $t \le R^2$ then we consider the contributions F_{61}, corresponding to $r \ge R/2$, and F_{62}, corresponding to $r \le R/2$. We have

(45) $$|F_{61}(x,t)| \le \beta^* C R^{-3}\int_0^t (1+\tau)^{2\varepsilon-2}d\tau \le \beta^* C R^{-2}.$$

Since $r \le R/2$, $R \ge m$ imply that $\tau \ge R/(4\beta^*)$, we find using (44) and $\beta^* \le 1$ that

(46) $$|F_{62}(x,t)| \le \beta^* C\int_{R/4}^t (1+\tau)^{2\varepsilon-2}d\tau\int_{\partial\Omega}(t-\tau+r^2)^{-3/2}d\sigma$$

$$\le \beta^* C R^{2\varepsilon-2} I \le \beta^* C R^{2\varepsilon-2}.$$

The bound (40) for $i = 6$ now follows from (43) - (46).

REFERENCES

1. Finn, R., Estimates at infinity for stationary solutions of the Navier-Stokes equations, *Bull. Math. Soc. Sci., Math. Phys. R.P. Roumaine, 3*(51)(1959), 387-418.

2. Finn, R., On the exterior stationary problem for the Navier-Stokes equations, and associated perturbation problems, *Arch. Rational Mech. Anal. 19*(1965), 363-406.

3. Finn, R., Mathematical questions relating to viscous fluid flow in an exterior domain, *Rocky Mt. J. Math. 3*(1973), 107-140.

4. Heywood, J., On stationary solutions of the Navier-Stokes equations as limits of nonstationary solutions, *Arch. Rational Mech. Anal. 37*(1970), 40-60.

5. Heywood, J., The exterior nonstationary problem for the Navier-Stokes equations, *Acta. Math. 129*(1972), 11-34.

6. Heywood, J., On nonstationary Stokes flow past an obstacle, *Indiana Univ. Math. J. 24*(1974), 271-284.

7. Knightly, G., On a class of global solutions of the Navier-Stokes equations, *Arch. Rational Mech. Anal. 21* (1966), 211-245.

8. Knightly, G., Stability of uniform solutions of the Navier-Stokes equations in n-dimensions, (Tech. Summary Rep. 1085, Mathematics Research Center, United States Army, University of Wisconsin, Madison, 1970).

9. Knightly, G., A Cauchy problem for the Navier-Stokes equations in R^n, *SIAM J. Math. Anal. 3*(1972), 506-511.

10. Oseen, C. U., *Neuere Methoden und Ergebnisse in der Hydrodynamik*, Akademische Verlagsgesellschaft, Leipzig, 1927.

*This research was supported in part by NSF Grant No. MPS 75-07579 and was completed while the author was a visitor at the University of Colorado during the academic year 1976-1977.

LOCAL CONTROLLABILITY IN NONLINEAR SYSTEMS

H. W. Knobloch
Mathematisches Institut
Universität Würzburg

1. INTRODUCTION

We deal with control systems which are defined by ordinary differential equations. Local controllability of the system along one of its trajectories means, roughly speaking, that one is able to steer the system from any point of the trajectory to a full neighborhood in arbitrarily small time. To find explicite conditions which guarantee that the system has this property is one of the objectives behind the approach described in this paper. Compared with what one can find in the literature our method is a straightforward one. We introduce formal Taylor series and relate them directly to the higher order effect, which local variations have on trajectories. The main contribution of Sections 2 - 4 is the development of an algorithm which allows to obtain the coefficients of these series from the data of the problem.

As an application we show then how the usage of the formal series provides a convenient and natural access to a higher order maximum principle. This result is closely related to the work of Krener [1]. Except for the method the difference concerns mainly two points. Locally our approach yields a class of variations which is larger and simpler to overlook than in [1]. Globally we need a countability hypothesis, which is not required in [1] (though the condition can probably easily be handled in concrete situations).

2. VARIABLE JUMP-POINTS: LOCAL TAYLOR EXPANSION

We first introduce the standard notation which will be used throughout the next three sections of this paper. t denotes a scalar variable, $\dot{}$ means differentiation with respect to t, $x = (x^1,\dots,x^n)^T$, $f = (f^1,\dots,f^n)^T$ are n-dimensional column vectors. Without stating this explicitly, we will always assume that a function $f(t,x) = (f^1(t,x),\dots,f^n(t,x))^T$ is defined and has continuous partial derivatives of any order with respect to t and the components of x on a set of the form

$$\{t,x \mid t_o-\varepsilon < t < \tilde{t}+\varepsilon,\ x \in X\} \tag{2.1}$$

for some $\varepsilon > 0$. X is a non-empty, open set in x-space, $[t_o,\tilde{t}]$ a given fixed t-interval.

Let N be a positive integer, which has to be regarded as fixed for the time being. We introduce the set

$$\tilde{W} = \{w=(w^1,\dots,w^N) \mid t_o<w^1\le w^2\le\dots\le w^N\le\tilde{t}\}. \tag{2.2}$$

Throughout the following considerations w will mostly be restricted to a neighborhood of the corner of $\tilde{W}$ which is farthest to the right, that is the point

$$\tilde{w} = (\tilde{t},\tilde{t},\dots,\tilde{t}). \tag{2.3}$$

Assume now that there are given N + 1 functions $f_i(t,x)$, $i = 0,\dots,N$, which are defined on the set (2.1) and which have the differentiability properties mentioned above. We define a function $f(t,x|w)$ as follows

$$(2.4)\quad f(t,x|w) = \begin{cases} f_o(t,x) \text{ if } t \le w^1, \\ f_i(t,x) \text{ if } w^i < t \le w^{i+1},\ i = 1,\dots,N-1, \\ f_N(t,x) \text{ if } t > w^N. \end{cases}$$

Except for the points on the hyperplanes $t = w^i$, $i = 1,\dots,N$, $f(t,x|w)$ is - for fixed w - a function of class C^∞ on the set (2.1).

Furthermore, assume that we have a fixed solution $\tilde{x}(t)$ of the diff. eq. $\dot{x} = f_o(t,x)$, which exists on $[t_o,\tilde{t}]$ and satisfies $\tilde{x}(t) \in X$. Since $f_o(t,x) = f(t,x|\tilde{w})$ (according to (2.3) and (2.4)), it follows then by standard arguments that the initial value problem

$$\dot{x} = f(t,x|w),\ x(t_o) = a$$

has a solution $x_w(t;a)$, which also exists on $[t_o,\tilde{t}]$ if $||w-\tilde{w}||$ and $||a-\tilde{a}||$ are sufficiently small, where $\tilde{a} = \tilde{x}(t_o)$. Furthermore this solution tends to $\tilde{x}(t) = x_{\tilde{w}}(t;\tilde{a})$ for $w \to \tilde{w}$ and $a \to \tilde{a}$. Hence the function

$$(2.5)\qquad c(w,a) = x_w(\tilde{t};a)$$

is defined on a set of the form $N \cap \tilde{W} \times A$, where N is a neighborhood of $\tilde{w}$ (in the $\mathbb{R}^N$) and A is a neighborhood of $\tilde{a}$ (in the $\mathbb{R}^n$). In this section we wish to study the analytic properties of the function $c(w,a)$.

2.1 Lemma. If N and A are sufficiently small one can extend the definition of $c(w,a)$ to the whole neighborhood $N \times A$ in such a way that $c(w,a)$ becomes a C^∞-function of (w,a).

Proof. We denote by $x_i(t;\sigma,a)$ the solution of the initial value problem

$$(2.6)\qquad \dot{x} = f_i(t,x),\quad x(\sigma) = a,\quad x(t) \in X$$

$i = 0,\dots,N$. $x_i(t;\sigma,a)$ is a C^∞-function of all its arguments on a neighborhood of $t = \tilde{t}$, $\sigma = \tilde{t}$, $a = \tilde{x}(\tilde{t})$ and satisfies

$x_i(t,t,a) = a$. Using this relation and standard arguments one can find positive numbers $\varepsilon_N, \varepsilon_{N-1},\ldots,\varepsilon_1$ and subsets $K_{N+1} = X, K_N,\ldots,K_1$ of X containing a neighborhood of $\tilde{x}(\tilde{t})$ such that the following statement holds. $x_i(t;\sigma,a)$ exists and satisfies

$x_i(t;\sigma,a) \in K_{i+1}$ if $|t-\tilde{t}| \leq \varepsilon_i$, $|\sigma-\tilde{t}| \leq \varepsilon_i$, $a \in K_i$, $i=1,\ldots,N$. Furthermore it is clear that $x_o(t;t_o,a)$ will exist as a function of t on the whole interval $[t_o,\tilde{t}]$ and stay close to $\tilde{x}(t)$ if $||a-\tilde{x}(t_o)||$ is sufficiently small. Hence one can find positive numbers ε_o,δ such that this result is true: If $||a-\tilde{x}(t_o)|| \leq \delta$ then $x_o(t;,t_o,a)$ exists on $[t_o,\tilde{t}]$ and satisfies

$$x_o(t;t_o,a) \in K_1 \text{ for } |t-\tilde{t}| \leq \varepsilon_o.$$

We now choose $\varepsilon = \text{Min}\,(\varepsilon_o,\varepsilon_1,\ldots,\varepsilon_N)$ and restrict for the remaining part of the proof w, t and a to the set

$$||w-\tilde{w}|| \leq \varepsilon,\ ||a-\tilde{a}|| \leq \delta,\ |t-\tilde{t}| \leq \varepsilon, \tag{2.7}$$

where $\tilde{a} = \tilde{x}(t_o)$ and $\tilde{w}$ is given by (2.3).

It can be seen immediately from the preceding considerations that for this choice of t, w and a the following functions $y_i(t;,w,a)$ are well defined and are C^∞-functions of all their arguments:

$$\begin{aligned} y_o(t;w,a) &= x_o(t;t_o,a), \\ y_i(t;w,a) &= x_i(t;w^i,y_{i-1}(w^i;w,a)),\ i = 1,\ldots,N. \end{aligned} \tag{2.8}$$

It should be noted that $y_i(t;w,a)$ actually depends on the components $w^1,\ldots,w^i$ of w only and that it satisfies as a function of t the equation $\dot{x} = f_i(t,x)$. Furthermore, it follows from the identity $x_i(t;t,a) = a$ that we have for $i = 1,\ldots,N$

$$y_i(w^i;w,a) = y_{i-1}(w^i;w,a) \text{ if } w = (w^1,\ldots,w^i,\ldots,w^N). \tag{2.9}$$

By putting $w = \tilde{w}$ we obtain in view of the first relation (2.8)

(2.10) $y_i(\tilde{t};\ \tilde{w},a) = x_o(\tilde{t};t_o,a),\ i = 0,\dots,N.$

Assume now, that $w \in \mathrm{int}(\tilde{W})$ (that is the set of all $w = (w^1,\dots,w^N)$ such that $t_o < w^1 < w^2 < \dots < w^N < \tilde{t}$). If one compares (2.9) with the definition of $x_w(t;a)$ the following statement becomes evident

$$y_o(t;w,a) = x_w(t;a) \text{ for } t \le w^1,$$
$$y_i(t;w,a) = x_w(t;a) \text{ for } w^i < t \le w^{i+1},\ i = 1,\dots,N-1,$$
$$y_N(t;w,a) = x_w(t;a) \text{ for } t > w^N.$$

Hence it follows from (2.5) that

(2.11) $c(w,a) = y_N(\tilde{t};w,a)$ if $w\in\mathrm{int}(\tilde{W}),\ ||w-\tilde{w}||\le\varepsilon,\ ||a-\tilde{a}||\le\delta.$

For reasons of continuity (2.11) holds also if $w \in \tilde{W}$. Thereby the lemma is proved, since, as we have noted before, the function on the right hand side of (2.11) is a C^∞-function of (w,a). The relations (2.8) - (2.11) will be of further use in obtaining some basic results about the partial derivatives of the function $c(w,a)$ which is our next objective. In the sequel, the symbol D will denote some differential operator acting on the components of w and a (but not on t) and $(Dc)(\tilde{w},\tilde{a})$ will be the value which the corresponding derivative of the function c assumes for $w = \tilde{w}$, $a = \tilde{a}$. In order to simplify our considerations we introduce for the moment the following notation:

(2.12) $w_j(w^1,\dots,w^j) = (w^1,\dots,w^j,\ \tilde{t},\dots\tilde{t}).$

Hence $w_j(w^1,\dots,w^j)$ is the N-tupel w with variable components $w^1,\dots,w^j$ and fixed $w^i = \tilde{t}$ for $i > j$. Now if one specializes w to $w_j(w^1,\dots,w^j)$ in the formula (2.9) then one obtains

$$y_i(\tilde{t};w_j(w^1,\dots,w^j),a) = y_{i-1}(\tilde{t};w_j(w^1,\dots,w^j),a),\ i > j,$$

and therefore

$$y_i(\tilde{t};w_j(w^1,\ldots,w^j),a) = y_j(\tilde{t};w,a) \text{ if } i > j$$

(note that $y_j(t;w,a)$ depends upon the components $w^1,\ldots,w^j$ of w only). It is now clear in view of (2.12) that differentiating the function on the left hand side with respect to $w^1,\ldots,w^j$ and then putting $w = \tilde{w}$ amounts to the same as if one would differentiate $y_i(\tilde{t};w,a)$ and then specialize $w \to \tilde{w}$. Hence we are arrived at this result.

2.2 Lemma. If the differential operator D involves no differentation with respect to any w^i with $i > j$ then

$$(Dc)(\tilde{w},\tilde{a}) = (Dy_N)(\tilde{t};\tilde{w},\tilde{a}) = (Dy_j)(\tilde{t};\tilde{w},\tilde{a}).$$

3. THE FORMULA FOR $(Dc)(\tilde{w},\tilde{a})$ IN CASE $D = \partial^\nu/(\partial w^i)^\nu$.

In this and the following sections we keep the initial value a fixed: $a = \tilde{a} = \tilde{x}(t_o)$. For shortness we write $y_i(t;w)$, $c(w)$ instead of $y_i(t;w,\tilde{a})$, $c(w,\tilde{a})$ (for the definition of these functions see (2.5) and (2.8)). In what follows we make frequently use of the relations (2.10), (2.11) which in the present notation assume the form

$$(3.1) \qquad c(\tilde{w}) = y_i(\tilde{t};\tilde{w}) = \tilde{x}(\tilde{t}),\ i = 0,\ldots,N.$$

Note that $x_o(t;t_o,\tilde{a}) = \tilde{x}(t)$ (= reference trajectory).

Our aim is to construct for every $i = 1,\ldots,N$ and every $\nu = 0,1,\ldots$ a function $h_i^{(\nu)}(t,x)$ of the two variables t and x for which the relation

$$(3.2) \qquad h_i^{(\nu)}(w^i,y_i(w^i;w)) = \left.\frac{\partial^\nu y_i}{(\partial w^i)^\nu}(t;w)\right|_{t=w^i} \quad {}^{*)}$$

holds identically in $w = (w^1,\ldots,w^i,\ldots,w^N)$. Before we proceed to the construction of such a function we note that by

---

*) that means: Regard y^i as function of the two variables t,w, perform differentiation with respect to w^i, then replace $t\to w^i$.

putting $w = \tilde{w}$ one obtains from (3.1), (3.2) and Lemma 2.2

$$\frac{\partial^{\nu} c}{(\partial w^{i})^{\nu}}(\tilde{w}) = h_i^{(\nu)}(\tilde{t},\tilde{x}(\tilde{t})), \tag{3.3}$$

which is the formula referred to in the heading of this section.

For notational convenience we introduce the following abbreviation. Let $g(y)$ be a function of y (infinitely often differentiable), where g and y are n-dimensional vectors. We then use the symbol

$$(\mathcal{D}^{\nu}g)(y^{o},\ldots,y^{\nu})$$

to denote a n-dimensional vector-valued function depending upon the n-dimensional vectors $y^{o},\ldots,y^{\nu}$, which is defined uniquely by the following property: If $y(\sigma)$ is any (n-dimensional vector-) function of the scalar variable σ and if $y(\sigma)$ is ν-times differentiable then

$$(\mathcal{D}^{\nu}g)(y(\sigma), \frac{dy}{d\sigma}(\sigma),\ldots, \frac{d^{\nu}y}{d\sigma^{\nu}}(\sigma)) = \frac{d^{\nu}}{d\sigma^{\nu}} g(y(\sigma)). \tag{3.4}$$

In other words: One obtains the ν-th derivative (with respect to σ) of the composite function $g(y(\sigma))$ from $\mathcal{D}^{\nu}g$ by means of the substitution $y^{i} \to \frac{d^{i}y}{d\sigma^{i}}(\sigma)$, $i = 0,\ldots,\nu$. If g depends on variables other than y, say on t, then the symbol

$$(\mathcal{D}_y^{\nu}g)(t,y^{o},\ldots,y^{\nu})$$

indicates that the operator $\mathcal{D}^{\nu}$ has to be applied with respect to the variable y only. Using (3.4) $\mathcal{D}^{\nu}g$ is easily calculated for $\nu = 0,1,2$:

$$\begin{aligned} &(\mathcal{D}^{o}g)(y^{o}) = g(y^{o}), \quad (\mathcal{D}^{1}g)(y^{o},y^{1}) = g_y(y^{o})y^{1}, \\ &(\mathcal{D}^{2}g)(y^{o},y^{1},y^{2}) = (g_y(y^{o})y^{1})_{y^{o}}\cdot y^{1}+g_y(y^{o})y^{2}. \end{aligned} \tag{3.5}$$

Here g_y denotes the functional matrix of g and $(\ldots)_{y^o}$ denotes the functional matrix with respect to y^{o} of the vector-valued function $g_y(y^{o})y^{1}$. - From the relation (3.4) one obtains immediately the following general recursion formula

$$(3.6) \qquad \mathcal{D}^{\nu+1}g = \sum_{\mu=0}^{\nu} (\mathcal{D}^{\nu}g)_{\mu}\cdot y^{\mu+1},$$

where $(\ldots)\mu$ denotes the Jacobian matrix of $\mathcal{D}^{\nu}g$ with respect to the variable y^{μ}.

Throughout the following considerations the components of w except w^i have to be regarded as fixed. For notational convenience we use for the moment the symbols in the first line of the list below instead of the corresponding symbols in the second line

ω	$y(t)$	$z(t,\omega)$	$k(t,x)$	$l(t,x)$
w^i	$y_{i-1}(t;w)$	$y_i(t;w)$	$f_{i-1}(t,x)$	$f_i(t,x)$

(Note that $y_{i-1}(t;w)$ actually does not depend upon $\omega = w^i$, see the remark following (2.8)). We now write down three relations which are satisfied identically in (t,ω). They are immediate consequences of (2.9) and of the remark following (2.8).

$$(3.7) \qquad \begin{aligned} &\frac{\partial}{\partial t} z(t,\omega) = l(t,z(t,\omega)), \\ &\dot{y}(t) = k(t,y(t)), \quad z(\omega,\omega) = y(\omega). \end{aligned}$$

If the first of these identities is differentiated ν times with respect to ω and if order of differentiation is interchanged we obtain

$$\frac{\partial}{\partial t}\left(\frac{\partial^{\nu}}{\partial \omega^{\nu}} z(t,\omega)\right) = (\mathcal{D}_x^{\nu} l)\left(t,z(t,\omega),\ldots,\frac{\partial^{\nu} z}{\partial \omega^{\nu}}(t,\omega)\right).$$

Here we used the notation just explained (cf. (3.4)). Next we introduce the functions

$$\begin{aligned} z_0(\omega) &= z(\omega,\omega) = y(\omega) \quad \text{(cf. (3.7))} \\ z_{\nu}(\omega) &= \frac{\partial^{\nu} z}{\partial \omega^{\nu}}(t,\omega)\Big|_{t=\omega}, \nu = 1,2,\ldots \end{aligned}$$

Using the preceding identity we obtain a recursion formula for $z_{\nu}(\omega)$ by differentiation of the defining relation with respect to ω:

$$(3.8)\quad \begin{aligned}\frac{dz_\nu}{d\omega}(\omega) &= \frac{\partial}{\partial t}\left(\frac{\partial^\nu z}{\partial \omega^\nu}(t,\omega)\right)\Big|_{t=\omega} + \frac{\partial^{\nu+1} z}{\partial \omega^{\nu+1}}(t,\omega)\Big|_{t=\omega} \\ &= (\mathcal{D}_x^\nu l)(\omega, z_o(\omega),\ldots,z_\nu(\omega)) + z_{\nu+1}(\omega).\end{aligned}$$

From this recursion formula one sees immediately that $z_\nu(\omega)$ admits a representation of the form $h^{(\nu)}(\omega, y(\omega))$, where $h^{(\nu)}(t,x)$ is a function of t and x, which can be expressed directly in terms of the functions $l(t,x)$ and $k(t,x)$. This is true for $\nu = 0$, since $z_o(\omega) = y(\omega)$. Furthermore, if we have a representation

$$z_\nu(\omega) = h^{(\nu)}(\omega, y(\omega)),$$

and if we use the fact that $y(t)$ satisfies the differential eq. $\dot{y} = k(t,y)$ we obtain

$$\frac{dz_\nu}{d\omega}(\omega) = h_t^{(\nu)}(\omega, y(\omega)) + (h^{(\nu)})_x(\omega, y(\omega))\cdot k(\omega, y(\omega)).$$

Hence we are arrived at the following result. Let the functions $h^{(\nu)}(t,x)$ be defined recursively as follows: $h^{(0)}(t,x) = x$, $h^{(\nu+1)}(t,x) = \frac{d}{dt}h^{(\nu)}(t,x) - (\mathcal{D}_x^\nu l)(t, h^{(0)}(t,x),\ldots,h^{(\nu)}(t,x))$, where $\frac{d}{dt}$ means differentiation with respect to the differential equation $\dot{y} = k(t,y)$. Then

$$(3.9)\qquad z_\nu(\omega) = h^{(\nu)}(\omega, y(\omega)).$$

We now return to the original notation and obtain from (3.9) the desired representation for the expression on the right hand side of (3.2) in terms of $\omega = w^i$ and $y(\omega) = z(\omega,\omega) = y_i(w^i;w)$.

<u>3.1</u> <u>Theorem</u>. Let the functions $h_i^{(\nu)}(t,x)$ be defined recursively as follows: $h_i^{(0)}(t,x) = x$,

$$h_i^{(\nu+1)}(t,x) = \left(\frac{d}{dt}\right)_{(i-1)} h_i^{(\nu)}(t,x) - (\mathcal{D}_x^\nu f_i)(t, h_i^{(0)}(t,x),\ldots, h_i^{(\nu)}(t,x)),$$

where $\left(\frac{d}{dt}\right)_{(i-1)}$ denotes differentiation with respect to the differential equation $\dot{x} = f_{i-1}(t,x)$ and where $(\mathcal{D}_x^\nu f_i)(\ldots)$ has to be understood according to the definition (3.4). Then the identity (3.2) and therefore also (3.3) holds.

Examples. We make use of (3.5) and omit occasionally the argument (t,x)

(3.10) $h_i^{(0)}(t,x) = x,\ h_i^{(1)} = f_{i-1} - f_i,$

(3.11)
$$\begin{aligned} h_i^{(2)} &= \left(h_i^{(1)}\right)_t + \left(h_i^{(1)}\right)_x f_{i-1} - (f_i)_x h_i^{(1)} \\ &= (f_{i-1})_t - (f_i)_t + (f_{i-1})_x f_{i-1} + (f_i)_x f_i - 2(f_i)_x f_{i-1} \end{aligned}$$

$$\begin{aligned} h_i^{(3)}(t,x) &= \left(h_i^{(2)}\right)_t(t,x) + \left(h_i^{(2)}\right)_x(t,x)\cdot f_{i-1}(t,x) \\ &\quad - \left[(f_i)_x(t,y)\cdot h_i^{(1)}(t,x)\right]_y \cdot h_i^{(1)}(t,x)\Big|_{y=x} \\ &\quad - (f_i)_x(t,x)h_i^{(2)}(t,x). \end{aligned}$$

4. THE FORMULA FOR GENERAL D.

In this section we are going to show how the calculation of $(Dc)(\tilde{w})$ can be reduced to the special case treated in the previous section.

Let us consider a differential operator D which can be represented as a formal product $D' \circ D''$, where D', D'' satisfy the following properties

(4.1) D' does not act on any w^j for $j \geq i$, $D'' = \partial^\nu/(\partial w^i)^\nu$.

Now the identity (3.2) can be written in this form

$$(D''y_i)(t;w)\Big|_{t=w^i} = h_i^{(\nu)}(w^i, y_i(w^i;w)).$$

If we apply the operator D' on both sides and take into account that D' does not act on w^i we obtain the relation

(4.2) $$(Dy_i)(t;w)\Big|_{t=w^i} = D'\left(h_i^{(\nu)}(w^i, y_i(w^i;w))\right).$$

The evaluation of this relation is based on the following remark. Let $D_1, D_2, \ldots, D_s$ be the different divisors of D', which of course also do not act on each w^j for $j \geq i$. The action of the operator D' on a composite function $h_i^{(\nu)}(t, y(w^1, \ldots, w^{i-1}))$, where y is an arbitrary function of

$w^1,\ldots,w^{i-1}$, can formally be described in this way:

$$D'\left(h_i^{(\nu)}(t,y(w^1,\ldots,w^{i-1}))\right) = q(t,(D_1y)(w^1,\ldots,w^{i-1}),\ldots,(D_sy)(w^1,\ldots,w^{i-1})). \tag{4.3}$$

This relation has to be understood in the following sense: One can associate with the operator D' a certain function $q = q(t,y^1,\ldots y^s)$ of the scalar variable t and s vector-valued variables y^μ such that (4.3) holds for any $y = y(w^1,\ldots,w^{i-1})$. Hence the expression on the right hand side of (4.2) can be obtained from this function q by means of the substitution $y^\mu \to D_\mu y_i(w^i;w)$. The latter term now turns into the derivative $(D_\mu c)(\tilde{w})$ if w is specialized to $\tilde{w}$, and the left hand side of (4.2) turns into $(Dc)(\tilde{w})$. All this follows from Lemma 2.2, since neither D nor $D_1,\ldots,D_s$ acts on any w^j for $j > i$. Hence we obtain the desired recursion formula $(Dc)(\tilde{w}) = q(\tilde{t},(D_1c)(\tilde{w}),\ldots,(D_sc)(\tilde{w}))$.

It is convenient to express this result in a similar way as we did it in the previous section, namely in terms of functions $h_D(t,x)$ which have the property

$$(Dc)(\tilde{w}) = h_D(\tilde{t},\tilde{x}(\tilde{t})). \tag{4.4}$$

It follows then from the preceding considerations that if functions h_{D_μ} with the property (4.4) (for $D = D_\mu$) have been constructed for all divisors D_μ of D', then one can write down also a function h_D with property (4.4) for $D = D' \circ D''$ (provided conditions (4.1) are satisfied), namely

$$h_D(t,x) = q(t,h_{D_1}(t,x),\ldots,h_{D_s}(t,x)). \tag{4.5}$$

We repeat the definition of $q(t,y^1,\ldots,y^s)$ (cf.(4.3)). This function can be obtained from $h_i^{(\nu)}$ by the following procedure. First apply D' to the composite function $h_i^{(\nu)}(t,y(w^1,\ldots,w^{i-1}))$, where y is an arbitrary, not specified function of $w^1,\dots,w^{i-1}$. Then express the result in terms of the derivatives $y^\mu = (D_\mu y)(w^1,\dots,w^{i-1}))$, $\mu = 1,\dots,s$. Note that y itself has to be counted among those derivatives. If say $D_1 y$ is identified with y then one has to take $h_{D_1}(t,x) = x$ (cf. also (3.1) and (3.10)). In particular, if $D' = \partial^\mu/(\partial w^j)^\mu$, then the divisors of D' are just the operators $\partial^\rho/(\partial w^j)^\rho$, $\rho = 0,1,\dots,\mu$, and the procedure described above is the same as the one which leads to the definition of the operator $\mathcal{D}^\nu$ (see (3.4)). Hence

$$h_D(t,x) = (\mathcal{D}_x^\mu h_i^{(\nu)})(t,h_j^{(0)}(t,x),\dots,h_j^{(\mu)}(t,x)) \quad \text{if } D = \frac{\partial^{\mu+\nu}}{(\partial w^i)^\nu(\partial w^j)^\mu} \text{ and } i > j. \tag{4.6}$$

In particular (cf. (3.5), (3.10))

$$h_D(t,x)=(h_i^{(\nu)})_x(t,x)h_j^{(1)}(t,x) \text{ if } D=\frac{\partial^{1+\nu}}{(\partial w^i)^\nu \partial w^j} \text{ and } i>j. \tag{4.7}$$

5. APPLICATION TO CONTROL PROBLEMS.

We consider a control system which is described by a differential equation

$$\dot{x} = f(t,x;u),\ x = (x^1,\dots,x^n)^T,\ u = (u^1,\dots,u^m)^T, \tag{5.1}$$

where the function f is defined and of class C^∞ on a neighborhood of $[t_o,t_e] \times X \times U$, X being an open set in x-space and U an arbitrary non-empty set in u-space. A piecewise C^∞ function $u(\cdot)$ with $u(t\pm 0) \in U$ for all t is called an admissible control function. Piecewise C^∞ means: There exists a subdivision of the interval $[t_o,t_e]$: $t_o < t_1 < t_2 \dots < t_k \le t_e$ such that $u(\cdot)$ coincides on (t_{j-1},t_j) with some function which is of class C^∞ on whole $\mathbb{R}$. At each point of discontinuity we agree to identify the values of $u(\cdot)$ and of its derivatives with their respective left-hand limits.

Let there be given N + 1 admissible control functions $u_o(\cdot),\ldots,u_N(\cdot)$ and let us consider the functions

$$f_i(t,x) = f(t,x;u_i(t)),\ i = 0,\ldots,N. \tag{5.2}$$

They are piecewise of class C^∞ on a neighborhood of the set $\{t,x|t_o \le t \le t_e,\ x \in X\}$, the discontinuities occuring at finitely many hyperplanes $t = t_j$. Each f_i can be extended as a C^∞ function of (t,x) into a set of the form $\{t,x|t_j \le t \le t_j+\varepsilon,\ x \in X\}$.

In the sequel the symbol D will again appear as a subscript and will have the same meaning as before, that is D is a formal differential operator of the form

$$\partial^{\nu 1}/(\partial w^1)^{\nu 1} \circ \ldots \circ \partial^{\nu N}/(\partial w^N)^{\nu N}. \tag{5.3}$$

We now associate with the given system of admissible control functions and with each D a function $h_D(t,x) = h_D(t,x;u_o(\cdot),\ldots,u_N(\cdot))$. The definition of h_D is the same as in the previous sections, the dependence from the control functions $u_i(\cdot)$ arises from the fact that f_i is now given by (5.2). For the reader's convenience we repeat the definition, first in case $D = \partial^\nu/(\partial w^i)^\nu$ - we write then again $h_i^{(\nu)}$ instead of h_D - then in case of a general $D = D' \circ D''$, where D', D" satisfy condition (4.1). Note that the operator $\mathcal{D}_x^\nu$ which appears in the formula for $h_i^{(\nu+1)}$ has to act on the function $f(t,x;u_i(t))$. The application of the operator according to the definition given in Section 3 is then followed by a replacement of all variables (except t of course) by the $h_i^{(\mu)}$, $\mu = 0,\ldots,\nu$.

$$h_i^{(o)}(t,x) = x,$$

$$\begin{aligned} h_i^{(\nu+1)}(t,x) &= \frac{\partial}{\partial t} h_i^{(\nu)}(t,x) + (h_i^{(\nu)})_x(t,x)\cdot f(t,x;u_{i-1}(t)) \\ &\quad - (\mathcal{D}_x^\nu f)(t,x,h_i^{(1)},\ldots,h_i^{(\nu)};u_i(t)),\ i=1,\ldots,N, \end{aligned} \tag{5.4}$$

$$h_D(t,x) = q(t,h_{D_1}(t,x),\ldots,h_{D_s}(t,x)),$$

where $D_1,\ldots,D_s$ are the different divisors of D'. For the definition of $q(t,y^1,\ldots,y^s)$ see the explanation following (4.5). In passing we note that

(5.5) $h_i^{(1)}(t,x;u_o(\cdot),\ldots,u_N(\cdot)) = f(t,x;u_{i-1}(t))-f(t,x;u_i(t))$

(cf. (3.10)).

As can be seen from (5.4) h_D is defined and piecewise of class C^∞ on the set $\{t,x|t_o \le t \le t_e, x \in X\}$.

For notational convenience we introduce some further abbreviations. If D is a differential operator of the form (5.3) we denote by $\nu(D)$ the multiindex $(\nu_1,\ldots,\nu_N)$, by $|\nu(D)|$ the number $\sum\nu_i$ and by $\phi(D)$ the number $(\nu_1!)\ldots(\nu_N!)$. Furthermore, given a N-dimensional vector $z = (z^1,\ldots,z^N)^T$, we denote by $z^{\nu(D)}$ the product of the $(z^i)^{\nu i}$. Using this notation we are now in the position to write down the formal Taylor-series which we associate with each system $u_o(\cdot),\ldots,u_N(\cdot)$ of admissible control functions and on which the main result of this section will be based. This formal series will be denoted by $P(t,x,z;u_o(\cdot),\ldots,u_N(\cdot))$ and it is given by

(5.6) $$\sum_D \frac{1}{\phi(D)} h_D(t,x;u_o(\cdot),\ldots,u_N(\cdot))z^{\nu(D)},$$

where the sum has to be extended over all operators of the form (5.3) (with $h_D(t,x;\ldots) = x$ in case $\nu(D) = 0$). Substituting $t \to \tilde{t}$, $x \to \tilde{x}(\tilde{t})$, $z \to w - \tilde{w}$ one obtains from (5.6) the Taylor-expansion (at $w=\tilde{w}$) of the function which was previously denoted by $x_w(\tilde{t})$. Our interest in the formal series (5.6) arises from the fact that the values of the function $x_w(\tilde{t})$ represent points on admissible trajectories, at least if $w \in \tilde{W}$(cf. (2.2)). This follows from (2.4) and (5.2). As one sees immediately the differential equation for $x_w(t)$ can be written in the form $\dot{x} = f(t,x;u_w(t))$, $u_w(t)$ being an admissible control function.

Finally we turn to the consideration of optimal solutions. So we assume from now on that there is given an admissible control function $u_o(\cdot)$ (reference control) and a solution $\tilde{x}(\cdot)$ of $\dot{x} = f(t,x;u_o(t))$ (reference trajectory), which exists on $[t_o,t_e]$ and which is optimal in the following sense. It minimizes a certain integral subject to boundary constraints of the form $x(t_o) \in M_o$, $x(t_e) \in M_e$, where M_o, M_e are manifolds. As usually we think of the value of the integral as being identical with the value of one component of the (augmented) state variable. Both the constraints and the optimality of $\tilde{x}(\cdot)$ will then be taken into account by the "transversality conditions" quoted in the theorem below. They are of the familiar type and need not be explained further.

5.1 Theorem. Assume that the reference trajectory is optimal in the sense explained above. Let T be a countable set of real numbers contained in (t_o,t_e) and let there be assigned to each $\tilde{t} \in T$ positive integers N,s, positive numbers λ_o, ε_o, a n-dimensional vector $p = p(\tilde{t})$ and vectors depending on a scalar parameter λ

$$z(\lambda) = (z^1(\lambda),\ldots,z^N(\lambda))^T, \quad u_i(t,\lambda), \quad i = 1,\ldots,N,$$

such that the following conditions hold.

(i) $z(\lambda)$ is of class C^∞ on a neighborhood of $[0,\lambda_o]$. We have $z(0) = 0$ and

$$z^1(\lambda) \le z^2(\lambda) \le \ldots \le z^N(\lambda) \le 0 \text{ if } 0 \le \lambda \le \lambda_o.$$

(ii) $u_i(t,\lambda)$ is of class C^∞ on a neighborhood of the set $\Lambda = \{t,\lambda \mid \tilde{t}-\varepsilon_o \le t \le \tilde{t},\ 0 \le \lambda \le \lambda_o\}$ and satisfies

$u_i(t,\lambda) \in U$ for all $(t,\lambda) \in \Lambda$ and $i = 1,\ldots,N$.

(iii) $P(\tilde{t},\tilde{x}(\tilde{t}),z(\lambda);u_o(\cdot), u_1(\cdot,\lambda),\ldots,u_N(\cdot,\lambda)) = \tilde{x}(\tilde{t}) + \lambda^s p + o(\lambda^s)$.

Conclusion. There exists a non-trivial solution $y(\cdot)$ of the adjoint variational equation which satisfies the transversality conditions at $t = t_o, t_e$ and the inequalities $y(t)^T p(t) \leq 0$ for all $t \in T$.

The proof of the theorem will appear elsewhere. Actually a stronger version of the theorem can be established. z, u_i and p are allowed to depend upon some (vector-valued) parameter q. The hypotheses of the theorem have to hold uniformly for all $q \in K$, K being a convex neighborhood of $q = 0$. The conclusion is then as follows: We have $y(t)^T p(t,0) \leq 0$ and in case $p(t,0) = 0$ also $y(t)^T p \leq 0$ for all p in the tangent cone of $p(t,K)$ at $q = 0$.

1. Corollary. The adjoint vector $y(\cdot)$ can always be chosen in such a way that the conclusion of the theorem and in addition the maximum-principle holds, that is

(5.7) $y(t)^T[f(t,\tilde{x}(t);u)-f(t,\tilde{x}(t);u_o(t)) \leq 0,\ u \in U.$

Proof. Choose a countable dense subset $\{u_\nu\}$ of U and for every ν countably many $t_{\nu,\mu}$ which form a dense subset of (t_o,t_e). This can be done in such a way that all $t_{\nu,\mu}$ are different from each other and from the elements of T. Take now the union of T and all these $t_{\nu,\mu}$ and call this set T^*. T^* consists of countably many numbers and it is easy to see that one can satisfy all hypotheses of the theorem by choosing $p(t)$ as before in case $t \in T$ and

$$p(t) = f(t,\tilde{x}(t);u_\nu) - f(t,\tilde{x}(t);u_o(t)$$

in case $t = t_{\nu,\mu}$ (Take $N=1$, $s=1, u_1(t,\lambda)=u_\nu$ and $z^1(\lambda)=-\lambda$. Use

(5.5)). It follows hence from the theorem that (5.7) holds for $t = t_{\nu,\mu}$ and $u = u_\nu$. For reasons of continuity it holds therefore in general.

A similar technique can be used in order to prove the following corollary which links our approach to results obtained via higher order control variations. Actually these variations correspond to special elements of the set E which we are going to introduce now. E consists of all pairs $\pi = (I,p(\cdot))$ which have the following properties.

(i) I is an open subinterval of (t_o,t_e),

(ii) $p(t) = (p^1(t),\ldots,p^n(t))^T$ is a continuous n-dimensional vector on I,

(iii) For each $\tilde{t} \in I$ one can find numbers $N,s,\lambda_o,\varepsilon_o$ and vectors z,u_i such that all hypotheses of Theorem 5.1 are satisfied (with $p = p(\tilde{t})$).

Note that in contrast to the hypotheses of the theorem the dependence of the vector p from t is supposed to be a continuous one.

2. Corollary. Let all hypotheses of Theorem 5.1 hold. Furthermore let there be associated with every $t \in (t_o,t_e)$ a subset Π_t of $\mathbb{R}^n$ and let there exist a countable subset E' of E which is dense in the family $\{\Pi_t\}$. This means: Given $t,p \in \Pi_t$ and $\varepsilon > 0$, one can find a $\pi = (I,p(\cdot) \in E'$ such that $t \in I$ and $||p-p(t)|| \leq \varepsilon$.

Conclusion. The adjoint state vector can be chosen in such a way that it meets all previous requirements and in addition satisfies $y(t)^T p \leq 0$ for all $p \in \Pi_t$ and all $t \in (t_o,t_e)$.

Addendum

The countability condition is unnecessary. The statement of Corollary 2 can be sharpened as follows.

Let the hypotheses of Theorem 5.1 be true. Then the adjoint state vector can be chosen such that it meets all previous requirements and in addition satisfies $y(t)^T p(t) \leq 0$ whenever $\pi = (I, p(\cdot)) \in E$ and $t \in I$.

REFERENCES

1. Krener, A. J., The high order maximal principle and its application to singular extremals, to appear.

2. Krener, A. J., The higher order maximal principle, in *Geometric Methods in System Theory*, Reidel, Dordrecht, 1973.

3. Hermes, H., Local controllability and sufficient conditions in singular problems, to appear.

4. Hermes, H., On local and global controllability, *SIAM J. Control* *12* (1974), 252-261.

5. Brockett, R., Lie algebras and Lie groups in control theory, in *Geometric Methods in System Theory*, Reidel, Dordrecht, 1973.

NEW STABILITY RESULTS FOR NONAUTONOMOUS SYSTEMS

J. P. LaSalle*
Lefschetz Center for Dynamical Systems
Brown University

1. INTRODUCTION

The new invariance properties that have been established by Artstein (see [1-4]) for the positive limit sets of solutions of nonautonomous ordinary differential equations greatly extend Liapunov's direct method for the study of the stability of such time-varying systems. (See [4] for a discussion of the most recent advances, and [4, 10, 11] for a brief history of these developments.) An important role here is played by results which relate Liapunov functions to the location of the positive limit sets of solutions. For autonomous systems results of this type go back to LaSalle in [8] and for nonautonomous systems to Yoshizawa in [16]. A variation of Yoshizawa's result was given by LaSalle in [9] (the two results overlap but neither is more general than the other). Another variation was given by Artstein in [4]. More recently, Onuchic et al in [15] generalize Yoshizawa's basic theorem. Here we will give a theorem (Theorem 1) that generalizes all of these and state some general results on asymptotic stability and instability. A more complete report on these results is to appear (see [12]).

For a generalization in a different direction see [5, 6] by Haddock (he places stronger conditions on $\dot{V}$ and weakens the restriction on f). All of this, the author believes, makes it possible to write the final chapter on the extension of Liapunov's direct method for ordinary differential equations (see [12]).

Consider

$$\dot{x} = f(t,x) \tag{1}$$

where $f:\Gamma^* = R_+ \times G^* \to R^n$; G^* is an open set in $R^n, R_+ = [0,\infty)$. and we always assume that f is continuous in x, measurable in t and satisfies Carathéodory's condition (see, e.g., [7,p.28]).

Our basic assumption on f will be

<u>Assumption B</u>. Given any compact set K in G*, any continuous $u:R_+ \to K$ and any $\alpha > 0$ there exist $T = T(\alpha,K,u) > 0$ and $\beta = \beta(\alpha,K,u) > 0$ such that $\left|\int_a^{a+t} f(s,u(s))ds\right| > \alpha$ for $a > T$ and $t > 0$ implies $t > \beta$.

Both Assumption A in [4] and the assumptions placed on the right-hand side of the differential equation in [15] imply Assumption B.

We are interested only in the future behavior of solutions and confine ourselves to solutions on maximal right-intervals. If we say that x(t) is a solution of (1) starting in a set $\Gamma \subset \Gamma^*$, we mean, for some $(t_0,x^0) \in \Gamma$, x(t) is a solution of (1) defined on a maximal right-interval $[t_0,\omega)$, $t_0 < \omega \leq \infty$, satisfying $x(t_0) = x^0$. If we say that x(t) starts in a set $G \subset G^*$, we mean $x(t_0) \in G$ for some $t_0 \geq 0$. Also x(t) in Γ means $(t,x(t)) \in \Gamma$, and so forth. We use Ω to denote the positive limit set of x(t) ($p \in \Omega$ if there is a sequence $t_n \in [t_0,\omega)$ such that $(t_n,x(t_n)) \to (\omega,p)$ as $n \to \infty$). We say that x(t) is <u>precompact</u> if x(t) is bounded on $[t_0,\omega)$ and $\Omega \subset G^*$ (so that $\omega = \infty$). If x(t) is precompact, then Ω

is nonempty, compact, connected, and is the smallest closed set that $x(t)$ approaches as $t \to \infty$.

In what follows we adopt the following notation: given $\Gamma \subset \Gamma^*$ define

$$G(t) = \{x; (t,x) \in \Gamma\},$$

$$G = \beta(\Gamma) = \bigcup_{t \geq 0} G(t),$$

$$\hat{G} = \alpha(\Gamma) = \bigcap_{t \geq 0} G(t),$$

and

$$G^{\infty} = \gamma(\Gamma) = \bigcap_{t_0 \geq 0} \overline{\bigcup_{t \geq t_0} G(t_0)}.$$

Note that $G(t)$ is the cross section of Γ at time t; G is the projection of Γ onto R^n and $R_+ \times G$ is the smallest cylinder contining Γ; $R_+ \times \hat{G}$ is the largest cylinder inside Γ. Also, $\hat{G} \subset G^{\infty} \subset \overline{G}$, and if $x(t)$ remains in $\Gamma, \Omega \subset G^{\infty}$; thus G^{∞} can give information about the location of Ω.

2. LIAPUNOV FUNCTIONS AND THE LOCATION OF POSITIVE LIMIT SETS.

We first generalize the definition of a Liapunov function given in [9]. Let $V: \Gamma^* \to R$. For $(t_0, x^0) \in \Gamma^*$ define

$$\dot{V}(t_0, x_0) = \sup \liminf_{\theta \to 0^+} \frac{1}{\theta} [V(t_0+\theta, x(t_0+\theta)) - V(t_0, x^0)],$$

where the sup is taken over all solutions $x(t)$ of (1) satisfying $x(t_0) = x^0$.

__1.__ __Definition.__ Let Γ be a subset of $\Gamma^* = R_+ \times G^*$. We say that $V: \Gamma^* \to R$ is a __liapunov function of__ (1) __on__ Γ if (i) V is continuous on Γ, (ii) given a compact set $K \subset G^*$ there is an $m(K)$ such that $m(K) \leq V(t,x)$ for all $(t,x) \in \Gamma$ and $x \in K$, and (iii) $\dot{V}(t,x) \leq 0$ for all $(t,x) \in \Gamma$.

It then follows (see [9]), if V is a liapunov function of (1) on Γ and $x(t)$ is a solution of (1) with $(t, x(t)) \in \Gamma$

for $t \in [\alpha,\beta]$, that $V(t,x(t))$ is differentiable almost everywhere on $[\alpha,\beta]$ and

$$V(t,x(t)) - V(\alpha,x(\alpha)) \le \int_\alpha^t \dot{V}(\tau,x(\tau))d\tau \le 0 \qquad (2)$$

for all $t \in [\alpha,\beta]$.

In [9] the definition of a liapunov function was confined to cylinders $\Gamma = R_+ \times G$, $G \subset G^*$. In this case we will say that V is a <u>liapunov function of</u> (1) <u>on</u> G.

Next we generalize the notion of the set E associated with a liapunov function V of (1) on Γ. The set E is defined by $x \in E$ if there is a sequence $(t_n,x^n) \in \Gamma$ such that $(t_n,x^n) \to (\infty,x)$ and $\dot{V}(t_n,x^n) \to 0$ as $n \to \infty$. Then E is closed relative to G^* and $E \subset G^\infty$. Define also $V^{-1}(c)$ by $x \in V^{-1}(c)$ if there is a sequence $(t_n,x^n) \in \Gamma$ such that $(t_n,x^n) \to (\infty,x)$ and $V(t_n,x^n) \to c$ as $n \to \infty$. We then have (the proof is an easy modification of the proof of Theorem 1 in [9]):

<u>1.</u> <u>Theorem</u>. Assume that Condition B is satisfied and that V is a liapunov function of (1) on Γ. If $x(t)$ is a precompact solution of (1) in Γ for all $t \ge t_0 \ge 0$, then $x(t)$ approaches, for some constant c, a component of $E \cap V^{-1}(c)$ as $t \to \infty$.

<u>1.</u> <u>Remark</u>. Under the same conditions as Theorem 1 it is also easy to see that <u>if</u> $x(t)$ <u>is a solution of</u> (1) <u>that remains in</u> Γ <u>for</u> $t \in [t_0,\omega)$ <u>then</u> $\Omega \cap G^* \subset E$. Thus, when E is empty, the result can give information about instability.

Let us now give in connection with Theorem 1 a result that is useful in applications. It shows how boundedness and precompactness of solutions can be determined by test functions V (which may or may not be liapunov functions).

<u>2.</u> <u>Definition</u>. A set $\Gamma \subset \Gamma^*$ is said to be <u>positively invariant</u> if each solution of (1) starting in Γ remains thereafter in Γ. If Γ is positively invariant and each solution

starting in Γ is precompact, we say that Γ is <u>strongly positively invariant</u>. A set $G \subset G^*$ is said to be (<u>strongly</u>) <u>positively invariant</u> if $R_+ \times G$ is (strongly) positively invariant.

If Γ is positively invariant and $x(t)$ starts in Γ at time t_0, then $x(t) \in G(t)$ for all $t \geq t_0$ and hence $x(t) \in G$. This does not imply that G is positively invariant but it does mean that solutions starting in $\hat{G}$ remain thereafter in G.

<u>1.</u> <u>Proposition</u>. Let $V:\Gamma^* \to R$ be continuous, and let Γ be a component of $\{(t,x) \in \Gamma^*; V(t,x) < c\}$. If $\dot{V}(t,x) \leq 0$ on Γ, Γ is positively invariant. If G is bounded and $\overline{G} \subset G^*$, then Γ is strongly positively invariant.

3. THEOREMS ON STABILITY AND INSTABILITY.

The first theorem of this section generalizes Liapunov's theorem on asymptotic stability. Let V be a liapunov function of (1) on Γ, and let W be a continuous function on $\overline{G} \subset G^*$ into R_+ such that

$$\dot{V}(t,x) \leq -W(x) \leq 0, \text{ all } (t,x) \in \Gamma.$$

Define

$$E(W) = \{x; W(x) = 0,\ x \in \overline{G} \cap G^*\}.$$

Note that

$$E \subset E(W).$$

<u>2.</u> <u>Theorem</u>. Let Γ be a strongly positively invariant set in Γ^*. If (i) V is a liapunov function of (1) on Γ, (ii) $E(W)$ is contained in the interior of $\hat{G}$, (iii) corresponding to each compact set $K \subset R^n$ there is a β_K such that $0 \leq V(t,x) \leq \beta_K$ for all $(t,x) \in \Gamma$ and $x \in K$ and (iv) $E(W)$ is uniformly stable, then $E(W)$ is uniformly asymptotically stable. If G is bounded,

then $x(t) \to E(W)$ as $t \to \infty$ uniformly for all solutions starting in Γ (and hence for all solutions starting in $\hat{G}$).

The next theorem is an immediate corollary of Theorem 1 and generalizes a result of Marachkov (see [13] or [17]).

3. Theorem. Let Γ be a strongly positively invariant set in Γ. If (i) f satisfies Assumption B and (ii) there is a Liapunov function of (1) on Γ, then solutions starting in Γ (or in $\hat{G}$) approach E as $t \to \infty$. If (iii) E is contained in the interior of $\hat{G}$, then E is an attractor and $\hat{G}$ is in its region of attraction. If (iv) E is stable, then E is asymptotically stable.

The differences between the two results is that in Theorem 2 there is a stronger condition on V whereas in Theorem 3 there is an additional restriction on f. Of interest here would be sufficient conditions for uniform asymptotic stabilty under Assumption B but with conditions on V weaker then those in Theorem 2. One such result can be found in [14] (the assumption on f is stronger but the proof holds under Assumption B).

Let $V:\Gamma^* \to R$, and let $\sum_0$ be a component of $\{(t,x) \in \Gamma^*;\ V(t,x) > 0\}$ and $\sum_c = \{(t,x) \in \sum_0;\ V(t,x) > c\}$. Define $\Gamma_c = \{(t,x) \in \sum_c;\ x \in N\}$, where N is a bounded open set of R^n and $\bar{N} \subset G^*$. Consider the following conditions ($p \in N$):

C_1. p is on the boundary of G_0.

C_2. p is on the boundary of $G_0(t_0)$ for some $t_0 \geq 0$.

C_3. p is on the boundary of $G_0(t_0)$ for each $t_0 \geq 0$.

C_4. p is on the boundary of $\hat{G}_0 = \alpha(\Gamma_0)$.

C_5. V depends only upon x, S_0 is a component of $\{x; V(x) > 0\}$, $G_0 = S_0 \cap N(\Gamma_0 = R_+ \times G_0)$, and p is on the boundary of G_0.

The next theorem follows from Theorem 1 and generalizes Cetaev's instability theorem.

4. Theorem. Assume (i) f satisfies Assumption B, (ii) $-V$ is a liapunov function of (1) on Γ_0, and (iii) $G_c^\infty \cap E$ is empty for each $c > 0$. If $p \in N$ is an equilibrium point in (1), then

(1). C_1 implies there are solutions starting arbitrarily near p that leave N and p is not uniformly stable.

(2). C_2 implies there is a t_0 such that there are solutions starting arbitrarily near p at time t_0 that leave N and p in unstable at time t_0.

(3). C_3 implies for each $t_0 \geq 0$ there are solutions starting arbitrarily near p at time t_0 that leave N and p is uniformly unstable.

(4). C_4 implies that each solution starting in $\hat{G}_0$ leaves N and p is uniformly unstable.

(5). C_5 implies that each solution starting in G_0 leaves N and p is uniformly unstable.

We have a variety of instabilities, but in each case N is a measure of the instability. If it is know that $S_0 = \beta(\Sigma_0)$ is positively invariant, then one can also conclude that p is not an attractor. Note that if C_5 is satisfied, then (iii) of Theorem 4 can be replaced by $G_0 \cap E$ is empty, but the conclusion then is, given a neighborhood N_0 of p with $N_0 \subset N$, each solution starting in $S_0 \cap N_0$ leaves N_0.

REFERENCES

1. Artstein, Z., Topological dynamics of an ordinary differential equation, *J. Diff. Eqs.* (to appear).

2. Artstein, Z., Topological dynamics of ordinary differential equations and Kurzweil equations, *J. Diff. Eqs.* (to appear).

3. Artstein, Z., The limiting equations of nonautonomous ordinary differential equations (preprint).

4. Artstein, Z., Limiting equations and stability of nonautonomous ordinary differential equations, Appendix A in [11].

5. Haddock, J., Liapunov functions and boundedness and global existence of solutions, *Appl. Anal. 1*(1972), 321-330.

6. Haddock, J., Stability theory for nonautonomous systems, Dynamical Systems, An International Symposium, ed. L. Cesari, J. Hale, and J. LaSalle, Academic Press, New York, 1976, Vol. II, 271-274.

7. Hale, J. K., *Ordinary Differential Equations*, Wiley-Interscience, New York, 1969.

8. LaSalle, J. P., The extent of asymptotic stability, *Proc. Nat. Acad. Sci., U.S.A. 46*(1960), 363-365.

9. LaSalle, J. P., Stability theory for ordinary differential equations, *J. Diff. Eqns. 4*(1968), 57-65.

10. LaSalle, J. P., Stability theory and invariance principles, Dynamical Systems, An International Symposium, ed. L. Cesari, J. Hale, and J. LaSalle, Academic Press, New York, 1976, Vol. I, 211-222.

11. LaSalle, J. P., *The Stability of Dynamical Systems*, CBMS Regional Conference Series in Applied Math. (to be published by SIAM).

12. LaSalle, J. P., Stability of nonautonomous systems, *J. Nonlinear Anal.: Theory, Methods, Applications* (to appear).

13. Marachkov, V., On a theorem of Liapunov, *Bull. Soc. Phys. Math. Kazan III 12*(1940), 171-174.

14. Morgan, A., and Narendra, K., On the uniform asymptotic stability of certain nonautonomous differential equations, *SIAM J. Control* (to appear).

15. Onuchic, N., Onuchic, L. R., and Taboas, P., Invariance properties in the theory of stability for ordinary differential systems and applications, *Appl. Anal. 5*(1975), 101-107.

16. Yoshizawa, T., Asymptotic behavior of solutions of a system of differential equations, *Contrib. Diff. Eqs.* *1*(1963), 371-387.

17. Yoshizawa, T., *Stability Theory by Liapunov's Second Method,* Publication No. 9, The Mathematical Society of Japan, Tokyo, 1966.

---

*This research was supported in part by the Air Force Office of Scientific Research Grant #AF-AFOSR-71-2078D and the United States Army Grant #AROD AAG 29-76-G-0052.

BRANCHING OF SOLUTIONS OF TWO-PARAMETER BOUNDARY VALUE PROBLEMS FOR SECOND ORDER DIFFERENTIAL EQUATIONS

W. S. Loud
Department of Mathematics
University of Minnesota

Dedicated to Professor Karl Klotter on his seventy-fifth birthday.

1. INTRODUCTION

This paper is a study of branching of solutions of two-point boundary-value problems for certain second-order nonlinear differential equations containing two parameters. The problem considered is

$$x'' + g(t,x,x',\lambda,\mu) = 0. \tag{1.1}$$

$$\begin{aligned} a_{11}x(0) + a_{12}x'(0) + b_{11}x(b) + b_{12}x'(b) &= k_1, \\ a_{21}x(0) + a_{22}x'(0) + b_{21}x(b) + b_{22}x'(b) &= k_2. \end{aligned} \tag{1.2}$$

It is assumed that $g(t,x,x',\lambda,\mu)$ is a sufficiently regular function of five real variables such that initial value problems for (1.1) have unique solutions, and that the 2×4 matrix of the a's and b's has rank 2.

The principal assumption of the paper is that there exists a solution $x_o(t)$ of (1.1), (1.2) when $\lambda = \mu = 0$. There are many recent results on the existence of solutions of nonlinear boundary-value problems, so that this is not an unreasonable assumption. We consider the problem of what solutions of (1.1), (1.2) exist for small nonzero values of the parameters λ and μ.

2. METHOD OF ATTACK

We consider the solutions of (1.1) for small λ and μ which have initial conditions close to those of $x_o(t)$. Let $x(t,\xi,\eta,\lambda,\mu)$ denote that solution of (1.1) for which at $t = 0$:

$$x(0,\xi,\eta,\lambda,\mu) = x_o(0) + \xi \ , \ x'(0,\xi,\eta,\lambda,\mu) = x_o'(0) + \eta.$$

The solution $x(t,\xi,\eta,\lambda,\mu)$ will satisfy the boundary conditions (1.2) if and only if $F(\xi,\eta,\lambda,\mu) = G(\xi,\eta,\lambda,\mu) = 0$ where $F(\xi,\eta,\lambda,\mu) = a_{11}\xi + a_{12}\eta + b_{11}(x(b,\xi,\eta,\lambda,\mu) - x_o(b)) + b_{12}(x'(b,\xi,\eta,\lambda,\mu) - x_o'(b))$ $G(\xi,\eta,\lambda,\mu) = a_{21}\xi + a_{22}\eta + b_{21}(x(b,\xi,\eta,\lambda,\mu) - x_o(b)) + b_{22}(x'(b,\xi,\eta,\lambda,\mu) - x_o'(b))$ Because $x(t,0,0,0,0) = x_o(t)$, $F(0,0,0,0) = G(0,0,0,0) = 0$.

The various partial derivatives of F and G at the origin can be computed by standard procedures. All such derivatives can be expressed as integrals involving various partial derivatives of g evaluated for $x = x_o(t)$, $x' = x_o'(t)$, $\lambda = \mu = 0$ together with solutions of the linear variational equation

(2.1) $$y''+g_{x'}(t,x_o(t),x_o'(t),0,0)y'+g_x(t,x_o(t),x_o'(t),0,0)y=0$$

and its adjoint. Let the matrix of first derivatives at the origin be

(2.2) $$M = \left\| \begin{matrix} F_\xi & F_\eta & F_\lambda & F_\mu \\ G_\xi & G_\eta & G_\lambda & G_\mu \end{matrix} \right\|,$$

and let J denote the submatrix which is the Jacobian of F and G with respect to ξ and η:

(2.3) $$J = \left\| \begin{matrix} F_\xi & F_\eta \\ G_\xi & G_\eta \end{matrix} \right\|.$$

We shall consider those cases for which M has rank 1 or rank 2. The rank of J will be 0 or 1 and, of course, Rank J $\le$ Rank M.

When the rank of J is 2, the system $F = G = 0$ can be solved uniquely for ξ and η as functions of λ and μ. This means that for small λ and μ there is a unique solution of (1.1), (1.2) near to $x_o(t)$. We shall call this the nonbranching case. It clearly corresponds to the variation equation (2.1) having no nontrivial solution which satisfies the boundary conditions (1.2) with $k_1 = k_2 = 0$.

3. BRANCHING RESULTS

The cases of branching will occur when the rank of J is less than 2. We consider four cases:

Case I: Rank M = 2, Rank J = 1;

Case II: Rank M = 2, Rank J = 0;

Case III: Rank M = 1, Rank J = 1;

Case IV: Rank M = 1, Rank J = 0.

In each of these cases the implicit function theorem allows the system F = G = 0 to be solved at least partially for some of the variables in terms of the others. To determine ξ and η as functions of λ and μ it is always necessary to impose further conditions, usually on second derivatives, but in some cases on higher derivatives. We call a case generic when the simplest such condition is satisfied, and we assume the generic case always. The detailed statements of the conditions on higher derivatives are quite combersome, and the reader is referred to the complete version of the paper for them. We shall give some of the details for Case I.

In Case I it is usually possible to solve the system F = G = 0 for ξ and λ as functions of η and μ:

$$\xi = H(\eta,\mu), \qquad \lambda = K(\eta,\mu)$$

where $H(0,0) = K(0,0) = K_\eta(0,0) = 0$. Assume that the second derivative $K_{\eta\eta}(0,0) \neq 0$. Then the equation $\lambda = K(\eta,\mu)$ can be solved for η as a function of λ and μ in the following sense.

There is a curve in the λ-μ plane, tangent at the origin to the line $\lambda = K_\mu(0,0)\mu$, such that for (λ,μ) on one side of this curve there are two solutions for η, while for (λ,μ) on the other side of the curve, there are none. Each such (λ,μ), when substituted into the relation $\xi = H(\eta,\mu)$ gives $\xi(\lambda,\mu)$, so that a solution of (1.1), (1.2) is determined.

1. Theorem. Let the rank of M be 2 *and let the rank* J *be* 1. *Let the critical second derivative be nonzero. Then there is a curve in the* λ-μ *plane passing through the origin with a well-defined tangent at the origin such that if* (λ,μ) *is near the origin and on one side of the curve, there are two solutions of the problem* (1.1) (1.2) *which converge to* $x_0(t)$ *as* $(\lambda,\mu) \to (0,0)$, *while if* (λ,μ) *is close to the origin and on the other side of the curve, there are no solutions of* (1.1), (1.2) *with this property*.

In Case II the condition for the problem to be generic is that a certain determinant not be zero.

2. Theorem. Let the rank of M *be* 2 *and let the rank of* J *be* 0. *Let the critical determinant* Δ *be nonzero. Then if* $\Delta > 0$, *for all points* (λ,μ) *sufficiently close to the origin there are solutions of* (1.1), (1.2) *which converge to* $x_0(t)$ *as* $(\lambda,\mu) \to (0,0)$. *If* $\Delta > 0$, *there are two nontangential curves through the origin in the* λ-μ *plane having well-defined tangents at the origin, which divide the neighborhood of the origin into four sector-like regions. For* (λ,μ) *in one of these four regions, there are four solutions of the problem* (1.1), (1.2) *which converge to* $x_0(t)$ *as* $(\lambda,\mu) \to (0,0)$ *in this region. For* (λ,μ) *in any one of the other three regions, there are no solutions of the problem* (1.1), (1.2) *with this property*.

In Case III it is usually possible to solve one of the equations $F = 0$ and $G = 0$ for ξ as a function of (η, λ, μ). The second equation will then give an equation $K(\eta, \lambda, \mu) = 0$. At $(0,0,0)$ K and its first partial derivatives are all zero. Then according to the algebraic signs of various second derivatives or combinations of second derivatives of K at the origin, there are no solutions for η for small λ and μ; two solutions for all small (λ, μ); two solutions for some (λ, μ) and none for others (λ, μ); or one solution for small (λ, μ) in a complete neighborhood of the origin with two diametrically opposite small sectors excluded. This leads to the following theorem.

3. *Theorem. Let the matrices* M *and* J *each have rank* 1. *Then according to the behavior of critical second derivatives the following possibilities for branching of solutions of* (1.1), (1.2) *near* $x_0(t)$ *exist.*

(i) *There may be no solutions of* (1.1), (1.2) *for small* $(\lambda, \mu) \neq (0,0)$ *near to* $x_0(t)$.

(ii) *For all small* (λ, μ) *close to* $(0,0)$ *there may exist two solutions of* (1.1), (1.2) *which converge to* $x_0(t)$ *as* $(\lambda, \mu) \to (0,0)$.

(iii) *There may exist two curves in the* (λ, μ)*-plane which pass through the origin with well-defined distinct tangents and which divide a neighborhood of the origin into four sector-like regions. For* (λ, μ) *in either of two opposite sectorial regions, there are two solutions of* (1.1), (1.2) *which converge to* $x_0(t)$ *as* $(\lambda, \mu) \to (0,0)$ *while for* (λ, μ) *in the remaining two sectorial regions there are no* solutions close to $x_o(t)$.

(iv) There may be a definite line through the origin such that if (λ,μ) is close to the origin and not included in small sectors containing the critical line, there is one solution of (1.1), (1.2) which converges to $x_o(t)$ as $(\lambda,\mu) \to (0,0)$ while remaining outside the excluded sectors.

The result for Case IV is the following.

4. Theorem. Let M of rank 1 and let J be of rank 0. Then branching of solutions of (1.1), (1.2) near $x_o(t)$ for small λ and μ can be described as follows. There is a definite straight line through the origin and a number of curves (at most four, and possibly none) which behave locally like parabolas with vertex at the origin and tangent there to the straight line. These curves divide a neighborhood of the origin into a number of subregions. Depending on the algebraic signs of critical quantites, for (λ,μ) in any one of these subregions there are either four, two, or no solutions of (1.1), (1.2) which converge to $x_o(t)$ as $(\lambda,\mu) \to (0,0)$.

Suppose finally that we are in Case I again but that for $\lambda = \mu = 0$, the problem (1.1), (1.2) has a one-parameter family of solutions. In this case the hypothesis of Theorem 1 will not be satisfied. Indeed the conclusion is quite different. Let the family of solutions be $x(t,\alpha)$ with $x_o(t) = x(t,0)$.

5. Theorem. Let M have rank 2, let J have rank 1, and let (1.1), (1.2) have a one-parameter family of solutions when $\lambda = \mu = 0$. Then if a critical second derivative is not zero, there are two lines in the λ-μ plane through the origin and small sectorial regions including the first of these lines

and excluding the second, such that if (λ,μ) is contained in one of these small sectorial regions there is a solution of (1.1), (1.2) near to $x_o(t)$ which converges to $x_o(t)$ if $(\lambda,\mu) \to (0,0)$ in a direction tangent to the first line. If $(\lambda,\mu) \to (0,0)$ in a direction tangent to some other line through the origin but within the small sector, then the solution tends to a member $x(t,\alpha)$ of the family of solutions with α small.

As an application of this let $x_o(t)$ be a nonconstant L-periodic solution of $x'' + g(x) = 0$, where $g(x)$ is sufficiently regular and satisfies $xg(x) > 0$ for $x \neq 0$. Then if $f(t)$ is also L-periodic the equation

$$x'' + \lambda x' + g(x) = \mu f(t)$$

will have a T-periodic solution not a translation of $x_o(t)$ if (λ,μ) is in a small sector which excludes the line $\mu = 0$. This solution will approach $x_o(t)$ or a translation of $x_o(t)$ depending on the direction of approach of (λ,μ) to $(0,0)$ within the sector.

REFERENCE

1. Loud, W. S., Branching of solutions of two-parameter boundary value problems for second order differential equations, *Ingenier Archiv* (to appear).

---

*The research for this paper was supported in part by Grant No. DA-ARO-D-31-124-73-G199, U. S. Army Research Office. The complete version of this paper will appear in Ingenieur-Archiv. What follows is a summary of the result.

PERIODIC SOLUTIONS OF NONLINEAR TELEGRAPH EQUATIONS

Jean Mawhin
Institut Mathématique
Université de Louvain

1. INTRODUCTION

This paper is devoted to the study of generalized periodic solutions of nonlinear telegraph equations of the form

(1.1) $$\beta u_t + u_{tt} - u_{xx} = g(u) + h(t,x)$$

where $\beta \neq 0$, g is a continuous real function and h is Lebesgue integrable over I^2 with $I = [0,2\pi]$. A generalized periodic solution of (1.1) (shortly GPS) is a real function $u \in H$, (where H will denote from now the Hilbert space $L^2(I^2)$ with the usual inner product (,)), such that, for all real C^2-functions v on I^2 which are 2π-periodic in both variables, one has

(1.2) $$(u, -\beta v_t + v_{tt} - v_{xx}) = (g(u),v) + (h,v)$$

Let us remark that a necessary condition for (1.2) to have a meaning is that g be such that $g(u) \in H$ when $u \in H$. Also the 2π-periodicity in x can be replaced by other boundary conditions like Dirichlet or Neumann ones. We consider it only for the sake of a greater symmetry in the treatment of equation (1.1). Also, many of the results obtained for (1.1) can be extended to equations of the form

$$\beta u_t + u_{tt} - u_{xx} = f(t,x,u)$$

or to systems of such equations but we restrict ourselve to (1.1) for the sake of simplicity. In contrast to previous results obtained for equations of type (1.1) by various authors (see [2-3, 18-20] for a bibliography), our results will not be restricted to small nonlinearities although g will have to satisfy a linear growth restriction in order to map H into itself. The obtained theorems will be more in the spirit of recent work on ordinary and elliptic partial differential equations generalizing to some nonlinear equations the nonresonance and resonance situations which occur in linear equations (see for example [6, 14] for recent surveys). Also, we shall be interested in getting results which do not depend upon the size of $|\beta|$, the nonvanishing of this number being only assumed to insure the compactness properties required by a treatment using degree theory. For existence and uniqueness results when $\beta = 0$ using the Banach fixed point theorem, see [15].

2. THE LINEAR CASE

Let us consider the equation

$$\beta u_t + u_{tt} - u_{xx} - \lambda u = h(t,x) \tag{2.1}$$

where $\beta \neq 0$ and λ are constants.

If u, h respectively have the Fourier series

$$u(t,x) = \sum_{(p,m)\in Z^2} u_{pm} \exp i\,(pt+mx)$$

$$h(t,x) = \sum_{(p,m)\in Z^2} h_{pm} \exp i\,(pt+mx)$$

with $\quad u_{-p,-m} = \overline{u}_{pm},\ h_{-p,-m} = \overline{h}_{pm},$

then the following result is easily proved.

<u>2.1</u> <u>Proposition</u>. $u \in H$ <u>is a</u> GPS <u>of</u> (2.1) <u>if and only if</u>, <u>for all</u> $(p,m) \in Z^2$,

$$[i\beta p + (m^2 - p^2 - \lambda)]u_{pm} = h_{pm}. \tag{2.2}$$

We can now prove the following Fredholm alternative for (2.1).

2.1 Theorem. Equation (2.1) has a GPS u for any $h \in H$ if and only if

(2.3) $$\lambda \neq m^2 \quad (m \in \mathbb{N})$$

in which case the solution is unique and defines a linear compact operator

$$T_\lambda : H \to H, \; h \mapsto u.$$

If

$$\lambda = q^2$$

for some $q \in \mathbb{N}$ then (2.1) has a solution if and only if

$$h \in H_q = \{h \in H : h_{oq} = 0\}$$

in which case there is a unique GPS $u \in H_q$ defining a linear compact operator

$$\tilde{T}_q : H_q \to H_q, \; h \to u.$$

Moreover one has, for any β,

$$|T_\lambda| \le [\text{dist}\ (\lambda, \Sigma)]^{-1}$$

$$|\tilde{T}_q| \le [\text{dist}\ (q^2, \Sigma \setminus \{q^2\})]^{-1}$$

where

$$\Sigma = \{m^2 - p^2 : m, p \in \mathbb{Z}\} = (2\mathbb{Z}+1) \cup 4\mathbb{Z}.$$

Proof. If (2.1) has a GPS for any $h \in H$, then (2.2) is solvable in u_{pm}, $(p,m) \in \mathbb{Z}^2$ for any h_{pm} and hence necessarily (2.3) holds. If (2.3) holds, then the Fourier series

(2.4) $$\sum_{(p,m)\in \mathbb{Z}^2} [i\beta p + m^2 - p^2 - \lambda]^{-1} h_{pm} \exp i\,(pt+mx)$$

is such that

(2.5) $$\sum_{(p,m)\in \mathbb{Z}^2} (1+p^2+m^2)\,|h_{pm}|^2 (\beta^2 p^2 + (m^2-p^2-\lambda)^2)^{-1} < \infty.$$

In fact, as is easily checked, one has for all $(p,m) \in \mathbb{Z}^2$,

$$\frac{1+p^2+m^2}{\beta^2 p^2 + (m^2-p^2-\lambda)^2} \le \min \left(\frac{1}{\sigma} + \frac{|\lambda+1|}{\sigma^2}, \; \frac{|2+\lambda|+2}{\beta^2}, \; \frac{2+\beta^2+|1+\lambda|}{\beta^2} \right)$$

where $\sigma = \text{dist}\ (\lambda, \{m^2 : m \in \mathbb{N}\})$, and hence the convergence of (2.5) follows from that of $\sum |h_{pm}|^2$. This implies moreover that the Fourier series (2.4) defines a function $u \in H$ which clearly satisfies (2.2) and hence is a GPS of (2.1). Its uniqueness follows from Proposition 2.1 and the compactness of T_λ follows from (2.5) and Rellich's theorem [1]. Now if $h \in H$,

$$|T_\lambda h|^2 = (2\pi)^2 \sum_{(p,m)\in Z^2} [\beta^2 p^2 + (m^2 - p^2 - \lambda)]^{-2} |h_{pm}|^2$$

$$\leq (2\pi)^2 [\text{dist } (\lambda, \Sigma)]^{-2} \sum_{(p,m)\in Z^2} |h_{pm}|^2$$

$$= [\text{dist } (\lambda, \Sigma)]^{-2} |h|^2$$

and similarly, if $h \in H_q$,

$$|\tilde{T}_q h|^2 = (2\pi)^2 \sum_{\substack{(p,m)\in Z^2 \\ (p,m)\neq(0,\pm q)}} [\beta^2 p^2 + (m^2 - p^2 - q^2]^{-2} |h_{pm}|^2$$

$$\leq [\text{dist } (q^2, \Sigma \setminus \{q^2\})]^{-2} |h|^2$$

Q. E. D.

In the sequel, $P_q : H \to H$ will denote the orthogonal projector such that $H = \text{Im } P_q \oplus H_q$. Hence, by Theorem 2.1, equation (2.1) with $\lambda = q^2$ will be solvable if and only if $P_q h = 0$, in which case the unique solution in H_q is given by

$$u = \tilde{T}_q h.$$

We shall also need the following result. Let $C(I^2)$ be the space of continuous functions on I^2 which are 2π-periodic in both variables and $||\cdot||$ the usual uniform norm on $C(I^2)$, and let

$$C_0 = C(I^2) \cap H_0 = \{h \in C(I^2) : P_0 h = 0\}.$$

<u>2.2 Proposition.</u> $\tilde{T}_0$ <u>maps</u> H_0 <u>continuously into</u> C_0 <u>and is a compact operator from</u> C_0 <u>into itself</u>.

<u>Proof</u>. Let $h \in H_0$. Then, if

$$h(t,x) = \sum_{|p|+|m|>0} h_{pm} \exp i\,(pt+mx)$$

and has

$$(\tilde{T}_0 h)(t,x) = \sum_{|p|+|m|>0} (i\beta p + m^2 - p^2)^{-1} h_{pm} \exp i\,(pt+mx)$$

and hence

$$4\pi^2 (\tilde{T}_0 h)(t,x) = (k*h)(t,x)$$

if the convolution product has a meaning, where k(t,x) is given by the Fourier series

$$\sum_{|p|+|m|>0} (i\beta p+m^2-p^2)^{-1} \exp i(pt+mx).$$

This series defines an element in H_0 because

$$\sum_{|p|+|m|>0} [\beta^2 p^2+(m^2-p^2)^2] = \sum_{m\neq 0} m^{-4} + \sum_{p\neq C} \beta^{-2} p^{-2} +$$

$$+ \sum_{\substack{|m|\neq|p| \\ p\neq 0}} [\beta^2 p^2+(m^2-p^2)^2]^{-1}.$$

and all the series are convergent because

$$[\beta^2 p^2+(m^2-p^2)^2]^{-1} \leq (m^2-p^2)^{-2}$$

and

$$\sum_{|m|\neq|p|} (m^2-p^2)^{-2}$$

converges. Therefore, by a classical argument of the convolution product theory [7], $k_* h$ is in $C_0(I^2)$. Now, if $\xi = (t,x)$,

$$|(\tilde{T}_0 h)(\xi+\delta) - (\tilde{T}_0 h)\xi| =$$

$$= |(2\pi)^{-2} \int_{I^2} [k(\xi+\delta-\eta)-k(\xi-\eta)]h(\eta)d\eta| \leq$$

$$\leq (2\pi)^{-2} [\int_{I^2} [k(\xi+\delta-\eta)-k(\xi-\eta)]^2 d\eta]^{1/2} |h|,$$

and because of

$$|h| \leq (2\pi)^2 ||h||,$$

and the fact that, k being in H,

$$\int_{I^2} [k(\xi+\delta-\eta)-k(\xi-\eta)]^2 d\eta \to 0$$

if $\delta \to 0$, the compactness of $\tilde{T}_0$ in C_0 follows from Arzela-Ascoli theorem.

Q.E.D.

3. A FIRST NONLINEAR NONRESONANCE CASE

The easiest nonresonance existence theorem for equation (1.1) is the following one, the notations and general assumptions, being the ones of sections 1 and 2.

3.1 Theorem. Assume that there exist

$$\mu < a \le b < \nu$$

with μ, ν consecutive elements of Σ, and that there exists $R > 0$ such that

$$a \le u^{-1} g(u) \le b \tag{3.1}$$

for $|u| > R$. Then, for each $h \in H$, equation (1.1) has at least one GPS.

Proof. Let $\lambda \in]\mu,\nu[$. Then equation (1.1) is equivalent to

$$\beta u_t + u_{tt} - u_{xx} - \lambda u = g(u) - \lambda u + h(t,x). \tag{3.2}$$

If we write, for $u \in H$,

$$G_\lambda u = g(u) - \lambda u + h,$$

then it follows from (3.1) and a result of Krasnosel'skii [9] that G_λ maps H continuously into itself and, as is easily checked,

$$|G_\lambda u| \le \max\ (|b-\lambda|,\ |a-\lambda|)\ |u| + c,$$

where $c \ge 0$ is a constant depending only upon $\max_{|u| \le R} |g(u)|$ and $|h|$. It follows now from Theorem 2.1 that equation (3.2), and then (1.1) is equivalent to the fixed point problem

$$u = T_\lambda G_\lambda u$$

in H and that $T_\lambda G_\lambda$ is compact on bounded sets of H. Moreover, for each $u \in H$,

$$|T_\lambda G_\lambda u| \le [\min(\lambda-\mu, \nu-\lambda)]^{-1} [\max(|b-\lambda|, |a=\lambda|)|u| + c]$$

and an elementary analysis shows that the coefficient of $|u|$ will be strictly smaller than one if λ is chosen in

$$]2^{-1}(\mu+b),\ 2^{-1}(\nu+a)[$$

which is always possible. The result follows then from Schauder [17] or Rothe [16] or Granas [8] fixed point theorem.

Q.E.D.

4. A SECOND NONLINEAR NONRESONANCE CASE

Theorem 3.1 shows that a GPS exists for equation (1.1) when $u^{-1}g(u)$ stays away from two consecutive elements of Σ for $|u|$ large. We shall now show that such a result essentially still holds if $u^{-1}g(u)$ "jumps" from one element of Σ to the preceding (or the following) one, in the terminology of Fucik [5,6]. See also Dancer [4] for the case of elliptic problems. If $u \in H$, we shall denote by u^+ (resp. u^-) the positive (resp. negative) part of u.

4.1 Lemma. *Let $\mu < \nu$ be two nonzero consecutive elements of Σ and a,b such that*

$$\mu \leq a \leq b \leq \nu$$

and

$$(a,b) \neq (\mu,\mu), \ (a,b) \neq (\nu,\nu).$$

Then the unique GPS of equation

$$\beta u_t + u_{tt} - u_{xx} = au^+ - bu^- \tag{4.1}$$

is $u = 0$.

Proof. If $u = u^+$, then

$$\beta u_t + u_{tt} - u_{xx} - au = 0$$

and hence, if $a \neq q^2$, $q \in Z\backslash\{0\}$ (note that necessarily $a \neq 0$), then $u = 0$ and if $a = q^2$, $q \in Z\backslash\{0\}$, then

$$u(t,x) = A \cos qx + B \sin qx,$$

and hence $u = u^+$ if and only if $u = 0$. Similarly if $u = u^-$ is a GPS of (4.1) then $u = 0$. If $a = b$, then $a = b \neq \mu$ and $a = b \neq \nu$ and hence

$$\beta u_t + u_{tt} - u_{xx} = au^+ - au^- = au$$

implies that $u = 0$. Thus we can assume that $a < b$. Let u be a GPS of (4.1) such that $u^+ \neq 0$, $u^- \neq 0$ and let

$$\rho = 2^{-1}(\nu+\mu), \ \sigma = 2^{-1}(\nu-\mu).$$

Then

$$\rho = \mu + \sigma = \nu - \sigma$$

and (4.1) is equivalent to

$$u = T_\rho[(a-\rho)u^+ - (b-\rho)u^-],$$

with the notations of Theorem 2.1. Thus

$$(4.2) \qquad |u| \leq \sigma^{-1}|(a-\rho)u^+ - (b-\rho)u^-|$$

with the equality sign if and only if one of the numbers λ, μ has the form q^2 with $q \in Z$ (that cannot be true for both) and $(a-\rho)u^+ - (b-\rho)u^-$ has the form

$$A \cos qx + B \sin qx,$$

i.e. satisfies

$$(4.3) \qquad \beta u_t + u_{tt} - u_{xx} - q^2 u = 0.$$

Now, using the orthogonality of u^+ and u^-,

$$|(a-\rho)u^+ - (b-\rho)u^-| =$$

$$(4.4) \qquad = [(a-\rho)^2|u^+|^2 + (b-\rho)^2|u^-|^2]^{1/2} \leq \sigma|u|,$$

with the equality sign if and only if

$$|a-\rho| = |b-\rho| = \sigma.$$

Therefore if one of the conditions for equality in (4.2) or (4.4) is not satisfied,

$$|u| < |u|,$$

a contradiction, and hence $u = 0$. If they are simultaneously satisfied, then either

$$\mu = q^2 = a, \; b = \nu$$

or

$$a = \mu, \quad \nu = q^2 = b$$

where in both cases $(a-\rho)u^+ - (b-\rho)u^-$ is a solution of (4.3). But then

$$(a-\rho)u^+ - (b-\rho)u^- = -\sigma u^+ - \sigma u^- = -\sigma|u| < 0$$

satisfies (4.3), which is impossible. Thus $u = 0$ and the

proof is complete.

Q.E.D.

The following result is a slight generalization of a theorem of Fučik [5]. We use the terminology of [13, 14].

4.2 *Lemma*. *Let* X, Z *be normed spaces*, $L : \operatorname{dom} L \subset X \to Z$ *be a linear Fredholm mapping of index zero with continuous right inverses*, $S : X \to Z$, *positive homogeneous (i.e.* $S(ax) = aS(x)$ *for all* $a \geq 0$ *and* $x \in X$), $R : X \to Z$ *quasibounded of quasi-norm zero, be mappings which are* L-*compact on bounded sets. Then, if for all* $k \in [0,1]$,

$$(4.5) \qquad Lx - (1+k)^{-1}Sx + k(1+k)^{-1}S(-x) = 0$$

implies $x = 0$, *one has*

$$(L-S-R)(\operatorname{dom} L) = Z.$$

Proof. By condition (4.5) and the homotopy invariance theorem of coincidence degree theory [10, 13] one has, for any open ball B(R) centered at the origin, with

$$S_k(x) = (1+k)^{-1}Sx - k(1+k)^{-1}S(-x),$$

$$d[(L,S),B(r)] = d[(L,S_0),B(R)] = d[(L,S_1),B(R)] = 1 \pmod 2$$

using the fact that

$$S_1(x) = 2^{-1}[Sx-S(-x)]$$

is odd and the generalized Borsuk theorem [10, 13].

Now it follows from Lemma III.2 in [13] that there exists $\mu > 0$ such that

$$|Lx-Sx| \geq \mu$$

when $|x| = 1$ and hence, using the positive homogeneity of S,

$$|Lx-Sx| \geq \mu|x|$$

when $x \in \operatorname{dom} L$. Now, N being quasibounded of quasinorm zero, there exists $\nu \geq 0$ such that

$$|Nx| \leq (\mu/2)|x| + \nu,$$

and hence, if $y \in Z$,

$$|Nx-y| \leq (\mu/2)|x| + \nu + |y|.$$

Therefore if $R > 2\mu^{-1}(\nu+|y|)$ and $|x| \leq R$ one has

$$|Nx-y| \leq (\mu/2)R + \nu + |y| < \mu R$$

and hence, using the generalized Rouché's theorem given in [13], Th. III.3, one gets

$$d[(L,S+R-y), B(R)] = d[(L,S), B(R)] = 1 \pmod 2)$$

and the proof is complete

Q.E.D.

4.1 Theorem. *Assume that* g *is such that there exist two consecutive nonzero elements* $\mu < \nu$ *of* Σ *and two reals* a *and* b *with*

$$\mu \leq a < b \leq \nu \tag{4.6}$$

such that either

$$\lim_{u\to\infty} u^{-1}g(u) = a, \quad \lim_{u\to-\infty} u^{-1}g(u) = b$$

or

$$\lim_{u\to-\infty} u^{-1}g(u) = a, \quad \lim_{u\to\infty} u^{-1}g(u) = b.$$

Then equation (1.1) *has at least one* GPS *for any* $h \in H$.

Proof. Let us consider the first case, the second one is similar. Equation (1.1) is equivalent to

$$\beta u_t + u_{tt} - u_{xx} - (au^+-bu^-) - g(u) - (au^+-bu^-)=h(t,x). \tag{4.7}$$

Let $P_0 : H \to H$ be the (orthogonal) projector defined in Section 2. It follows from Theorem 2.1 and the equivalence theorem of [10] that the GPS of (4.7), and hence of (1.1) are the solutions of the equation in H

$$u - P_0u - [P_0+\tilde{T}_0(I-P_0)](Au+Bu+h) \tag{4.8}$$

where $\tilde{T}_0$ is defined in Theorem 2.1,

$$A : H \to H, \; u \mapsto au^+ - bu^-$$

$$B : H \to H, \; u \mapsto g(u) - (au^+-bu^-)$$

If

$$L = I - P_0,\ S = [P_0+\tilde{T}_0(I-P_0)]Au,\ R = [P_0+\tilde{T}_0(I-P_0)]B,$$
$$y = [P_0+\tilde{T}_0(I-P_0)]h,$$

then (4.8) has the form

$$Lu - Su - Ru = y,$$

and $L : H \to H$ is Fredholm of index zero with continuous right inverses, S is positive homogeneous, R is quasibounded of quasinormzero and both S and R are L-compact on bounded sets because of the compactness of P_0 and $\tilde{T}_0(I-P_0)$. Thus, to apply Lemma 4.2, we only have to verify (4.5), i.e. to verify that for all $k \in [0,1]$,

$$Lu - (1+k)^{-1}Su + k(1+k)^{-1}S(-u) = 0 \Rightarrow u = 0,$$

which is equivalent to u = 0 being the unique GPS of

(4.9) $\quad \beta u_t + u_{tt} = u_{xx} - (1+k)^{-1}[(a+bk)u^+-(b+ak)u^-] = 0.$

But, if (4.6) holds and $k \in [0,1]$, then

$$\mu \le (1+k)^{-1}(a+kb) < (1+k)^{-1}(b+ba) \le \nu,\ k \in [0,1[,$$

and for k = 1, (4.9) is equivalent to

$$\beta u_t + u_{tt} - u_{xx} - 2^{-1}(a+b)u = 0.$$

Using then Lemma 4.1 the proof is complete.

Q.E.D.

5. A THIRD NONLINEAR NONRESONANCE CASE

Theorem 4.1 covers the case where $u^{-1}g(u)$ "jumps" between two consecutive elements of Σ when u goes from $-\infty$ to $+\infty$ or vice versa. We shall now consider cases where $u^{-1}g(u)$ has for limit the same element of Σ when $|u| \to \infty$. In this section we shall assume that this limit is an element of Σ which is not a perfect square. By analogy to the case of a linear function g, this is still a nonresonance case.

5.1 Theorem. Assume that

$$u^{-1}g(u) \to \infty \text{ as } |u| \to \infty,$$

where $\mu \in \Sigma$ but is not a perfect square. Then, for each $h \in H$, equation (1.1) has at least one GPS.

Proof. Equation (1.1) is clearly equivalent to

$$\beta u_t + u_{tt} - u_{xx} - \mu u = g(u) - \mu u + h(t,x)$$

and hence to the equation in H

$$u = T_\mu G_\mu u$$

if

$$G_\mu u = g(u) - \mu u + h.$$

By the assumptions made, G_μ maps H continuously into itself and is quasibounded of quasi-norm zero. The same is therefore true for the map $T_\mu G_\mu$ which is moreover compact on bounded sets. The result then follows from Granas theorem [8].

Q.E.D.

6. A FIRST NONLINEAR RESONANCE CASE

Theorem 5.1 covers the case where $u^{-1}g(u)$ tends to a non perfect square element of Σ when $|u| \to \infty$. We shall now assume that this limit is a perfect square and, by analogy to the case where the function g is linear, this will be called a resonance case.

We shall assume in this section that

(6.1) $$u^{-1}g(u) \to 0 \text{ as } |u| \to \infty.$$

Like in the linear case, supplementary conditions upon h are required to insure the existence of a GPS. Moreover here supplementary requirements will be made upon g.

6.1 Theorem. Assume that (6.1) holds and that

$$\lim_{u \to \pm\infty} g(u) = g_\pm$$

with g_+ *and* g_- *finite or not. Then if*

(6.2) $$g_- < -(2\pi)^{-2} \int_{I^2} h < g_+$$

or

$$g_+ < -(2\pi)^{-2} \int_{I^2} h < g_-,$$

equation (1.1) *has at least one* GPS. *If moreover, for all* $u \in R$,

$$g_- < g(u) < g_+$$

or

$$g_+ < g(u) < g_-$$

then condition (6.2) *or* (6.3) *is necessary.*

Proof. Like in Theorem 4.1 and with the same definition for P_0 we reduce our problem to the equation in H

(6.4) $$u - P_0 u = [P_0 + \tilde{T}_0(I-P_0)](Gu+h)$$

where

$$G : H \to H, \ u \mapsto g(u).$$

That G maps H continuously into itself follows from the condinuity of g, (6.1) and Krasnosel'skii's result [9]. If

$$L = I - P_0, \ N = [P_0 + \tilde{T}_0(I-P_0)](G+h)$$

it is easy to check that $L : C(I^2) \to C(I^2)$ is a continuous Fredholm operator of index zero and, using Proposition 2.2, $N : C(I^2) \to C(I^2)$ is compact on bounded sets and quasibounded of quasinorm zero. We then apply Theorem 6.1 of [11] to the equation

$$Lu = Nu$$

in $C(I^2)$. The argument is exactly similar to the one given in [12] and will not be reproduced here.

Q.E.D.

7. A SECOND NONLINEAR RESONANCE CASE

We shall now assume that

(7.1) $$u^{-1}g(u) \to q^2$$

when $|u| \to \infty$, with $q \in Z\backslash\{0\}$. Thus g can be written

(7.2) $$g(u) = q^2 u + r(u)$$

where

(7.3) $$u^{-1}r(u) \to 0 \text{ if } |u| \to \infty.$$

The following result gives the existence of a GPS for (1.1) if (7.3) is satisfied in a restricted way.

7.1 Theorem. Assume that $g : R \to R$ has the form (7.2) where $r : R \to R$ is such that, for some $\delta \in [0,1[$, $r_+, r_- \in R$, one has

(7.4) $$|u|^{-\delta} r(u) \to r_{\pm} \text{ as } u \to \pm\infty.$$

Then, if $\delta > 0$ and

(7.5) $$r_+ > r_-,$$

equation (1.1) has at least one GPS for any $h \in H$. If $\delta = 0$ and if $h \in H$ is such that

(7.6) $$4\pi(r_+ - r_-) + \int_{I^2} hw > 0$$

for all $w(s) = \sin(qx+\phi)$, then equation (1.1) has at least one GPS.

Proof. Equation (1.1) can be written as

$$\beta u_t + u_{tt} - u_{xx} - q^2 u = r(u) + h(t,x)$$

and hence is equivalent to the abstract equation in H (see e.g. [10])

(7.7) $$(I-P_q)u = [P_q + \tilde{T}_q(I-P_q)]Ru,$$

where P_q and $\tilde{T}_q$ are defined in section 2 and

$$(Ru)(t,x) = r(u(t,x)) + h(t,x).$$

By our assumptions, $L = I - P_q$ is continuous Fredholm of index zero and $N = [P_q + \tilde{T}_q(I-P_q)]R$ is L-compact on bounded subsets of H. Moreover, for some $\mu \geq 0$, $\nu \geq 0$,

$$|Nu| \leq \mu |u|^{\delta} + \nu$$

for all $u \in H$. We shall apply Theorem 3.3 of [14] to

$$Lu = Nu$$

and show first that

$$(\forall \text{ bounded } V \subset \text{Im } L)(\exists\, t_0 > 0)(\forall\, t \geq t_0)(\forall\, v \in V)$$
$$(\forall\, w \in \ker L \cap \partial B(1)) : (N(tw + t^{\delta} v), w) > 0.$$

If it is not the case, then

$$(\exists \text{ bounded } V \subset \text{Im } L)(\exists \text{ sequence } (t_n),\ t_n > 0,\ t_n \to \infty \text{ if}$$
$$n \to \infty)(\exists \text{ sequence } (v_n),\ v_n \in V)(\exists \text{ sequence } (w_n),$$
$$w_n \in \ker L \cap \partial B(1)) : (N(t_n w_n + t_n^{\delta} v_n), w_n) \leq 0,$$

i.e.

$$\int_{I^2} [t_n^{-\delta} r(t_n w_n + t_n^{\delta} v_n) w_n + h w_n] \leq 0. \tag{7.9}$$

By going if necessary to subsequences we can assume that $w_n \to w \in \ker L \cap \partial B(1)$ and using an argument strictly similar to the one used in the proof of Theorem 5.3 of [14], one shows that the left-hand side of (7.9) goes, when $n \to \infty$, to

$$\int_{I_+^2} r_+ |w|^{1+\delta} - \int_{I_-^2} r_- |w|^{1+\delta} \tag{7.10}$$

if $\delta > 0$ and to

$$\int_{I_+^2} r_+ |w| - \int_{I_-^2} r_- |w| + \int_{I^2} hw \tag{7.11}$$

if $\delta = 0$, where

$$I_{\pm}^2 = \{(t,x) \in I^2 : w(x) \gtrless 0 = I \times \{x \in I : w(x) \gtrless 0\} = I \times I_{\pm}.$$

But, $w(x)$ being of the form $c \sin(qx+D)$, $c \geq 0$, one verifies easily that, for any $\delta \in [0,1[$,

$$\int_{I_+^2} |w|^{1+\delta} = \int_{I_-^2} |w|^{1+\delta} = (1/2) \int_{I^2} |w|^{1+\delta}.$$

Therefore (7.9) and (7.10) imply, if $\delta > 0$,

$$r_+ - r_- \leq 0,$$

a contradiction to (7.5), and, if $\delta = 0$, because of

$$(1/2) \int_{I^2} |w| = 4\pi c,$$

$$4\pi(r_+ - r_-) + \int_{I^2} hc^{-1} w \leq 0,$$

a contradiction to (7.6). Now, using (7.8) with $V = \{0\}$ and the orthogonality of P_q one gets, for some $t_0 > 0$ and all $w \in \ker L \cap \partial B(1)$,

$$(P_q N(t_0 w), w) = (N(t_0 w), w) > 0$$

and hence

$$P_q N(z) \neq 0$$

for $z \in \ker L \cap \partial B(t_0)$, and, using the Poincaré-Bohl theorem,

$$d_B[P_q N \ker L, B(t_0), 0] = 1,$$

which completes the proof.

Q.E.D.

Remark. An elementary but somewhat tedious computation shows that (7.6) is equivalent to

$$r_+ - r_- - 4\pi|h_{0,q}| > 0.$$

8. REMARKS ON THE GPS OF NONLINEAR HEAT EQUATIONS

The study of the GPS of the nonlinear heat equation

$$u_t - u_{xx} = g(u) + h(t,x) \tag{8.1}$$

where g and h are like in section 1 can be made in a completely similar way as for the telegraph equation (1.1). If Σ denotes now the set of perfect squares, one can easily verify that Theorems 3.1, 4.1 hold for equation (8.1), as well as Theorem 6.1 and 7.1, the proofs being completely similar.

REFERENCES

1. Agmon, S., *Lectures on Elliptic Boundary Value Problems*, Van Nostrand, Princeton, 1965.

2. Cesari, L., Functional analysis and differential equations, *SIAM Studies in Appl. Math 5* (1969), 143-155.

3. Cesari, L., Nonlinear analysis, in *Nonlinear Mechanics*, (CIME Lecture Notes, Cremonese, Roma, 1973), 3-95.

4. Dancer, E. N., On the Dirichlet problem for weakly non-linear elliptic partial differential equations, to appear.

5. Fučik, S., Boundary value problems with jumping nonlinearities, *Casopis pestov. matem.* *101*(1976), in press.

6. Fučik, S., Open problems in the solvability of nonlinear equations, *Tagung über Nichtlineare Funktionalanalysis und ihre Anwendungen, Oberwolfach*, (January, 1976).

7. Garnir, H. G., *Fonctions de Variables Réelles, Vol. II*, Vander, Louvain, 1967.

8. Granas, A., The theory of compact vector fields and some of its applications, *Rozprawy Matem.* *30*(1962), 1-91.

9. Krasnosel'skii, M. A., *Topological Methods in the Theory of Nonlinear Integral Equations*, Pergamon, New York, 1963.

10. Mawhin, J., Equivalence theorems for nonlinear operator equations and coincidence degree theory for some mappings in locally convex topological vector spaces, *J. Differential Equations* *12*(1972), 610-636.

11. Mawhin, J., The solvability of some operator equations with a quasibounded nonlinearity in normed spaces, *J. Math. Anal. Appl.* *45*(1974), 455-467.

12. Mawhin, J., Problèmes aux limites du type de Neumann pour certaines équations différentielles ou aux dérivées partielles non linéaires, in *Equations Differentielles et Fonctionnelles non Linéaires*, Hermann, Paris, 1973, 123-134.

13. Mawhin, J., *Nonlinear Perturbations of Fredholm Mappings in Normed Spaces and Applications to Differential Equations*, Univ. de Brasilia, Trabalho de Matem. no. 61, May, 1974.

14. Mawhin, J., Topology and nonlinear boundary value problems, in *Intern. Symposium on Dynamical Systems, Vol. I*, Academic Press, New York, 1976, 51-82.

15. Mawhin, J., Solutions periodiques d'équations aux dérivées partielles hyperboliques non linéaires, to appear.

16. Rothe, E., Zur theorie der topologischen Ordnung und der Vektorfelder in Banach Räumen, *Compositio Math.* *5*(1937), 177-197.

17. Schauder, J., Der fixpunktsatz in functionalräumen, *Studia Math.* *2*(1930), 171-180.

18. Vejvoda, O., Nonlinear boundary value problems for differential equations, in *Differential Equations and Their Applications*, Praha, Publ. House, Czech. Acad. Sci., 1963, 199-215.

19. Vejvoda, O., Periodic solutions of nonlinear partial differential equations of evolution, in *Differential Equations and their Applications*, Bratislava, Acta Fac.

Rer. Natur. Univ. Comenianae, Math., 17, 1967, 293-300.

20. Vejvoda, O., Periodic solutions of partial differential equations, in *Proceed. IVth Intern. Confer. Nonlinear Oscillations*, Praha, Academia, 1968, 277-283.

NON-SELFADJOINT SEMILINEAR EQUATIONS AT SIMPLE RESONANCE IN THE ALTERNATIVE METHOD[1]

P. J. McKenna
Department of Mathematics
University of Michigan

1. INTRODUCTION

Since the introduction of methods of functional analysis (Cesari [2], [3]) into the classical bifurcation process of Poincaré, Lyapunov, and Schmidt, the theory of alternative methods has developed. The central idea is the splitting of a given operational equation

(1.1) $$Ex = Nx$$

in suitable spaces, into a system of auxiliary and bifurcation equations which are then studied by a variety of different methods (Banach's and Schauder's fixed point theorem, topological degree, Schauder's invariance of domain, etc., c.f. [4]).

In this paper we assume E to be a non-selfadjoint linear operator in a Hilbert space S, and by systematic consideration of the adjoint operator E^* and of the selfadjoint products E^*E and EE^*, we present a new method of splitting equation (1.1) into auxiliary and bifurcation equations, which may then be studied by the aforementioned methods.

[1]The results presented here are contained in the author's Ph.D. thesis written at the University of Michigan under the direction of Professor Lamberto Cesari.

In particular, we discuss the case where N is Lipschitzian, and then we show that the splitting can always be achieved in such a way that the auxiliary equation can be solved by Banach's fixed point theorem. The solution of the bifurcation equation is then a finite-dimensional problem.

This extends in a natural way the results of Cesari for selfadjoint operators [3], [4] and the subsequent work of Knobloch [7], [8]. Our analysis gives a constructive process for a choice of subspaces and projection operators, similar to those whose existence had been proved by Hale [6] and by Rothe [13] under different hypotheses.

Futhermore, the reduction to the finite-dimensional problem is achieved under far less restrictive hypotheses on L than those used by Osborn and Sather. The method may also be used in cases where the operator L has Freholm indices (p,q) where $p \neq q$.

2. PRELIMINARIES

Let S be a real Hilbert space with inner product (,) and norm || ||. We shall assume that the operator $E:\mathcal{D}|E| \to S$, $D(E) \subset S$, satisfies:

(2.1) The non-negative selfadjoint operator E^*E possesses an orthonormal sequence of eigenvectors ϕ_i associated eigenvalues α_i^2, where $\alpha_i \geq 0$ and $\alpha_i \to +\infty$. Let J be the smallest integer such that $\alpha_i > 0$ for $i \geq J$.

(2.2) The range of E is of finite codimension.

We remark that conditions (2.1) and (2.2) are satisfied if E is any real uniformly elliptic partial differential operator of order 2m with smooth coefficients and coercive boundary conditions [1].

Note that EE^* is also a nonnegative selfadjoint operator, and we shall explicitly describe below an orthonormal sequence of eigenvectors of EE^* which form a basis of RanE.

<u>Lemma</u>. Let $\psi_i = E\phi_i/(||E\phi_i||)$, $i \geq J$. Then ψ_i are eigenvectors of EE^* and are an orthonormal basis for RanE. First, the ψ_i's are eigenvectors of EE^*. Indeed,

$$EE^*(E\phi_i) = E(E^*E)\phi_i = \alpha_i^2 E\phi_i$$

and thus

$$EE^*\phi_i = \alpha_i^2\phi_i.$$

Also,

$$(E\phi_i, E\phi_i) = (E^*E\phi_i, \phi_i) = \alpha_i^2(\phi_i, \phi_i) = \alpha_i^2,$$

and thus

$$||E\phi_i|| = \alpha_i, \quad E\phi_i = \alpha_i\psi_i, \quad i = 0,1,2,\ldots$$

Also, the ψ_i's are orthonormal. Indeed, for $i \neq j$ we have

$$(\psi_i, \psi_j) = \frac{1}{||E\phi_i||}\frac{1}{||E\phi_j||}(E\phi_i, E\phi_j)$$

$$= \frac{1}{||E\phi_i||}\frac{1}{||E\phi_j||}(E^*E\phi_i, \phi_j)$$

$$= \frac{\alpha_i^2\ \delta_{ij}}{||E\phi_i||\ ||E\phi_j||} = \delta_{ij}$$

Finally, the ψ_j's are a basis for RanE. Indeed, let $y \in$ RanE, and assume that $(y, \psi_i) = 0$, $i \geq J$. Then $(y, \sum_{i\geq J} c_i\psi_i) = 0$ for all c_i with $\Sigma_{i\geq J}\, c_i^2 > \infty$. Thus for every $x \in \mathcal{D}(E)$ we have $x = \sum_{1=0}^{\infty} a_i\phi_i$, $Ex = \Sigma_i a_i(E\phi_i) = \Sigma a_i\alpha_i\psi_i$ with $\Sigma a_i^2\alpha_i^2 < \infty$. For $c_i = a_i\alpha_i$, we then have $(y, Ex) = 0$ and this is proved for all $y \in \mathcal{D}(E)$. Thus if $(y, \psi_i) = 0$ for all $i \geq J$, we have $y \in (\mathrm{RanE})\perp = \ker E^*$. Thus $y = 0$ and our lemma is proved.

We will require that N satisfy:

(2.3) There exists $A > 0$ so that we have $||Nx-NY|| \leq A||x-y||$ for all $x,y \in S$.

Let k be an integer $k \geq J$ and let us define the projection operator $P_k: S \to S$. For every $x \in S$ or $x = \Sigma_0^\infty c_i\phi_i$ with $\Sigma_0^\infty c_i^2 < \infty$, we take

$$P_k x = \sum_{i=0}^{k} c_i\phi_i, \quad (I-P_k)x = \Sigma_{k+1}^\infty c_i\phi_i.$$

Now, we define the projection $Q_k: S \to S$. For every $x \in S$ we define $I - Q_k$ by

$$(I-Q_i) = \Sigma_{k+1}^\infty (x,\psi_i)\psi_i.$$

Thus for $x = y + z$, $y \in \ker E^*$, $z \in (\ker E^*)\perp = \operatorname{Ran} E$ $z = \Sigma_{i\geq J}\, c_i\psi_i$, we take

$$Q_k x = y + \sum_{i=J}^{i=k} c_i\psi_i.$$

3. THE MAIN CONSTRUCTION

(3.i) <u>Theorem</u>. Under the assumptions (2.1) and (2.2) on E and the assumption (2.3) on N, the problem $Ex = Nx$ can be reduced to a finite-dimensional problem.

In other words, the original equation $Ex = Nx$ can be split into an auxiliary and bifurcation equation and we do this in such a way that the auxiliary equation is solvable by Banach's fixed point theorem.

<u>Proof</u>. We define a partial inverse H_k of E,

$$H_k : (I-Q_k)S \to (I-P_k)S_p, \quad p < 1$$

by

$$H_k(\Sigma_{k+1}^\infty c_i\psi_i) = \Sigma_{k+1}^\infty (1/\alpha_i)c_i\phi_i.$$

Then for $x \in \Sigma_{k+1}^\infty c_i\psi_i$, we have

$$\begin{aligned}||H_kx||^2 &= ||H_k(\sum_{k+1}^{\infty} c_i\psi_i)||^2 \\ &= \Sigma_{k+1}^{\infty} \alpha_i^{-2}c_i^2 \\ &\le \alpha_{k+1}^{-2} (\Sigma_{k+1}^{\infty} c_i^2)^{\theta} \\ &\le \alpha_{k+1}^{-2} ||x||^2.\end{aligned}$$

Then the norm of the operator $H_k:(I-Q_k)S \to (I-P_k)S$ is α_{k+1}^{-1}. Thus $||H_k||$ can be made as small as we want by taking k sufficiently large and the following basis axioms of the alternative methods hold.

(2.4)
(i) $EH_k(I-Q_k)u = (I-Q_k)u, \forall\, u \in H$.
(ii) $H_k(I-Q_k)Eu = (I-P_k)u, \forall\, u \in \mathcal{D}(E)$.
(iii) $EPu = QEu, \forall\, u \in \mathcal{D}(E)$.

Thus, under these conditions, equation (1.1) is equivalent to the following system of auxiliary and bifurcation equations:

(2.5)
(i) $x = Px + H(I-Q)Nx$
(ii) $Q(Ex-Nx) = 0.$

To solve (i) for any given $x_o \in P_kS_p$, we search for a fixed point of the map $T(x_o+x_1) = x_o + H(I-Q_k)N(x_o+x_1)$ where T is regarded as mapping the fibre $x_o + (I-P_k)S_p$ into itself. If k is chosen so that

$$A\alpha_{k+1}^{-1} < a < 1,$$

then T is a contraction since

$$\begin{aligned}||Tx_1-Tx_1'|| &= ||H_k(I-Q_k)(N(x_o+x_1) - N(x_o+x_1'))|| \\ &\le ||H_k||A||x_1-x_1'|| \le a||x_1-x_1'||.\end{aligned}$$

Therefore for each $x_o \in P_{kp}$, we can find $\tau(x_o) \in (I-P_k)p$ such that $x_o + \tau(x_o)$ satisfies the auxiliary equation.

$\tau(x_o)$ depends continuously on x_o. The problem is reduced to finding an $x_o \in P_kS_p$ which satisfies

$$Q_k(Ex_o - N(x_o + \tau(x_o))) = 0.$$

This is just the question of looking for zeros of a continuous map from the finite dimensional space $P_k S_p$ into the space $Q_k S$. Theorem (3.i) is thereby proved.

4. BOUNDED NONLINEARITIES FOR FREDHOLD INDICES (1,1).

In this section, we discuss the bifurcation equation. Namely we search for conditions under which the bifurcation map $\Phi: P_k S_p \to Q_k S$, defined by $\Phi(x_o) = Q_k(Ex_o - N(x_o + \tau(x_o)))$, possesses zeros in $P_k S_p$. Our results will depend on the restrictions we place on N and also on the nature of the linear problem $Eu = f$.

We consider the case where E is a differential operator of some order 2m, not necessarily selfadjoint, satisfying (2.1) in a bounded domain G of R^n, $n \geq 1$, where $S = L_2(G)$ and

(4.1) $$Nx = g(x) - f(t), \quad t \in G,$$

and $f \in L_2(G)$, $g: R' \to R'$ is a continuous bounded Lipschitzian real-valued function with

$$|g(s)| \leq M, \quad |g(x) - g(x')| \leq A|s - s'|$$

for all s', $s \in R$ (without loss of generality, means $G = 1$).

Moreover, we assume that both the kernel and cokernel of the linear operator E are equal to one.

In this case $S = L^2$. Then $J = 1$ and for any $k \geq 1$, $P_k S$ and $Q_k S$ have the same dimension $k + 1$. Let $\{\psi_o\}$ be a basis for $\ker E^*$. Thus every element $y \in Q_k S$ is of the form $y = \Sigma_o^k \, d_i \psi_i$.

We define the map $R: Q_k S \to P_k S$ by $R(\Sigma_o^k \, d_i \psi_i) = \Sigma_o^k \, d_i \psi_i$. Then $R\Phi: P_k S \to P_k S$. Thus, the bifurcation equation takes the form

$$RQ_k(Ex_o - N(x_o + \tau(x_o))) = 0$$

with $x_o = \Sigma_o^k c_i\phi_i \in P_k S$.

For the study of this equation, we shall make use of the assumptions above on E and N and we shall need the further hypothesis of the same kind studies in Chapter 2.

(4.2) $\forall R_1 > 0\ \exists R_o > 0$ so that if $c \in R'y \in (\ker E)\perp$, $|c| \geq R_o$, $||y|| \leq R_1$, then $(N(c\phi_o + y), c\psi_o) \geq 0$ (or ≤ 0).

(ii) Theorem. Under assumptions (2.1) and E = (1,1) on E and hypotheses (4.1), (4.2) on N, then the equation Ex = Nx has at least a solution $x \in S(E) \subset S$.

Proof. We can regard R as a map from R^{k+1} into R^{k+1}. Let us consider the hypercube

$$\Pi = \{c = (c_o,\ldots,c_k) \,|\, |c_i| \leq A_i,\ i = 1,\ldots,k,\ |c_o| \leq A_o\}$$

(or the corresponding set of elements $x = \Sigma_o^k c_i\phi_i$ in $P_k S$), where $A_i > 2\alpha_i^{-1} J$, $i = 1\ldots k$, and $A_o > 0$ will be chosen later. We shall ensure that on the (k+1) sides $F_j = \{|c_i| \leq A_i$, $i = 0,\ldots,k$, $i \neq j$, $c_j = \pm A_j\}$, the corresponding coordinate functions $(R\Phi x_o, \phi_j)$ have opposite signs. We take $R_1 = \Sigma_1^k A_i^2 + \alpha_{k+1}^{-1} M$ and take A_o to be the corresponding R_o in (4.2).

First we deal with the face F_j, $j = 1\ldots k$. Let

$$x_o = \Sigma_o^k{}_{,\ i \neq j}\ a_i\phi_i \pm A_i\phi_i$$

$$R\Phi(x_o) = \Sigma_{\substack{i=1\\ i\neq j}}^k a_i\alpha_i\phi_i \pm A_j\alpha_j\phi_j - RQ_k N(x_o + \tau(x_o))$$

$$R\Phi(x_o), \phi_j) = \pm A_j\alpha_i - (RQ_k N(x_o + \tau(x_o), \phi_j)$$

and by the choice of A_j, and the fact that

$$|(RQ_k N(x_o + \tau(x_o)), \phi_j)| \leq M$$

we see that $(R\Phi(x_o))_i$ has the same sign as $\pm A_i$ for $i = 1\ldots k$.

On the zeroth face of the hypercube Π we have

$$x_o = \pm A_o \phi_o + \Sigma_1^k a_i \phi_i, \text{ where } \Sigma_1^k a_i^2 \leq \Sigma_1^k A_i^2$$

and since $||\tau(x_o)|| = ||H(I-P)N(x_o+\tau(x_o))|| \leq ||H||M$, we have ensured that

$$(R\Phi(x_o)\phi_o) = (QN(\pm R\phi_o+\Sigma_1^k a_i\phi_i+\tau(x_o)), \pm R_o\phi_o)$$

is greater than or equal to zero by (4.2), and thus $(R\phi)_o$ takes opposite signs on the opposite faces $F_o, -F_o$ of Π.

Thus $(R\Phi)_j$ takes opposite signs on the opposite sides $\pm F_j, j = 0,1...k$, of the hypercube Π. By the well known Miranda's version [11] of Brouwer's fixed point theorem, we have solved the bifurcation equation of Theorem (3.ii) is thereby proved.

In the situation of this section, with $N(x) = g(x) - f(t)$, if the limits $g(+\infty)$ and $g(-\infty)$ exist and are finite, the following set of relations (a), (b)

(a) $$g_+ \int_{G+} \psi_o dx - g_- \int_{G-} \psi_o dx - \int h\psi_o dx > 0$$

(b) $$g_- \int_{G+} \psi_o dx - g_+ \int_{G-} \psi_o dx - \int h\psi_o > 0$$

$$G+ = \{t \in G, \phi_o(t) > 0\}$$

$$G- = \{t \in G, \phi_o(t) < 0\},$$

imply (4.2) under an additional assumption that $\{t \in G:\phi_o(t) = 0\} \subseteq \{t \in G:\psi_o(t) = 0\}$. The proof that (a) and (b) imply (4.2) is a routine measure theoretic result, similar to that given in [15]. It is implicit in the proof of a theorem of Shaw [15].

REFERENCES

1. Agmon, S., Douglis, A., and Nirenberg, L., Estimates near the boundary for solutions of elliptic partial differential equations satisfying general boundary conditions, *Comm. Pure Appl. Math.* *12* (1959), 623-727.

2. Cesari, L., Functional analysis and periodic solutions of nonlinear differential equations, *Contributions to Differential Equation 1*, Wiley, 1963, (149-187).

3. Cesari, L., Functional analysis and Galerkin's method, *Mich. Math. J. 11*(1964), 385-414.

4. Cesari, L., Alternative methods in nonlinear analysis, *International Conference on Differential Equations*, Academic Press, 1974.

5. Hale, J., *Applications of Alternative Problems*, Brown University Lecture Notes, 1971.

6. Hale, J., Periodic solutions of a class of hyperbolic equations containing a small parameter, *Arch. Rat. Mech. Anal. 23*(1967), 380-398.

7. Knobloch, H. W., Eine neue methode zuer approximation periodischer Lösungen nicht-linearer differentialgleichungen zweiter Ordnung, *Math. Zeitschr. 82*(1963), 177-197.

8. Knobloch, H. W., Remarks on a paper of L. Cesari on functional analysis and nonlinear differential equations, *Mich. Math. J. 10*(1963), 417-430.

9. Landesman, E. M., and Lazer, A. C., Nonlinear perturbations of linear elliptic boundary value problems at resonance, *J. Math. Mech. 19*(1970), 609-623.

10. McKenna, P. J., Non-selfadjoint semilinear problems in the alternative method, PhD thesis, University of Michigan (1976).

11. Miranda, C., Un'osservazione su un teorema di Brouwer, *Boll. Un. Mat. Ital. 3*(2)(1940), 5-7.

12. Osborn, J., and Sather, D., Alternative problems for nonlinear equations, *J. Differential Equations 17*(1975), 12-31.

13. Rothe, E., On the Cesari index and Browder-Petryshim degree, to appear.

14. Schmidt, E., Entwicklung willkurlicher functionen nach systemen vorgeschriebener, *Math. Ann. LXIII*(1926), 433-476.

15. Shaw, H., A nonlinear elliptic boundary value problem at resonance, *Trans. Amer. Math. Soc.*, 1976, to appear.

16. Williams, S., A sharp sufficient condition for solutions of a nonlinear elliptic boundary value problem, *Journ. Diff. Equations 8*(1970), 580-586.

PERIODIC SOLUTIONS OF SOME NONLINEAR INTEGRAL EQUATIONS

Roger D. Nussbaum*
Department of Mathematics
Rutgers University

Introduction. Recently K. Cooke and J. Kaplan [3] have suggested a model for the spread of a disease or (under different assumptions on the function f below) the growth of a population. Specifically, if τ is a positive constant and $f(t,x)$ is a nonnegative, continuous real-valued function which is periodic of period ω in the t variable, they consider

$$x(t) = \int_{t-\tau}^{t} f(s,x(s))ds \tag{1}$$

and look for positive solutions $x(t)$ which are periodic of period ω. Note that the dependence of f on t in (1) distinguishes equation (1) from those studied by Cooke and Yorke in [4], and the results we shall describe have little interest if f has no t dependence. Cooke and Kaplan establish the existence of a positive number $\beta = \beta(f)$ such that equation (1) has positive, ω-periodic solutions for $\tau > \beta$. However, numerical studies for the special case $f(s,x) = (1+\frac{1}{2}\sin 2\pi s)\, x\, (1-x)$ for $0 \le x \le 1$, $f(s,x) = 0$ otherwise, suggested the existence of positive solutions of (1) for $\tau > 1$, although the number β is 2 in this case.

In [11] we have studied a more general class of integral equations, namely

$$x(t) = \int_{t-\tau}^{t} P(t-s,\tau)f(s,x(s))ds = (F_\tau x)(t) \tag{2}$$

We have proved by means of a global bifurcation theorem the existence of a positive number τ_0 such that (2) has a positive periodic solution of period ω for $\tau > \tau_0$ and will, in general, have none for $0 \leq \tau < \tau_0$. For the previously mentioned example, τ_0 is easily shown to be 1.

If, however, one considers more general functions P than are allowed in [11] or if one considers a generalization of equation (1) of the form

$$x(t) = \int_{t-\beta(\tau)}^{t-\alpha(\tau)} f(s,x(s))ds \qquad (1)'$$

(discussed in Section 2 below), then the abstract global bifurcation theorem of [11] is not easily applicable. For this reason we prove in Section 1 a new global bifurcation theorem (Theorem 3 below) more suitable for applications to equations like (1)' or (2). In Section 2 we briefly sketch how this bifurcation theorem can be applied to (1)'. In Section 3 we discuss the question of uniqueness of positive, ω-periodic solutions of (1). A corollary of our work is that the example numerically studied by Cooke and Kaplan has a unique, positive periodic solutions of period one for each $\tau > 1$.

1. A NEW GLOBAL BIFURCATION THEOREM.

Let X be a Banach space. By a wedge $C \subset X$ we shall mean a closed, convex subset of X such that if $x \in C$, then $tx \in C$ for all $t \geq 0$; by a cone (with vertex at 0) we shall mean a wedge such that $x \in C$ and $x \neq 0$ implies that $-x \notin C$. Let $J = (a,\infty)$ denote an open interval of reals with $-\infty \leq a < \infty$ and write $Y = X \times \mathbb{R}$ with $||(x,\lambda)|| = ||x||+|\lambda|$ for $(x,\lambda) \in Y$. Suppose that $C \subset X$ is a wedge and $F: C\times J \rightarrow C$ is a continuous map and make the following assumptions on F:

H1. $F\colon C\times J \to C$ is a continuous map such that $F(0,\alpha) = 0$ for all $\alpha \in J$. It is a local strict-set-contraction with respect to the norms on Y and X (see [9] for definitions). If B is any bounded subset of $C \times J$, the set $\{(x,\alpha) \in B\colon F(x,\alpha) = x\}$ has compact closure in Y.

H2. If $a > -\infty$ and $F(x_k,\lambda_k) = x_k$ for some sequence $(x_k,\lambda_k) \in C \times J$ with $x_k \neq 0$ and $\lambda_k \longrightarrow a$, then it follows that $\lim_{k\to\infty} ||x_k|| = \infty$. (Here $J = (a,\infty)$.)

H3. There exists a subset Λ of J with no accumulation points in J such that if J_0 is any compact interval for which $J_0 \cap \Lambda$ is empty, one has $F(x,\alpha) \neq x$ for $\alpha \in J_0$ and $x \in C$ with $0 < ||x|| \leq \varepsilon(J_0)$, $\varepsilon(J_0)$ a positive number.

For notational convenience, define $F_\alpha\colon C \longrightarrow C$ by $F_\alpha(x) = F(x,\alpha)$ and define $B_\rho = \{x\in C\colon ||x||<\rho\}$. If $\lambda \in \Lambda$, there exists a positive number δ such that $\Lambda \cap [\lambda-\delta,\lambda+\delta] = \{\lambda\}$ and there exists a positive number ρ such that $F_\alpha(x) = x$ for $0 < ||x|| \leq \rho$ and for $\alpha = \lambda + \delta$ or $\alpha = \lambda - \delta$. Recalling notation for the fixed point index of F_α on a subset of C open in the relative topology (see [9]) we define

$$\Delta(\lambda) = i_C(F_{\lambda+\delta},B_\rho) - i_C(F_{\lambda-\delta},B_\rho) \tag{3}$$

Our next theorem is stated in [10]; however, only a hint of the proof (in the notation below) that $\sum_{\lambda\in\Lambda_0} \Delta(\lambda) = 0$ is given there. Since we shall need this fact, we present a more detailed proof here.

<u>1. Theorem</u>. Let notation be as above and suppose that C is a wedge in a Banach space X, $J = (a,\infty)$ is an open interval of reals, and $F\colon C\times J \longrightarrow C$ is a continuous map for which H1, H2 and H3 hold. Let S denote the closure in $C \times J$ of $\{(x,\alpha)\in C\times J\colon x\neq 0 \text{ and } F(x,\alpha)=x\}$. Suppose that $\lambda_0 \in \Lambda$ (Λ as in

H3) and that $\Delta(\lambda_0) \neq 0$ and let S_0 denote the connected component of S which contains $(0,\lambda_0)$. Then S_0 is nonempty and if $\Lambda_0 = \{\lambda\in\Lambda: (0,\lambda) \in S_0\}$ either (a) S_0 is not compact or (b) S_0 is compact (so Λ_0 is finite) and $\sum_{\lambda\in\Lambda_0} \Delta(\lambda) = 0$.

<u>Proof</u>. The proof that S_0 is nonempty follows exactly as in [10]. Assume S_0 is compact and let $\Lambda_0 = \{\mu_1,\mu_2,\ldots,\mu_n\}$ where $\mu_1 < \mu_2 < \ldots < \mu_n$. Take $\varepsilon > 0$ such that $\inf\{|\mu_j-\lambda|:\lambda\in\Lambda, \lambda\neq\mu_j, 1\leq j\leq n\} \geq 2\varepsilon$ and such that $|\lambda-a| \geq 2\varepsilon$ for $(x,\lambda) \in S_0$. Now standard arguments using a theorem of Whyburn [15, Theorem 9.3, Chapter 1] shows that there exists a bounded open neighborhood Ω of S_0 in $C \times J$ (the topology is always the relative topology on $C \times J$ and "open" means open as a subset of $C \times J$ in this topology) such that $(0,\lambda)\notin\bar{\Omega}$ if $|\lambda-\mu_j| \geq \varepsilon$ for some j with $1 \leq j \leq n$, $(x,\lambda) \notin \bar{\Omega}$ for $|\lambda-a|\leq\varepsilon$, and such that if $F(x,\lambda) = x$ for some $(x,\lambda) \in \bar{\Omega}-\Omega$, then $x = 0$.

Now let $\Omega_\lambda = \{x\in C: (x,\lambda)\in\Omega\}$, a relatively open subset of C. The generalized homotopy property [9, Theorem 2, Section E] for local strict-set-contractions shows that $i_C(F_\lambda,\Omega_\lambda)$ is constant for $\mu_j+\varepsilon \leq \lambda \leq \mu_{j+1}-\varepsilon$ for $1\leq j\leq n-1$ for $\lambda\leq\mu_1-\varepsilon$ and for $\lambda \geq \mu_n+\varepsilon$. Since Ω_λ is empty for $\lambda = a+\varepsilon$ and for λ large, we find that $i_C(F_\lambda,\Omega_\lambda) = 0$ for $\lambda = a+\varepsilon$ and for $\lambda = \mu_n+\varepsilon$. Thus to complete the proof it suffices to show that

$$i_C(F_{\mu_j+\varepsilon},\Omega_{\mu_j+\varepsilon}) = i_C(F_{\mu_j-\varepsilon},\Omega_{\mu_j-\varepsilon}) + \Delta(\mu_j) \qquad (4)$$

To prove (4), first select positive numbers ρ and η such that

$$\{(x,\lambda)\in C\times J: ||x||\leq\rho \text{ and } |\lambda-\mu_j|\leq\eta\} \subset \Omega$$

By applying the generalized homotopy property we find

$$i_C(F_{\mu_j-\eta},\Omega_{\mu_j-\eta}) = i_C(F_{\mu_j+\eta},\Omega_{\mu_j+\eta}) \qquad (5)$$

By our assumptions there exists a positive number $r < \rho$ such that if $0 < ||x|| \leq r$ and $\eta \leq |\mu_j - \lambda| \leq \varepsilon$ then $F(x,\lambda) \neq x$. Define $\tilde{\Omega} = \{(x,\lambda) \in C \times [\mu_j+\eta, \mu_j+\varepsilon] : (x,\lambda) \in \Omega$ or $||x|| < r\}$ and $\tilde{\Omega}_\lambda = \{x : (x,\lambda) \in \tilde{\Omega}\}$. One can check that the hypotheses of the homotopy property are satisfied so that $i_C(F_\lambda, \tilde{\Omega}_\lambda)$ is constant for $\mu_j+\lambda \leq \lambda \leq \mu_j+\varepsilon$. For $\lambda = \mu_j+\varepsilon$, the additivity property of the fixed point index gives that

$$i_C(F_{\mu_j+\varepsilon}, \tilde{\Omega}_{\mu_j+\varepsilon}) = i_C(F_{\mu_j+\varepsilon}, \Omega_{\mu_j+\varepsilon}) + i_C(F_{\mu_j+\varepsilon}, B) \tag{6}$$

where $B = \{x \in C : ||x|| < r\}$. Since $\tilde{\Omega}_\lambda = \Omega_\lambda$ for $\lambda = \mu_j+\eta$, the right hand side of (6) equals $i_C(F_{\mu_j+\eta}, \Omega_{\mu_j+\eta})$. Similarly, we find that

$$i_C(F_{\mu_j-\eta}, \Omega_{\mu_j-\eta}) = i_C(F_{\mu_j-\varepsilon}, \Omega_{\mu_j-\varepsilon}) + i_C(F_{\mu_j-\varepsilon} B) \tag{7}$$

Combining (5), (6) and (7) yields equations (4).

Results related to the observation that $\sum_{\lambda \in \Lambda_0} \Delta(\lambda) = 0$ have been obtained independently by several authors: see, for example, [5], [6], [10] and [14]. Note that [10] was submitted slightly before [6] despite the different publication dates.

We wish to apply Theorem 1 to study bifurcation in cones. First we need some lemmas. Recall that a cone K in a B-space X is "total" if X equals closed linear span of K; K is "reproducing" if $X = \{x-y : x,y \in K\}$.

1. Lemma. Let K be a total cone in a B-space X and $L: X \to X$ a bounded linear operator such that $L(K) \subset K$. Assume that $Lx \neq x$ for all $x \in K$ such that $x \neq 0$. Let $B = \{x \in K : ||x|| < 1\}$. If the spectral radius $r(L)$ of L is strictly greater than one, then $i_K(L,B) = 0$ and if $r(L) < 1$, then $i_K(L,B) = 1$.

Proof. If $r(L) < 1$, consider the homotopy $L_t = tL$ for $0 \le t \le 1$ and observe that $L_t(x) \ne x$ for $||x|| = 1$ and $0 \le t \le 1$. It follows that $i_K(L,B) = i_K(L_0,B) = 1$.

If $r(L) = r > 1$, the Krein-Rutman theorem implies that there exists $x_0 \in K - \{0\}$ with $Lx_0 = rx_0$. Consider the homotopy $L_s(x) = Lx+sx_0$ for all $s \ge 0$. The argument used in the final paragraph on p. 329 of [10] shows that $L_s(x) \ne x$ for $||x|| = 1$, $x \in K$ and $s \ge 0$. Since $L_s(x) = x$ has no solutions $x \in B$ for s large enough, it follows that $i_K(L,B)=i_K(L_s,B)=0$ □

Lemma 1 also appears in [2]; however, since the result is a simple consequence of a result communicated privately by this author to Amann in 1971 and mentioned in [1, p. 348] we have included the proof.

Our next lemma presumably appears somewhere in the literature on compact linear operators, but we do not know of a reference.

2. Lemma. Let X denote a real Banach space and [a,b] a compact interval of reals and assume that for each $\lambda \in [a,b]$ there is a compact bounded linear operator L_λ from X to X and the map $\lambda \longrightarrow L_\lambda$ is continuous in the uniform operator topology. Then it follows that the map $\lambda \longrightarrow r(L_\lambda)$ = spectral radius of L is continuous. If in addition there is a total cone $K \subset X$ such that $L_\lambda(K) \subset K$ for all $\lambda \in [a,b]$ and if $L_\lambda(x) \ne x$ for all $\lambda \in [a,b]$ and $x \in K - \{0\}$, then it follows that either $r(L_\lambda) > 1$ for all $\lambda \in [a,b]$ or $r(L_\lambda) < 1$ for all $\lambda \in [a,b]$ and $\inf\{||L_\lambda(x)-x||: x \in K, ||x|| = 1, \lambda \in [a,b]\} = c > 0$.

Proof. Assume we have shown $\lambda \longrightarrow r(L_\lambda)$ is continuous. If a total cone K exists and $r(L_\lambda) - 1$ is not of constant sign on [a,b], then by continuity we have $r(L_\lambda) = 1$ for some λ. For this λ, the Krein-Rutman theorem implies the existence of $x \in K - \{0\}$ with $L_\lambda(x) = x$, a contradiction.

To show that the infimum above is positive, suppose not and select $x_n \in K$ with $||x_n|| = 1$ and $\lambda_n \in [a,b]$ such that $L_{\lambda_n}(x_n)-x_n \longrightarrow 0$. By taking a subsequence we can suppose $\lambda_n \rightarrow \lambda$ and $L_\lambda(x_n)-x_n \longrightarrow 0$. Since L_λ is compact, by taking a further subsequence we can suppose that $L_\lambda(x_n)$ converges and hence that x_n converges to x. For this x we have $L_\lambda(x)=x$, a contradiction.

It remains to show $\lambda \longrightarrow r(L_\lambda)$ is continuous. Take $\lambda_0 \in [a,b]$. Since we know in general that

$$\limsup_{\lambda\to\lambda_0} r(L_\lambda) \le r(L_{\lambda_0})$$

it suffices to show that

$$\liminf_{\lambda\to\lambda_0} r(L_\lambda) \ge r(L_{\lambda_0}).$$

If $r(L_{\lambda_0}) = 0$, this is immediate, so we assume $r(L_{\lambda_0}) > 0$. By extending L_λ in the natural way to the complexification of X all hypotheses are preserved, so we can assume X is a complex Banach space. According to the spectral theory of compact linear operators, there exists an isolated point $z_0 \in C$ in the point spectrum of L_{λ_0} with $|z_0| = r(L_{\lambda_0})$. Furthermore, if Γ is a circle in C about z_0 which contains no other points of the spectrum of L_{λ_0} and we define a projection P_{λ_0} by

$$P_{\lambda_0} = \frac{1}{2\pi i}\int_\Gamma (z-L_{\lambda_0})^{-1}dz$$

then $X_{\lambda_0} = R(P_{\lambda_0})$ = the range of P_{λ_0} is finite dimensional, $L_{\lambda_0}P_{\lambda_0} = P_{\lambda_0}L_{\lambda_0}$ and if $A_{\lambda_0} = L_{\lambda_0}|X_{\lambda_0}$, the spectrum of A_{λ_0} is

$\{z_0\}$ (see [7, p. 579]). Define a projection P_λ by

$$P_\lambda = \frac{1}{2\pi i} \int_\Gamma (z-L_\lambda)^{-1} dz$$

It is not hard to show that there exists $\delta > 0$ such that P_λ is defined for $|\lambda-\lambda_0| < \delta$ and $||P_\lambda - P_{\lambda_0}|| < 1$; furthermore one has $||P_\lambda - P_{\lambda_0}|| \to 0$ as $\lambda \to \lambda_0$. Let X_λ denote the range of P_λ and write $B_\lambda = I - P_{\lambda_0} + P_\lambda$ and $\tilde{B}_\lambda = I - P_\lambda + P_{\lambda_0}$. Since $||P_\lambda - P_{\lambda_0}|| < 1$, both B_λ and $\tilde{B}_\lambda$ are one-one maps of X onto X; and since $B_\lambda(X_{\lambda_0}) \subset X_\lambda$ and $\tilde{B}_{\lambda_0}(X_\lambda) \subset X_{\lambda_0}$, it follows that X_{λ_0} and X_λ are isomorphic finite dimensional spaces and that $B_\lambda^{-1}(X_\lambda) \subset X_{\lambda_0}$. If we define $A_\lambda : X_{\lambda_0} \to X_{\lambda_0}$ by $A_\lambda = B_\lambda^{-1} L_\lambda B_\lambda | X_{\lambda 0}$, $A_\lambda \to A_{\lambda_0}$ as $\lambda \to \lambda_0$. Spectral theory for finite dimensional spaces implies that for $|\lambda-\lambda_0|$ small enough, A_λ will have an eigenvalue z close to z_0 and a corresponding eigenvector $x \in X_{\lambda_0}$. Since $B_\lambda x$ will then be an eigenvector of L_λ corresponding to the same eigenvalue z, this proves that $\liminf_{\lambda\to\lambda_0} r(L_\lambda) \geq r(L_{\lambda_0})$. □

We now suppose that C is a total cone in a Banach space X and that $F: C\times J \to C$ is a map which satisfies H1 and H2. We strengthen H3 to H4:

H4. For each $\lambda \in J$ there exists a compact bounded linear operator L_λ such that $L_\lambda(C) \subset C$ and such that for any compact interval $[c,d] \subset J$ $\lim_{||x||\to 0} (||x||^{-1})(||F(x,\lambda) - L_\lambda(x)||) = 0$ uniformly in $\lambda \in [c,d]$. The map $\lambda \to L_\lambda$ is continuous with respect to the norm topology on bounded linear operators. There exists a countable set $\Lambda \subset J$ with no accumulation points in J such that $x \neq L_\lambda(x)$ for $x \in K-\{0\}$ and $\lambda \in \Lambda$.

The reader can verify (using Lemma 2) that H4 implies H3.

3. Lemma. Assume that H1, H2 and H4 hold and let $\Delta(\lambda)$ be defined as in equation (3). If $\lambda_0 \in \Lambda$ and $r(L_\lambda) > 1$ for

$\lambda_0 < \lambda < \lambda_0+\varepsilon$ and $r(L_\lambda) < 1$ for $\lambda_0-\varepsilon < \lambda < \lambda_0$ $(\varepsilon>0)$, then $\Delta(\lambda_0) = -1$. If $r(L_\lambda) < 1$ for $\lambda_0 < \lambda < \lambda_0+\varepsilon$ and $r(L_\lambda) > 1$ for $\lambda_0-\varepsilon < \lambda < \lambda_0$ $(\varepsilon>0)$, then $\Delta(\lambda_0) = 1$. Finally, if $r(L_\lambda)-1$ is of constant sign for $0 < |\lambda-\lambda_0| < \varepsilon$, then $\Delta(\lambda_0) = 0$. (Note that according to H4 and Lemma 2, there must be an $\varepsilon > 0$ for which one of these possibilities hold).

Proof. If $B_\rho = \{x\in K: ||x||<\rho\}$ and $\lambda \notin \Lambda$, then the homotopy property implies that for ρ small enough,

$$i_C(F_\lambda, B_\rho) = i_C(L_\lambda, B_\rho)$$

The lemma now follows from Lemma 1. □

With the aid of Lemmas 1, 2 and 3, our next theorem follows directly from Theorem 1.

2. Theorem. Suppose that C is a total cone in a B-space X and that H1, H2 and H4 holds; for some $\lambda_0 \in \Lambda$ suppose that $r(L_\lambda) > 1$ for $\lambda > \lambda_0$ and λ near λ_0 and $r(L_\lambda) < 1$ for $\lambda < \lambda_0$ and λ near λ_0 or vice-versa ($r(L)$ denotes the spectral radius of L). Let S and S_0 be as in Theorem 1. Then it follows that S_0 is nonempty and either a) S_0 is unbounded or b) S_0 is bounded (hence compact) and if Λ_0 denotes $\{\lambda\in J:(0,\lambda)\in S_0\}$,

$$\sum_{\lambda\in\Lambda_0} \Delta(\lambda) = 0.$$

One difficulty in applying Theorem 2 is in verifying that there exists a set Λ as in H4 with no accumulation points in J. As we shall see, this may not be obvious even for relatively simple looking equations.

With this in mind we introduce the following weakening of H4:

H5. For each $\lambda \in J = (a,\infty)$ there exists a compact, bounded linear operator L_λ such that $L_\lambda(C) \subset C$ and such that for any compact interval $[c,d] \subset J$

$\lim_{||x||\to 0} (||x||^{-1})(||F(x,\lambda)-L_\lambda(x)||) = 0$ uniformly in $\lambda \in [c,d]$. The map $\lambda \longrightarrow L_\lambda$ is continuous with respect to the norm topology on bounded linear operators. If $\Lambda = \{\lambda\in J: L_\lambda(x) = x$ for some $x \in K-\{0\}\}$, then there exist numbers $\alpha > a$ and $\beta < \infty$ such that $\Lambda \subset [\alpha,\beta]$, $r(L_\lambda) > 1$ for $\lambda \in (a,\alpha]$ and $r(L_\lambda) > 1$ for $\lambda \in [\beta,\infty)$, where $r(L)$ denotes the spectral radius of L.

Our next result is actually a lemma preparatory to the proof of Theorem 3, but it may have some independent interest.

1. Proposition. Let C be a total cone in a Banach space X, $J = (a,\infty)$ an interval of reals, and $F: C\times J \longrightarrow C$ a map which satisfies H1, H2 and H5. If Λ is as in H5, suppose that $\Lambda \subset \bigcup_{k=1}^{n} I_k \subset J$, where $I_k = [a_k,b_k]$ is a compact interval and $b_j < a_k$ if $1 \le j \le k \le n$. Define a function $s(\lambda)$ by $s(\lambda) = 1$ for $a < \lambda < a_1$, $s(\lambda) = 0$ for $\lambda > b_n$ and $s(\lambda) = 1$ on (b_k,a_{k+1}) if $r(L_\lambda) < 1$ for all $\lambda \in (b_k,a_{k+1})$ and $s(\lambda) = 0$ if $r(L_\lambda) > 1$ for all $\lambda \in (b_k,a_{k+1})$. (Note that $s(\lambda)$ is well-defined by Lemma 2). For each interval I_k, define $\Delta(I_k) = \lim_{\delta\to 0^+} s(b_k+\delta)-s(a_k-\delta)$. Finally define $Q = \{(0,\lambda):\lambda\in \bigcup_{k=1}^{n} I_k\}$, $Q_k = \{(0,\lambda):\lambda\in I_k\}$ and T = closure $\{(x,\lambda):F(x,\lambda) = x$ and $x \neq 0\} \cup Q$. Then if $(0,\lambda_0) \in T$ and T_0 is the connected component of T which contains $(0,\lambda_0)$, it follows that either T_0 is unbounded or T_0 is bounded and $\sum_{Q_k\cap T_0\neq\phi} \Delta(I_k) = 0$, where the summation is over k such that $Q_k\cap T_0$ is nonempty. Furthermore, there is a $(0,\lambda_0) \in Q$ such that the connected component T_0 of T containing $(0,\lambda_0)$ is unbounded.

Proof. The proof is a slight modification of the argument used in Theorem 1. Assume that T_0 is bounded, so that it is compact and let

$Q^* = \{\lambda : \lambda \in Q_k$ for some Q_k such that $Q_k \cap T_0$ is nonempty$\}$.

Take $\varepsilon > 0$ such that $d(\lambda,Q^*) \geq 2\varepsilon$ for $\lambda \in Q-Q^*$ and $|\lambda-a| \geq 2\varepsilon$ for $(x,\lambda) \in T_0$; here $d(\lambda,Q^*)$ denotes the distance of λ to Q^*. Just as in the proof of Theorem 1, there exists an open (in the relative topology on $C \times J$), bounded neighborhood Ω of T_0 in $C \times J$ with the following properties:

1) $(0,\lambda) \notin \overline{\Omega}$ if $d(\lambda,Q^*) \geq \varepsilon$,

2) $(x,\lambda) \notin \overline{\Omega}$ if $|\lambda-a| \leq \varepsilon$ and

3) if $F(x,\lambda) = x$ for some $x \in \overline{\Omega}-\Omega$, then $x = 0$.

Just as before, let $\Omega_\lambda = \{x \in C : (x,\lambda) \in \Omega\}$ and note that the homotopy property shows that

$$i_C(F_{b_k+\varepsilon},\Omega_{b_k+\varepsilon}) = i_C(F_{a_{k+1}-\varepsilon},\Omega_{a_{k+1}-\varepsilon}) \text{ for } 1 \leq k \leq n-1$$

$$i_C(F_{b_n+\varepsilon},\Omega_{b_n+\varepsilon}) = 0 \tag{8}$$

$$i_C(F_{a_1-\varepsilon},\Omega_{a_1-\varepsilon}) = 0$$

Thus to complete the proof of the first half of the theorem, it suffices to show that

$$i_C(F_{b_k+\varepsilon},\Omega_{b_k+\varepsilon}) = i_C(F_{a_k-\varepsilon},\Omega_{a_k-\varepsilon}) + \Delta(I_k) \tag{9}$$

for all k such that $Q_k \cap T_0$ is nonempty.

To prove (9), select positive numbers ρ and η, $\eta < \varepsilon$, such that

$$\{(x,\lambda) \in C \times J : ||x|| \leq \rho \text{ and } d(\lambda,I_k) \leq \eta\} \subset \Omega$$

This can be done because Ω is an open set in $C \times J$ which contains Q_k. By applying the generalized homotopy property on the interval $[a_k-\eta,b_k+\eta]$ we find that

$$i_C(F_{a_k-\eta},\Omega_{a_k-\eta}) = i_C(F_{b_k+\eta},\Omega_{b_k+\eta}) \tag{10}$$

By using Lemma 2 there is a positive number $r \leq \rho$ such that if

$B = \{x \in C: ||x|| < r\}$, then $F_\lambda(x) \neq x$ for $(x,\lambda) \in (B-\{0\}) \times [a_k-\varepsilon, a_k-\eta]$ and for $(x,\lambda) \in (B-\{0\}) \times [b_k+\eta, b_k+\varepsilon]$. Furthermore, by using the homotopy $tF_\lambda + (1-t)L_\lambda$ for small r and for $0 \le t \le 1$, we obtain that

$$i_C(F_\lambda, B) = i_C(L_\lambda, B) \tag{11}$$

for $\lambda \in [a_k-\varepsilon, a_k-\eta]$ or $\lambda \in [b_k+\eta, b_k+\varepsilon]$.

By using the same argument as at the end of Theorem 1 we obtain that

$$\begin{aligned} i_C(F_{b_k+\eta}, \Omega_{b_k+\eta}) &= i_C(F_{b_k+\varepsilon}, \Omega_{b_k+\varepsilon}) + i_C(F_{b_k+\varepsilon}, B) \\ i_C(F_{a_k-\eta}, \Omega_{a_k-\eta}) &= i_C(F_{a_k-\varepsilon}, \Omega_{a_k-\varepsilon}) + i(F_{a_k-\varepsilon}, B) \end{aligned} \tag{12}$$

By combining equations (10), (11) and (12) and the formula for $i_C(L_\lambda, B)$ in Lemma 1, we obtain equation (9).

It remains to prove that for some interval I_k, the connected component of T containing Q_k is unbounded. Suppose not. Then if T_k denotes the connected component of T containing Q_k one has

$$\sum_{Q_j \cap T_k \neq \phi} \Delta(I_j) = 0$$

and this easily implies that

$$\sum_{j=1}^{n} \Delta(I_j) = 0 \tag{13}$$

On the other hand, if we select $\lambda_0 \in (a, a_1)$, $\lambda_j \in (b_j, a_{j+1})$ for $1 \le j < n$ and $\lambda_n \in (b_n, \infty)$ and define $s_j = s(\lambda_j)$, we have that $\Delta(I_j) = s_j - s_{j-1}$ and

$$\sum_{j=1}^{n} \Delta(I_j) = \sum_{j=1}^{n} (s_j - s_{j-1}) = s_n - s_0 = -1$$

This contradiction proves that an unbounded T_k exists. □

The proof of Proposition 1 is essentially the same as that of Theorem 1. The main point is that it is necessary to adjoin Q to closure $\{(x,\lambda): F(x,\lambda)=x, x\neq 0\}$ rather than working directly with the latter set.

We are now in a position to prove our main abstract bifurcation theorem, which can be viewed as a limiting case of Proposition 1.

3. Theorem. Let C be a total cone in a Banach space X, $J = (a,\infty)$ an interval of real numbers and $F: C\times J \to C$ a map which satisfies H1, H2 and H5. If Λ is as in H5, denote by Q the set $\{(0,\lambda):\lambda\in\Lambda\}$ and by S the set closure $\{(x,\lambda)\in C\times J: F(x,\lambda)=x \text{ and } x\neq 0\} \cup Q$. Then there exists $\lambda_0 \in \Lambda$ such that if S_0 is the connected component of S which contains $(0,\lambda_0)$, S_0 is unbounded.

Proof. A simple argument using the compactness of the L_λ shows that Λ is closed, and since we assume Λ is contained in a bounded interval $[\alpha,\beta] \subset J$, it follows that Λ is compact. Thus we can write $J-\Lambda = \bigcup_{n=1}^{\infty} J_n$, where the J_n are pairwise disjoint open intervals and we can suppose that length $(J_k) \geq$ length (J_m) if $k \leq m$. For each $\varepsilon > 0$ let $m(\varepsilon)$ denote the largest integer such that length $(J_k) \geq \varepsilon$ for $k \leq m(\varepsilon)$ and note that $m(\varepsilon) < \infty$ because Λ is compact. Select a sequence of positive reals ε_n such that $\varepsilon_n \to 0$ as $n \to \infty$ and $\varepsilon_n < \alpha - a$ for all n. For any fixed n consider $\bigcup_{j=1}^{m(\varepsilon_n)} J_j$; by reindexing we can assume that $J_j = (a_j, b_j)$ and that $b_j < a_{j+1}$ for $1 \leq j < m(\varepsilon_n)$. Define $I_j = [b_j, a_{j+1}]$ for $1 \leq j < m(\varepsilon_n)$ and note that $\Lambda \subset \bigcup_{j=1}^{m(\varepsilon_n)} I_j$ and that the hypothesis of Proposition 1 are met. Thus if we define $\Lambda^{(n)}$ and $Q^{(n)}$ by the formulas

$$\Lambda^{(n)} = J - \bigcup_{j=1}^{m(\varepsilon_n)} J_j \text{ and } Q^{(n)} = \{(0,\lambda):\lambda\in\Lambda^{(n)}\} \qquad (14)$$

and if we write $S^{(n)} = Q^{(n)} \cup$ closure $\{(x,\lambda):F(x,\lambda) = x, x\neq 0\}$, then there is an unbounded component $S_0^{(n)}$ of $S^{(n)}$.

We shall now let $n \to \infty$ and give a kind of limit argument to obtain S_0. First, recall that if $\varepsilon > 0$ and $M > 0$ and if $F(x,\lambda) = x$ for any (x,λ) with $x \neq 0$ and $\varepsilon \leq d(\lambda,\Lambda) \leq M$, then there exist $\delta = \delta(\varepsilon,M) > 0$ such that $||x|| \geq \delta$; this is a consequence of Lemma 2 and H5. For each n select $(0,\lambda_n)\in S_0^{(n)}$. Our construction ensures that $d(\lambda,\Lambda) < \varepsilon_n$ for all $\lambda \in \Lambda^{(n)}$, so $d(\lambda_k,\Lambda) \to 0$ and by taking a subsequence we can assume that $\lambda_n \to \lambda\in\Lambda$. We claim that the connected component of S containing $(0,\lambda)$ is unbounded. In order to prove this, we shall first show that if $\mathcal{O}$ is any bounded, open (in the relative topology on $C\times J$) neighborhood of $(0,\lambda)$, then $S \cap \partial\mathcal{O}$ is nonempty. Certainly, the connectedness and unboundedness of $S_0^{(n)}$ implies that there exists $(x_n,\lambda_n) \in S_0^{(n)} \cap \partial\mathcal{O}$. Since $F(x_n,\lambda_n) = x_n$, H1 implies that by taking a subsequence we can assume $(x_n,\lambda_n) \to (x,\mu)$ and $F(x,\mu) = x$. If $x \neq 0$ we have $(x,\mu) \in S \cap \partial\mathcal{O}$. If $x = 0$, we must show that $\mu \in \Lambda$. If $\mu \notin \Lambda$, we must have $d(\mu,\Lambda^{(n)}) \geq \varepsilon$ for some positive ε and for all large n. It follows that there exists $\delta > 0$ such that $||x_n|| \geq \delta$ for all large n. This contradiction shows that $\mu \in \Lambda$ if $x = 0$.

We can now show S_0 is unbounded. The proof is by contradiction. Suppose not. The usual argument using Whyburn's lemma [15, Chapter 1] shows that there is a bounded open neighborhood $\mathcal{O}$ of S_0 such that $S \cap \partial\mathcal{O}$ is empty. But we have seen that $S \cap \partial\mathcal{O}$ is nonempty for every bounded open neighborhood of $(0,\lambda)$. □

To conclude this section we recall some standard results and notation which we shall need later. If K is a cone in a Banach space X, K induces a partial ording on X by $x \le y$ iff $y-x \in K$. If $L:X \to X$ is a bounded linear operator such that $L(K) \subset K$, L is "positive" (with respect to K). If there exists a fixed $u \in K-\{0\}$ such that for every $x \in K-\{0\}$ there exist positive constants α and β and an integer $n \ge 1$ (all depending on x) such that $\alpha u \le L^n(x) \le \beta u$, L is called "u-positive." If r is an eigenvalue of a bounded linear operator $L:Z \to Z$, Z a Banach space, r is an "(algebraically) simple eigenvalue" if $\{x \in Z : (rI-L)^n(x) = 0 \text{ for some } n \ge 1\}$ is one dimensional.

Our next proposition follows directly from results in [8, Chapter 2] and from the Krein-Rutman theorem.

<u>2.</u> <u>Proposition</u>. (See [8, Chapter 2]). Let C be a reproducing cone in a real Banach space Z and $L:Z \to Z$ a compact, positive linear operator. Assume that L is u-positive for some $u \in C-\{0\}$. Then $r = r(L)$ = the spectral radius of L is positive and r is a simple eigenvalue of L with a corresponding eigenvector $x \in C-\{0\}$. If $Ly = \lambda y$ for some $y \in C-\{0\}$, it follows that $\lambda = r$.

2. PERIODIC SOLUTIONS OF SOME NONLINEAR INTEGRAL EQUATIONS

The results of the previous section are motivated by two integral equations (each parametrized by τ):

$$x(t) = \int_{t-\tau}^{t} f(s,x(s)ds \qquad (15)$$

$$x(t) = \int_{t-\beta(\tau)}^{t-\alpha(\tau)} f(s,x(s))ds \qquad (16)$$

We shall always assume about f that

<u>H6</u>. $f:\mathbb{R}\times\mathbb{R}^+ \to \mathbb{R}^+$ is a continuous function such that $f(s,0) = 0$ for all s and $f(s+\omega,x) = f(s,x)$ for all $(s,x) \in \mathbb{R}\times\mathbb{R}^+$,

where ω is a fixed positive number. There is a continuous, strictly positive periodic function $a(s)$ of period ω such that $\lim_{x\to 0^+} \frac{f(s,x)}{x} = a(s)$ uniformly in s. Finally we assume $\lim_{x\to\infty} \frac{f(s,x)}{x} = 0$ uniformly in s.

We shall always assume about the functions $\alpha(\tau)$ and $\beta(\tau)$ that

H7. $\alpha,\beta:\mathbb{R}^+ \longrightarrow \mathbb{R}^+$ are continuous nonnegative functions such that $\beta(\tau)-\alpha(\tau) > 0$ for all $\tau > 0$, $\lim_{\tau\to 0} \beta(\tau)-\alpha(\tau) = 0$ and $\lim_{\tau\to\infty} \beta(\tau)-\alpha(\tau) = \infty$.

We shall always denote by X the Banach space of continuous, real-valued functions which are periodic of period ω; the norm is the sup norm. If $x \in X$, $x(t)$ is thought of as defined for all $t \in \mathbb{R}$. If we write $K = \{x \in X : x(t) \geq 0 \text{ for all } t\}$, we are seeking solutions $x \in K-\{0\}$ of equation (15) or (16). If L_τ denotes the Frechet derivative of the operator $F_\tau : K \longrightarrow K$ defined by $(F_\tau x)(t) = \int_{t-\beta(\tau)}^{t-\alpha(\tau)} f(s,x(s)ds$, then

$$(L_\tau u)(t) = \int_{t-\beta(\tau)}^{t-\alpha(\tau)} a(s)u(s)ds \tag{17}$$

If e denotes the function identically equal to one, H6 and H7 imply that L_τ is e-positive and hence (by Proposition 2) that $\Delta = \{\tau \in (0,\infty) : L_\tau(x) = x \text{ for some } x \in K-\{0\}\} = \{\tau \in (0,\infty) : r(L_\tau) = 1\}$ It is also easy to verify that

$$\inf_t \int_{t-\beta(\tau)}^{t-\alpha(\tau)} a(s)dx \leq r(L_\tau) \leq \sup_t \int_{t-\beta(\tau)}^{t-\alpha(\tau)} a(s)ds \tag{18}$$

so that by using H7 we find that Λ lies inside some compact interval $J_0 \subset J$.

If $\alpha(\tau) = 0$ and $\beta(\tau) = \tau$ (the case considered in [11]) is it easy to verify (see [11]) that $r(L_\tau)$ is a strictly increasing, continuous (by Lemma 2) function of τ and that there is a unique value τ_0 such that $r(L_\tau) = 1$. However, even for the case $\alpha(\tau) = \frac{1}{2}\tau$, $\beta(\tau) = \tau$, it is unclear exactly what Λ looks like. One approach to this problem is to attempt a more detailed study of Λ, perhaps making more detailed assumptions on α and β; another is to prove a result like Theorem 3.

Our next theorem is a direct consequence of Theorem 3. The verifications necessary to show that the hypotheses of Theorem 3 are satisfied are the same as those in [11] and are routine in nature; we omit them.

4. Theorem. Assume that H6 and H7 hold and let $L_\tau : X \longrightarrow X$ and $F_\tau : X \to X$ be defined by equation 17 and Λ by $\Lambda = \{\tau \in (0,\infty) : r(L_\tau) = \text{spectral radius of } L_\tau = 1\}$. Then Λ is a compact subset of $(0,\infty)$ and if $S = \{(0,\lambda):\lambda\in\Lambda\} \cup \text{closure}\ \{(x,\tau)\in K\times(0,\infty) : F_\tau(x) = x \text{ and } x \neq 0\}$ there is an unbounded connected component S_0 of S. If M is any finite constant, then $\{(x,\tau)\in S : \tau\le M\}$ is bounded.

We remark that Theorem 3 can also be used to study

$$x(t) = \int_{t-\beta(\tau)}^{t-\alpha(\tau)} P(t-s,\tau)f(s,x(s))ds \tag{19}$$

under less restrictive assumptions on the function P than those in [11].

3. UNIQUENESS OF SOLUTIONS

Theorem 4 provides little detailed information about the set S_0. In this section we wish to investigate uniqueness of solutions $x \in K-\{0\}$ of $F_\tau(x) = x$. For simplicity we shall restrict ourselves to equations of the form

$$x(t) = \int_{t-\tau}^{t} a(s)g(x(s))ds = (F_\tau x)(t) \tag{20}$$

and while the techniques we shall give apply to more general equations (eg., equation 16), the uniqueness question is already nontrivial for the example mentioned in the introduction, namely $a(s) = 1+\frac{1}{2}\sin 2\pi s$, $g(x) = x(1-x)$ for $0 \le x \le 1$, $g(x) = 0$ otherwise.

We shall always assume about the a and g in (20) that

H8. The function $a(s)$ is strictly positive, continuous and periodic of period ω. The function $g:[0,\infty) \longrightarrow [0,\infty)$ is a continuous map such that $g(0) = 0$, $g(x) = 0$ for $x \ge M$, M a positive constant, and $g(x) > 0$ for $0 < x < M$. Furthermore, $g|[0,M]$ is C^1 and $g'(0) > 0$; and there exists a number $M_0 < M$ such that $g'(M_0) = 0$, $g'(x)$ is strictly decreasing for $0 \le x \le M_0$ and $g'(x) \le 0$ for $M_0 \le x \le M$.

For definiteness we shall always write K for the cone of nonnegative functions in X and L_τ for the Frechet derivative (with respect to K) of F_τ at the origin, so

$$(L_\tau u)(t) = \int_{t-\tau}^{t} a(s)g'(0)u(s)ds \tag{21}$$

Since $L_{\tau_1} \ge L_{\tau_2}$ if $0 < \tau_1 \le \tau_2$, it is easy to see that $\tau \longrightarrow r(L_\tau)$ is an increasing function; and since e (the function identically equal to one) is an eigenvector of L_τ for $\tau = j\omega$ with eigenvalue $\int_{t-j\omega}^{t} a(s)g'(0)$, it follows that $r(L_\tau) \longrightarrow \infty$ as $\tau \longrightarrow \infty$

It will also be convenient to have a definition.

1. Definition. $h(x) = \inf_{0<y\le x} (y^{-1})g(y)$. Notice that $h(x) > 0$ for $0 < x < M$ and that $h(x)$ is a decreasing function. Our first few lemmas provide general qualitative information about the solutions of (20).

4. Lemma. Assume that a and g satisfy H8 and that $x \in K-\{0\}$ is a solution of (20). Then it follows that $0 < x(t) < M$ (M as in H8) for all t. Furthermore, if h is as in Definition 1, one must have

$$(g'(0))^{-1}h(||x||)r(L_\tau) \le 1 \tag{22}$$

If $r(L_\tau) < 1$, equation (20) has no solution $x \in K-\{0\}$.

Proof. Since $x \in K-\{0\}$, we must have $x(t) > 0$ for t in some interval of length $d > 0$. It follows from the form of (20) that x must be strictly positive in an interval of length $d_1 = d+\tau$. Repeating the argument n times gives that x is strictly positive on an interval of length $d + n\tau$, and if $n\tau \ge \omega$, we have $x(t) > 0$ everywhere.

Since $g(u) = 0$ for $u \ge M$, it can't be true that $x(t) \ge M$ for all t. If $x(t) = M$ for some t, there will be a positive δ and a t_1 such that $x(t_1) = M$ and $x(t) < M$ for $t_1-\delta \le t < t_1$. For t on this interval we have

$$\begin{aligned} x'(t) &= a(t)g(x(t))-a(t-\tau)g(x(t-\tau)) \\ &\le a(t)g(x(t)) \le C(M-x(t)) \end{aligned}$$

where C is some constant. This gives that

$$\frac{x'(t)}{M-x(t)} \le C$$

and integrating both sides from $t_1 - \delta$ to t_1, we obtain a contradiction.

If $x \in K-\{0\}$ is a solution of (20) we have

$$x(t) \ge (g'(0))^{-1}h(||x||)L_\tau(x) = L(x) \tag{23}$$

The spectral radius ρ of the linear operator L in (23) is

$$\rho = (g'(0))^{-1}h(||x||)r(L_\tau)$$

Suppose that, contrary to the lemma, $\rho > 1$. Since L is a positive linear operator, the Krein-Rutman theorem implies there is a positive eigenvector v with corresponding eigenvalue ρ, and since $x(t) > 0$ for all t, we can assume $v \le x$. Then we have

$$x \ge Lx \ge Lv = \rho v \tag{24}$$

Applying L^n to (24) we find that

$$x \geq \rho^n v \tag{25}$$

which is a contradiction, since $\rho > 1$.

To prove the last assertion in the lemma, observe that $F_\tau(x) \leq L_\tau(x)$ for all $x \in K$, so that $F_\tau^n(x) \leq L_\tau^n(x)$ for every integer $n \geq 1$. If $F_\tau(x) = x$ for some $x \in K-\{0\}$, we get

$$x \leq L_\tau^n x$$

and this is impossible if $r(L_\tau) < 1$. □

It follows from Lemma 4 and Theorem 4 that if τ_0 is the unique value of τ for which $r(L_\tau) = 1$, equation (20) has no positive solutions for $\tau < \tau_0$ and at least one positive solution for each $\tau > \tau_0$.

5. Lemma. If H8 is satisfies and $x \in K$ is a solution of (20) for some τ such that $j\omega \leq \tau \leq (j+1)\omega$, j an integer, then

$$x(t) \geq \frac{j}{j+1} ||x||$$

for all t.

Proof. If $B = \int_{t-\omega}^{t} a(s)g(x(s))ds$ = a constant independent of t, then it follows from (20) that for any t

$$jB \leq \int_{t-j\omega}^{t} a(s)g(x(s))ds + \int_{t-\tau}^{t-j\omega} a(s)g(x(s))ds \leq (j+1)B. \tag{26}$$

Equation (26) gives the lemma. □

Our next few lemmas are more directly related to the question of uniqueness. First we state a simple abstract result.

6. Lemma. Let C be a total cone in a Banach space Z and $F: C \to C$ a compact map. Assume that $F(0) = 0$ and that there exists a compact, positive linear operator L such that $\lim_{||x|| \to 0} (||x||^{-1})(F(x) - L(x)) = 0$ and such that the spectral radius of L is strictly greater than 1 and $Lx \neq x$ for $x \in K-\{0\}$. Suppose that $\lim_{||x|| \to \infty} (||x||^{-1})F(x) = 0$. Finally, assume that

each nonzero fixed point x of F is isolated and that if U is a relatively open neighborhood of x which contains no other fixed points of F, $i_K(F,U) = 1$. Then it follows that F has precisely one nonzero fixed point.

Proof. For $\rho > 0$ let $B_\rho = \{x \in K: ||x|| < \rho\}$. By our assumptions 0 is an isolated fixed point of F, and according to Lemma 1, if ρ is chosen small enough, 0 is the only fixed point of F in $\bar{B}_\rho$ and

$$i_K(F,B_\rho) = i_K(L,B_\rho) = 0$$

Select R so large that $(||x||^{-1})||F(x)|| < 1$ for $||x|| \geq R$. The homotopy $F_t(x) = tF(x)$ on B_R shows that

$$i_K(F,B_R) = 1.$$

By compactness and the assumption that fixed points are isolated, there are only finitely many fixed points x such that $\rho < ||x|| < R$. Label these fixed points $x_1,\ldots,x_n$ and take pairwise dispoint open neighborhoods $U_1,U_2,\ldots,U_n$. By the additivity property and our assumptions, if we set $U = \{x \in K: \rho < ||x|| < R\}$ we have

$$1 = i_K(F,U) = \sum_{j=1}^{n} i_K(F,U_j) = n$$

This completes the proof. □

7. Lemma. Let K be a cone in a Banach space Z and suppose that K has nonempty interior $\mathring{K}$. Suppose that $F:K \to K$ is a compact map, that F has a fixed point $x_0 \in \mathring{K}$, and that F has a Frechet derivative A at x_0 such that I - A is one-one and onto. Assume that there is a homotopy A_t, $0 \leq t \leq 1$, such that A_t is a compact linear map for each t, $t \to A_t$ is continuous, $I-A_t$ is one-one for $0 \leq t \leq 1$, and $A_1 = A$ and $A_0 = 0$. Then x_0 is an isolated fixed point of F, and if U is any

relatively open neighborhood of x_0 in K which contains no other fixed points, $i_K(F,U) = 1$.

Proof. The fixed point is isolated because I - A is one-one. Select $\rho > 0$ such that $x_0+A_t(x-x_0) \in K$ for $0 \le t \le 1$ and for all x such that $||x-x_0|| \le \rho$ and define $V = \{x\in K: ||x-x_0||>\rho\}$. Then if we define $F_t(x) = x_0+A_t(x-x_0)$ and assume that ρ is also chosen so small that x_0 is the only fixed point of F in V and $i_K(F,V) = i_K(F_1,V)$, the homotopy property implies that

$$i_K(F,U) = i_K(F,V) = i_K(F_t,V) = 1. \quad \square$$

Our next two propositions show that the hypotheses of Lemma 7 are frequently satisfied for the function F_τ in equation (20).

3. Proposition. Assume that H8 is satisfied and that $x\in K-\{0\}$ is a solution of equation (20) such that $x(t) \le M_0$ for all t, where M_0 is as in H8. Then x is an isolated fixed point of F_τ and if U is a relatively open neighborhood of x in K which contains no other fixed points of F_τ, it follows that $i_K(F_\tau,U) = 1$.

Proof. By Lemma 4 we know that $x \in \mathring{K}$, so it suffices to show that the conditions of Lemma 7 hold. It is easy to show that F_τ has a Frechet derivative A at x given by the formula

$$(Au)(t) = \int_{t-\tau}^{t} a(s)g'(x(s))u(s)ds \tag{26}$$

Because of Lemma 7, it suffices to show that the spectral radius of A is less than 1; and because A is a compact positive linear operator, to prove that $r(A) < 1$ it suffices to prove that $Ay \le cBy$ for all $y \in K$, where $c < 1$ and B is a compact, positive linear operator with spectral radius 1. Define B by the formula

$$(Bu)(t) = \int_{t-\tau}^{t} a(s) \left(\frac{g(x(s))}{x(s)}\right) u(s)ds$$

First observe that because $0 < x(s) \le M_0$ and $g'(y)$ is assumed strictly decreasing for $0 \le y \le M_0$, it follows that there exists a positive constant $c < 1$ such that

$$g'(x(s)) \le c \frac{g(x(s))}{x(s)} \tag{27}$$

Equation (27) implies that $Ay \le cBy$ for all $y \in K$. Next notice that $Bx = x$ and that (as one can easily show) B is e-positive. It follows from Proposition 2 that the spectral radius of B is one. □

Our next proposition treats the case in which a solution $x(t)$ of (20) satisfies $x(t) \ge M_0$ for all t.

4. Proposition. Assume that H8 holds and that $x \in K$ is a solution of equation (20) such that $x(t) \ge M_0$ for all t, where M_0 is as in H8. In addition suppose that

$$\sup_t \int_t^{t+\frac{\omega}{2}} a(s)|g'(x(s))|ds < 1 \tag{28}$$

Then x is an isolated fixed point of F_τ and if U is any relatively open neighborhood of x in K which contains no other fixed points $i_K(F,U) = 1$.

Proof. As in Proposition 3, F_τ has a Frechet derivative A at x given by the formula

$$(Au)(t) = \int_{t-\tau}^{t} a(s)g'(x(s))u(s)ds \tag{29}$$

Define a compact, linear operator A_λ, $0 \le \lambda \le 1$, by the formula

$$(A_\lambda u)(t) = \lambda \int_{t-\tau}^{t} a(s)g'(x(s))u(s)ds \tag{30}$$

Because of Lemma 7, it suffices to show that $I - A_\lambda$ is one-one for $0 \le \lambda \le 1$. If we define $b(s) = \lambda a(s)g'(x(s))$ and define a bounded linear operator B by

$$(Bu)(t) = \int_{t-\tau}^{t} b(s)u(s)ds \tag{31}$$

it suffices to prove that I - B is one-one whenever $b \in X$ is a function such that

$$b(s) \leq 0 \text{ for all } s \text{ and}$$
$$\sup_t \int_t^{t+\frac{\omega}{2}} |b(s)|dx < 1 \tag{32}$$

In order to prove I - B one-one, we suppose not and try for a contradiction. Select an integer j such that $j\omega \leq \tau \leq (j+1)\omega$ and denote by u a nonzero element of X such that $(I-B)(u) = 0$. There are two cases to consider.

Case 1. $j\omega \leq \tau \leq j\omega+\frac{1}{2}\omega$. We have the question $u(t) - \int_{t-\tau}^{t-\tau+j\omega} b(s)u(s))ds - \int_{t-\tau+j\omega}^{t} b(s)u(s)ds = 0$. Write $\sigma = \tau-j\omega$ and define a bounded linear operator B_σ by

$$(B_\sigma v)(t) = \int_{t-\sigma}^{t} b(s)v(s)ds \tag{33}$$

Notice that $||B_\sigma|| = \sup_t \int_{t-\sigma}^{t} |b(s)|ds < 1$, so that $I - B_\sigma$ is one-one, and we have

$$(I-B_\sigma)^{-1} = \sum_{k=0}^{\infty} B_\sigma^k \tag{34}$$

The integral $\int_{t-\tau}^{t-\tau+j\omega} b(s)u(s)ds$ is a constant independent of t; it is nonzero, because otherwise $(I-B_\sigma)(u) = 0$ for a nonzero u. By multiplying u by a nonzero constant we can assume

$$\int_{t-\tau}^{t-\tau+j\omega} b(s)u(s)ds = 1 \tag{35}$$

and it follows that

$$u = (I-B_\sigma)^{-1}(1) = \sum_{k=0}^{\infty} B_\sigma^{2k}(I+B_\sigma)(1) \tag{36}$$

Notice that B_σ^{2k} is a positive linear operator and that $(I+B_\sigma)(1)$ is a positve function, since $||B_\sigma|| < 1$. It follows that u is a nonnegative function and hence that -Bu is nonnegative, which contradicts the assumption that u-Bu = 0.

Case 2. $j\omega+\frac{1}{2}\omega \le \tau \le (j+1)\omega$. We obtain the equation

$$u(t) - \int_{t-\tau}^{t-\tau+(j+1)\omega} b(s)u(s)ds + \int_{t}^{t+(j+1)\omega-\tau} b(s)u(s)ds = 0 \quad (37)$$

If we write $\sigma = (j+1)\omega-\tau$ and define a linear operator C_σ by

$$(C_\sigma v)(t) = \int_{t}^{t+\sigma} b(s)v(s)ds \quad (38)$$

then just as before we can assume $\int_{t-\tau}^{t-\tau+(j+1)\omega} b(s)u(s)ds = 1$ and we find

$$u+C_\sigma u = 1 \quad (39)$$

Equation (39) implies that

$$u = \sum_{k=0}^{\infty} (-C_\sigma)^k(1) \quad (40)$$

and since $-C_\sigma$ is a positive linear operator, u is a nonnegative function. Just as before this is impossible. □

1. Remark. If $b \in X$ is a nonpositive function and B is defined by (31), the argument used in Proposition 4 shows that I - B is one-one for τ near some $j\omega$, j an integer. In general the question of what additional assumptions on b will ensure that I - B is 1-1 for all τ remains open; certainly much better results than those given in Proposition 4 are true.

With the aid of Lemmas 4-7 and Propositions 3 and 4 we can obtain the following crude uniqueness result for solutions of equation (20).

5. Theorem. Assume that H8 holds, and if M_0 and M are as in H8 define a constant C by $C = \sup\{|g'(x)|:M_0 \le x \le M\}$. Assume that $\sup_t C\int_t^{t+\frac{1}{2}\omega} a(s)ds < 1$. If τ_0 is the unique value of τ such that $r(L_\tau) = 1$ and $S = \{(x,\tau):x\neq 0,\ F_\tau(x)=x\}$, let $\tau_1 \ge \tau_0$ be a number such that $||x|| \le M_0$ for all $(x,\sigma) \in S$ such that $\tau_0 \le \sigma \le \tau_1$ and let τ_2 be a number such that $x(t) \ge M_0$ for all $(x,\sigma) \in S$ such that $\sigma \ge \tau_2$. Then it follows that equation

(20) has a unique solution $x = x_\tau \epsilon K-\{0\}$ for $\tau_0 < \tau \le \tau_1$ and for $\tau \ge \tau_2$.

Proof. Note that Lemmas 4 and 5 imply that $\tau_2 < \infty$ and give a crude estimate for τ_2. If $\tau_0 < \tau \le \tau_1$, Proposition 3 implies that every positive solutions of $F_\tau(x) = x$ is isolated and has local fixed point index one, so Lemma 6 implies there is precisely one fixed point. If $\tau \ge \tau_2$, every solution $x \in K-\{0\}$ of $F_\tau(x) = x$ satisfies $x(t) \ge M_0$ for all t. Furthermore one has

$$\sup_t \int_t^{t+\frac{1}{2}} a(s)|g'(x(s)|ds \le C \sup_t \int_t^{t+\frac{1}{2}\omega} a(s)ds < 1$$

It follows from Proposition 4 and Lemma 6 that uniqueness holds for $\tau \ge \tau_2$. □

As an example, we shall now completely answer the uniqueness question for the model equation of Cooke and Kaplan.

1. Corollary. Consider the equation

$$x(t) = \int_{t-\tau}^{t} a(s)g(x(s))ds \tag{41}$$

where $a(s) = 1+\frac{1}{2}\sin 2\pi s$ and $g(x) = x(1-x)$ for $0 \le x \le 1$ and $g(x) = 0$ otherwise. For each $\tau > 1$ the equation has precisely one positive, periodic solution $x_\tau(t)$ of period 1. For $0 \le \tau \le 1$, the equation has no positive, periodic solutions $x \in X$.

Proof. $\tau \longrightarrow r(L_\tau)$ is strictly increasing, and if $\tau = j =$ integer, the function identically one is an eigenvector with eigenvalue $j = r(L_j)$. Thus τ_0 in Theorem 5 is one and by Lemma 4 there are no positive solutions for $\tau < 1$.

In the notation of assumption H8, $M_0 = \frac{1}{2}$ and $M = 1$; and if $h(x)$ is as in Definition 1, $h(x) = 1-x$ for $0 \le x \le 1$. It follows from Lemma 4 that if x is a solution of (41), then

$$||x|| \ge 1-(r(L_\tau))^{-1} \tag{42}$$

In particular $||x|| \geq 2/3$ for $\tau \geq 3$, and it follows from Lemma 5 that $x(t) \geq \frac{1}{2}$ for all t if $\tau \geq 3$. Thus Theorem 5 implies uniqueness of solutions for $\tau \geq 3$.

On the other hand, since $g(x) \leq \frac{1}{4}$ for all x, it follows that for any $x \in K$,

$$||F_\tau(x)|| \leq \frac{1}{4} \sup_t \int_{t-\tau}^{t} a(s)ds \tag{43}$$

In particular, any solution x of (41) for $1 < \tau \leq 2$ must satisfy $||x|| \leq \frac{1}{2}$, so Theorem 5 gives uniqueness for $1 < \tau \leq 2$.

It remains to consider the case $2 < \tau < 3$, and here we use an ad hoc argument in the spirit of Proposition 4. If x is any positive solution of (41) for τ with $2 < \tau < 3$, (43) implies that $||x|| < 3/4$, and (42) implies that $\frac{1}{2} < ||x||$, so that according to Lemma 5, $x(t) > \frac{1}{3}$ for all t. If follows that

$$-\frac{1}{2} \leq g'(x(t)) = 1-2x(t) \leq \frac{1}{3} \tag{44}$$

for any solution x as above. Define a bounded linear operator $A_\lambda : X \longrightarrow X$ for $0 \leq \lambda \leq 1$ by the formula

$$(A_\lambda u)(t) = \int_{t-\tau}^{t} \lambda a(s) g'(x(s)) u(s) ds$$

To prove uniqueness of positive solutions for $2 < \tau < 3$, it suffices (because of Lemmas 6 and 7) to show that $I - A_\lambda$ is one-one. For notational convenience, write $b(s)=\lambda g'(x(s)) \in X$ and define $B: X \longrightarrow X$ by

$$(Bu)(t) = \int_{t-\tau}^{t} b(s)a(s)u(s)ds$$

It suffices to show $I - B$ is one-one. Suppose not and let $u \in X$ be a nonzero solution of the equation $(I-B)(u) = 0$. Proceeding as in Proposition 4 we can write

$$u(t) - \int_{t-\tau}^{t-\tau+3} b(s)a(s)u(s)ds + \int_{t}^{t+\sigma} b(s)a(s)u(s)ds = 0 \tag{45}$$

where $\sigma = 3-\tau$. If we define a bounded linear operator $B_\sigma : X \longrightarrow X$ by

$$(B_\sigma v)(t) = \int_t^{t+\sigma} b(s)a(s)v(s)ds$$

then under our assumptions on b we have for $0 \le \sigma < 1$

$$||B_\sigma|| = \sup_t \int_t^{t+\sigma} |b(s)a(s)|ds < \frac{1}{2} \sup_t \int_t^{t+\sigma} a(s)ds < \frac{1}{2} \qquad (46)$$

It follows that $I + B_\sigma$ is one-one, and we must have $\int_{t-\tau}^{t-\tau+3} b(s)a(s)u(s)ds$ equal to a nonzero constant; by multiplying u by a constant we can assume that

$$\int_{t-\tau}^{t-\tau+3} b(s)a(s)u(s)ds = 1$$

Thus we obtain from (45) that

$$u = \sum_{k=0}^{\infty} (-B_\sigma)^k(1) = 1+v \qquad (47)$$

Since $||B_\sigma|| < \frac{1}{2}$ we find that

$$||v|| < \sum_{k=1}^{\infty} (\tfrac{1}{2})^k = 1$$

and consequently $u(t) > 0$ for all t. On the other hand, if t_0 is chosen such that $u(t_0) = ||u||$, the equation (45) cannot be satisfied: we find (using the positivity of u) that

$$||u|| - \int_{t_0-\tau}^{t_0} a(s)b(s)u(s)ds \ge ||u|| - \frac{1}{3}\int_{t_0-\tau}^{t_0} a(s)||u||ds$$
$$= ||u||[1-\frac{1}{3}\int_{t_0-\tau}^{t_0} a(s)ds] > 0 \qquad (48)$$

This contradiction proves that I - B is one-one and completes the proof of the corollary. □

2. Remark. Hal Smith [13] has proved uniqueness of positive solution of (41) for $1 < \tau \le 2$ by using ideas of Krasnoselskii [8], and the same observation was independently made by this author in [11, p. 19]. However the Krasnoselskii approach completely fails for $\tau > 2$, and for this reason we have used a very different set of ideas here.

REFERENCES

1. Amann, H., On the number of solutions of nonlinear equations in ordered Banach spaces, *J. Functional Analysis 11*(1972), 346-384.

2. Amann, H., Fixed points of asymptotically linear maps in ordered Banach spaces, *J. Functional Analysis 14*(1973), 162-171.

3. Cooke, K. L., and Kaplan, J. L., A periodicity threshold theorem for epidemics and population growth, to appear in *Math. Biosciences*.

4. Cooke, K. L., and Yorke, J. A., Equations modelling population growth and gonorrhea epidemiology, *Math. Biosciences 16*(1973), 75-101.

5. Dancer, E. N., On the structure of solutions of nonlinear eigenvalue problems, *Indiana U. Math. J. 23*(1974), 1069-1076.

6. Dancer, E. N., Solution branches for mappings in cones, and applications, *Bull. Australian Math. Soc. 11*(1974), 131-143.

7. Dunford, N. and Schwartz, J. T., *Linear Operators Part I*, Interscience Publishers Inc., New York, 1967, (fourth printing).

8. Krasnosel'skii, M. A., *Positive Solutions of Operator Equations*, P. Noordhoff Ltd., Groningen, The Netherlands, 1964.

9. Nussbaum, R. D., The fixed point index for local condensing maps, *Ann. Mat. Pura Appl. 89*(1971), 217-258.

10. Nussbaum, R. D., A global bifurcation theorem with applications to functional differential equations, *J. Functional Analysis 19*(1975), 319-339.

11. Nussbaum, R. D., A periodicity threshold theorem for some nonlinear integral equations, to appear.

12. Rabinowitz, P. H., Some global results for nonlinear eigenvalue problems, *J. Functional Analysis 7*(1971), 487-513.

13. Smith, H., preprint, to appear.

14. Turner, R. E. L., Transversality and cone maps, *Arch. Rat. Mech. Anal. 58*(1975), 151-179.

15. Whyburn, G. T., *Topological Analysis*, Princeton U. Press, Princeton, N.J., 1958.

---

*Partially supported by a National Science Foundation Grant.

UNITARY TREATMENT OF VARIOUS TYPES OF SYSTEMS IN STABILITY-THEORY

V. M. Popov*
Department of Mathematics
University of Florida

1. INTRODUCTION

The purpose of this paper is to make more flexible the technique of finding criteria of stability - especially of the frequency-domain type. The paper is basically self-contained, although it develops some ideas from a previous work of the author [1]. The class of problems which can be investigated by this approach is very large. As a prototype, one can take the nonlinear Volterra equation

$$\underline{1.1} \qquad \sigma(t) + \int_0^t g(t-\tau)\phi(\sigma(\tau))d\tau = f(t), \; t \geqq 0.$$

Basic information about this equation, from the stability viewpoint, can be found, for instance, in C. Corduneau [2] or R. Miller [3]. Considerable attention has been devoted to obtaining sharper results and to extend the theory to more general types of systems (see, for instance, J. A. Nohel and D. F. Shea [4], V. M. Popov [1], O. J. Staffans [5], etc.). A natural counterpart of the equation 1.1 is given by the "discrete" case

$$\underline{1.2} \qquad \sigma_k + \sum_{j=0}^{k} g_{k-j}\,\phi(\sigma_j) = f_k, \; k = 0, 1, 2, \ldots$$

(See, for instance, K. S. Narendra and Yo-Sung Cho [6] or V. A. Yakubovich [7], where one can also find references to previous basic work of Ya. Z. Tsypkin, G. Szegö, R. E. Kalman, J. B. Pearson, E. J. Jury, B. W. Lee, Yu. A. Dmitriev, etc.

The problem has been intensively studied and our list is very incomplete.)

It has been recognized that the theory of stability for the equations 1.1 and 1.2 can be presented in a unitary way, in the framework of Abstract Harmonic Analysis (A. Halanay [8], R. Datko [9]). A theory of stability under a very general setting has been given by M. I. Freedman, P. L. Falb, and G. Zames [10]. Extensions of equation 1.1, involving maximal monotone operators, have been studied by V. Barby [11]. A detailed study of the problem of stability for nonlinear delay differential equations which describe control systems can be found in a recent book of Vl. Răsvan [12].

In view of this great diversity of problems, it is obviously desirable to present every new element of technique in a sufficiently general form, to facilitate its application to new problems and to contribute to the unification of the field. Sections 2 - 6 of this paper are written with this purpose in mind. On this basis, the applications consist in giving appropriate meanings to the symbols involved.

To illustrate (and test) a new technique, it is often advisable to show how it works in the simplest and best studied cases. Such a purpose is served by the applications in sections 7 and 8, where we obtain new stability results for problems of the form 1.2. We hope that these applications will also serve to point out how one can obtain similar results in amny other problems.

2. SYSTEMS

We must adopt a more flexible meaning of the word "system." Let F be a normed space and denote the norm by $|.|$. Let T and H be two sets and denote, as usual, by H^T

the set of all mappings from T into H and by $P(H^T)$ the set of all subsets of H^T. A "system" is, by definition, a mapping

$$S : F \to P(H^T).$$
$$f \mapsto S_f$$

Thus, for every f in F, S_f will denote a set of functions from T into H - and we do not exclude the case in which this set is empty for some f. Concrete examples of systems are given in sections 7 and 8.

3. STABILITY

Let Λ^2 be a set of functions from T into H and let $|.|_2$ be a function from Λ^2 into R_+ (as usual, $R_+ = [0,\infty[$). (In most cases, Λ^2 will be a space of square integrable functions and $|.|_2$ will be the usual L^2-norm.)

3.1 Definition. A system S is said to be stable if 1: for every $f \in F$, one has the inclusion $S_f \subset \Lambda^2$ and 2: there exists a mapping χ: $R_+ \longrightarrow R_+$, continuous at the origin and vanishing there, such that, for every $f \in F$ and every $u \in S_f$, one has the inequality

$$|u|_2 \leqq \chi(|f|).$$

We remark that the above definition agrees with the Liapunov prototype. If Λ^2 is an L^2 space, one obtains the usual concept of L^2-stability.

4. SATURABILITY

Let Λ^∞ be a set of functions from T into H. (In most applications, Λ^∞ is taken as a set of bounded functions.) For every $f \in F$, define a sequence of mappings T_n: $S_f \longrightarrow \Lambda^2$, n = 1,2,..., with the properties:

4.1 $|T_n u|_2 \leqq |T_{n+1} u|_2$, for n = 1,2,... and $\forall u \in S_f$,

$$\underline{4.2} \qquad \sup_n |T_n u|_2 = \begin{cases} |u|_2 & \text{if } u \in S_f \cap \Lambda^2 \\ \infty & \text{if } u \notin \Lambda^2. \end{cases}$$

(Our notations recall the most usual interpretation of the symbols. Thus T, which denotes the domain of our functions, recalls the word "time" whereas the T_n's just introduced recall the word "truncation".)

Definition. We say that the system S is Λ^∞-saturable if, for every $f \in F$ and every $u \in S_f$, there exists a sequence $u^m \in S_f \cap \Lambda^\infty$, m = 1,2,..., with the property

$$\underline{4.3} \qquad \overline{\lim_{m\to\infty}} |T_n(u^m)|_2 = |T_n u|_2, \text{ for } n = 1,2,\ldots .$$

The terminology will appear more natural after reading section 7.

Observe that every system is Λ^2-saturable. However, this particular case is less useful.

5. INDEX FUNCTIONS

Given a Λ^∞-saturable system S, we introduce a class of functions which serve to investigate its stability. We denote by D_0 the set of all triples (f,u,n), where $f \in F$, $u \in (S_f \cap \Lambda^2) \cup (S_f \cap \Lambda^\infty)$ and n is a strictly positive integer. A function J of the form

$$J : D_0 \longrightarrow R$$
$$(f,u,n) \longmapsto J(f,u,n)$$

will be called an index function.

We shall need index functions which are bounded and positive, in the following sense:

5.1 Definition. An index function J is said to be bounded if there exist: (1) a function $\nu : R_+ \longrightarrow R_+$, continuous at the origin and vanishing there; and (2) a function $\rho : D_0 \longrightarrow R$, such that

$$\overline{\lim_{n\to\infty}}\ \rho(f,u,n) = 0,\ \forall\ f \in F \text{ and } \forall\ u \in S_f \cap \Lambda^2$$

$$\sup_n \rho(f,u,n) < \infty,\ \forall\ f \in F \text{ and } \forall\ u \in S_f \cap \Lambda^\infty$$

$$J(f,u,n) \leqq \nu(|f|) + \rho(f,u,n),\ \forall (f,u,n) \in D_0.$$

5.2 Definition. An index function J is said to be positive if there exist: an increasing mapping η, from R_+ onto R_+, and a mapping μ: $R_+ \longrightarrow R_+$, such that: (1) η and μ are continuous on R_+ and vanish at the origin; and (2) one has the inequality

$$J(f,u,n) \geqq \eta^{-1}(|T_n u|_2) - \mu(|f|),\ \forall (f,u,n) \in D_0.$$

6. THE CORE OF THE SATURABILITY-TECHNIQUE

The concepts introduced above are motivated by the following result:

6.1 Theorem. A system is stable if and only if there exists a set Λ^∞ such that: (1) the system is Λ^∞-saturable; and (2) there exists a bounded and positive index function.

Proof. If the system is stable, then, by definition, $S_f \subset \Lambda^2$. Choosing Λ^∞ such that $\Lambda^\infty \supset S_f$ and taking $u^m = u$, $m = 1,2,\ldots$, one sees immediately that the system is Λ^∞-saturable, no matter how one chooses the T_n's from 4.1 - 2. Taking χ as in the property of stability 3.1, we define $J(f,u,n) = \chi(|f|)$ and find immediately that J is bounded and positive. (This part of the proof serves only to show that, if a system is stable, then, in principle, one can always establish this fact by finding a suitable set Λ^∞ and a bounded and positive index function.)

Suppose now that one has found a set Λ^∞ such that the system is Λ^∞-saturable and that there exists a bounded and positive index function J. From 5.1 and 5.2 it follows that, if $u \in S_f \cap \Lambda^\infty$, then

$$|T_n u|_2 \leqq \eta(\mu(|f|) + \nu(|f|) + \rho(f,u,n)) \leqq K < \infty,$$

where K can be chosen independent of n. Using now 4.2 one finds that:

6.2 If $u \in S_f \cap \Lambda^\infty$, then $u \in S_f \cap \Lambda^2$.

From 5.1 and 5.2 it also follows that, if $u \in S_f \cap \Lambda^2$, then

$$\overline{\lim_{n \to \infty}} |T_n u|_2 \leqq \eta(\nu(|f|) + \mu(|f|)).$$

Hence, using 4.1 - 2 and defining the function χ by

$$\chi(r) = \eta(\nu(r) + \mu(r)), \forall r \geqq 0,$$

one obtains the conclusion

6.3 If $u \in S_f \cap \Lambda^2$, then $|u|_2 \leqq \chi(|f|)$.

Obviously, the function χ, defined above, is continuous at the origin and vanishes there. Therefore, it only remains to show that $S_f \subset \Lambda^2$. To prove this, choose $u \in S_f$ and consider a sequence $u^m \in S_f \cap \Lambda^\infty$, $m = 1,2,\ldots$, with the properties from section 4. By 6.2, $u^m \in S_f \cap \Lambda^2$. Hence, by 6.3, $|u^m|_2 \leqq \chi(|f|)$. Using now 4.1 and 4.2, we find that $|T_n u^m|_2 \leqq \chi(|f|)$, for $n = 1,2,\ldots$, and $m = 1,2,\ldots$. Letting m go to infinity, for fixed n, and using 4.3, one obtains $|T_n u|_2 \leqq \chi(|f|)$. Letting here n tend to infinity and using 4.2 one finds that $u \in S_f \cap \Lambda^2$ - and this ends the proof, by 6.3.

7. AN EXAMPLE

Consider Eq. 1.2. Let κ be a number in $]0,\infty[$. We assume that $\phi: R \longrightarrow R$ is nondecreasing, continuous at the origin, vanishes at the origin and satisfies the inequalities

7.1 $0 \leqq \phi(\sigma)\sigma \leqq \kappa\sigma^2, \forall \sigma \in R.$

We denote by Φ the set of all functions ϕ with the above properties. We shall use the spaces l_1, l_2 and l_∞, with the corresponding norms (as in N. Dunford and J. T. Schwartz [13]). A sequence will be denoted in the form $\{u_k\}$, or simply by u. We assume that

7.2 $\{f_k\} \in l_1$, $\{g_k\} \in l_1$ and $g_0 \geqq 0$

The last condition secures the fact that, given $\{f_k\} \in l_1$ and $\phi \in \Phi$, Eq. 1.2 has a unique solution.

In order to apply the present method, we have to give appropriate meanings to the symbols $F, T, H, \Lambda^2, \Lambda^\infty, T_n, S_f, J$ etc. We take $F = l_2$ and denote the corresponding norm by $|.|$. We define $T = \{0,1,2,...\}$ and $H = R$. For a given $f = \{f_k\} \in F$, the set S_f will be defined as follows:

7.3 S_f is the set of all sequences $\{u_k\}$ with the following property: There exists a function $\phi \in \Phi$ such that, if by $\{\sigma_k\}$ one denotes the corresponding solution of 1.2, then one has the equalities $u_k = -\phi(\sigma_k)$, $k = 0,1,2,....$

Observe that, according to the above definition, to a given $\{u_k\} \in S_f$ there corresponds one and only one sequence $\{\sigma_k\}$ with the property from 7.3. Moreover, this sequence satisfies the equation

7.4 $$\sigma_k = f_k + \sum_{j=0}^{k} g_{k-j} u_j, \quad k = 0,1,2,.....$$

Now our system is defined and we turn to the concept of stability. We take $\Lambda^2 = l_2$ and take $|.|_2$ as the corresponding norm. Then the concept of stability from section 3 takes a concrete meaning (and corresponds to the usual concept of absolute stability). Next, we take $\Lambda^\infty = l_\infty$ and show that our system is Λ^∞-saturable. Given $\{u_k\}$ in S_f and a strictly positive integer n, we define $T_n\{u_k\} = \{u_0, u_1, ..., u_n, 0, 0, ...\}$ (the truncated sequence). Then properties 4.1 - 2 are obviously satisfied. Let now m be a strictly positive integer. Given $\{u_k\}$ in S_f, we know, by 7.3, that there exists a function ϕ with the property stated in 7.3. We define a new function ϕ_m as follows:

$$\underline{7.5} \quad \phi_m(\rho) = \begin{cases} \phi(-m) & \text{if } \rho \leqq -m \\ \phi(\rho) & \text{if } |\rho| \leqq m \\ \phi(m) & \text{if } \rho > m \end{cases}$$

Clearly $\phi_m \in \Phi$. For every fixed m, consider now the sequence $\{\sigma_0^m, \sigma_1^m, \sigma_2^m, \ldots\}$ which satisfies the equation

$$\underline{7.6} \quad \sigma_k^m + \sum_{j=0}^{k} g_{k-j}\, \phi_m(\sigma_j^m) = f_k,\ k = 0,1,2,\ldots.$$

Now we define $u_k^m = -\phi_m(\sigma_j^m)$, $k = 0,1,2,\ldots.$ For convenience, we use the abbreviations $u^m = \{u_k^m\}$, $u = \{u_k\}$. By comparing Eqs. 7.6 and 1.2 and by using 7.5, one easily sees that $T_n u^m = T_n u$, for every m such that $m > \max_{1\leqq k\leqq n} |\sigma_k|$. Therefore condition 4.3 is satisfied.

Thus our system is Λ^∞-saturable. According to section 6, if one can find a bounded and positive index function, then the system is stable.

From our assumptions concerning ϕ, it follows that

$$\underline{7.7} \quad -\phi(\sigma_k)[\sigma_k - \tfrac{1}{\kappa}\phi(\sigma_k)] \leqq 0,\ k = 0,1,2,\ldots .$$

Moreover, if one defines the function $\Psi : R \longrightarrow R$ by $\Psi(\sigma) = \int_0^\sigma \phi(\rho)d\rho$ (which is nonnegative, vanishes at the origin and satisfies the inequality $\phi(\sigma)(\sigma'-\sigma) \leqq \Psi(\sigma') - \Psi(\sigma), \forall \sigma, \sigma' \in R$), then one obtains the inequalities (where, for convenience, we put $\sigma_{-1} = 0$)

$$\underline{7.8} \quad -\phi(\sigma_k)(\sigma_k - \sigma_{k-1}) \leqq -\Psi(\sigma_k) + \Psi(\sigma_{k-1}),\ k = 0,1,2,\ldots$$

$$\underline{7.9} \quad -\phi(\sigma_k)(\sigma_k - \sigma_{k+1}) \leqq \Psi(\sigma_{k+1}) - \Psi(\sigma_k),\ k = 0,1,2,\ldots.$$

We mention that such inequalities are commonly used in almost any study of our problem. The present method leads to improvements of the existing criteria because it can handle inequalities of the form 7.9 as easily as the inequalities 7.7 - 8 (which can be handled by any of the older methods).

Using these inequalities we can define a bounded index function as follows. Choose $f \in F$, $u \in (S_f \cap \Lambda^2) \cup (S_f \cap \Lambda^\infty)$ and n: a strictly positive integer. Define $\{\sigma_k\}$ by 7.4. According to 7.3, one can write $u_k = -\phi(\sigma_k)$. We introduce now the sequence $\{\tilde{\sigma}_k\}$, defined by

$$\underline{7.10} \quad \tilde{\sigma}_k = f_k + \sum_{j=0}^{\min(k,n)} g_{k-j} u_j, \quad k = 0,1,2,\ldots.$$

Then the above definitions imply that $\tilde{\sigma}_k = \sigma_k$ for k = 0,1,2,...n, and

$$\underline{7.11} \quad \sigma_{n+1} = \tilde{\sigma}_{n+1} - g_0\phi(\sigma_{n+1}).$$

Let now α, β and γ be nonnegative constants and define

$$\underline{7.12} \quad J(f,u,n) = \sum_{k=0}^{n} u_k[\alpha(\tilde{\sigma}_k + \tfrac{1}{\kappa} u_k) + \beta(\tilde{\sigma}_k - \tilde{\sigma}_{k+1}) + \gamma(\tilde{\sigma}_k - \tilde{\sigma}_{k-1})]. \quad (\tilde{\sigma}_{-1} = 0).$$

Using the above relations between σ_k and $\tilde{\sigma}_k$ and the inequalities 7.7 - 9, one easily finds the inequality

$$J(f,u,n) \leqq \beta\Psi(\sigma_{n+1}) + \gamma\Psi(\sigma_0) + \beta\phi(\sigma_n)\, g_0\phi(\sigma_{n+1}).$$

This can be rewritten in the form from 5.1 if one defines ρ by

$$\rho(f,u,n) = \beta[\Psi(\sigma_{n+1}) + \phi(\sigma_n)\, g_0\, \phi(\sigma_{n+1})]$$

and one chooses a suitable function ν such that $\nu(|f|) \geqq \gamma\Psi(\sigma_0)$.

It is easy to see that ρ has the properties required in 5.1: If $u \in S_f \cap \Lambda^2$ then, by 7.4 and our assumption that f and g belong to l_1, we find that σ_k approaches zero as k tends to infinity; therefore, $\rho(f,u,n) \to 0$ as $n \to \infty$. If $u \in S_f \cap \Lambda^\infty$ then, by similar arguments, $\{\sigma_k\}$ is bounded and, therefore, $\sup_n \rho(f,u,n) < \infty$. This proves that J is bounded in the sense of 5.1.

It remains to find conditions for J to be positive. First, we define the "transfer function"

7.13 $$\hat{g}(z) = \sum_{k=0}^{\infty} g_k z^{-k}$$

for every complex z with absolute value $|z| = 1$. In terms of $\hat{g}$ we introduce the frequency-domain condition

7.14 $$\text{Re } [\hat{g}(z)(\alpha + (\frac{\gamma}{z} - \beta)(z-1)] > -\frac{\alpha}{K}, \quad |z| = 1$$

From 7.10 - 12 and 7.13 - 14, by using Parseval formula, one finds that there exist $\varepsilon > 0$ and $K > 0$ such that

7.15 $$J(f,u,n) \geqq \varepsilon |T_n u|_2^2 - K |f|_2^2.$$

(The procedure of obtaining such estimates is the same as in the older methods and, therefore, we omit the details.) This shows that J is positive in the sense 5.2. We can now use Th. 6.1 and get the following stability criterion: If there exist $\alpha \geqq 0$, $\beta \geqq 0$ and $\gamma \geqq 0$ such that 7.14 holds, then our system (under the general assumptions mentioned in this section) is stable.

The same technique can be applied in connection with almost any existing frequency-domain condition of stability from literature. Thus one shows, for instance, that Theorem 6 from V. A. Yakubovich [7] remains true even if one removes the restriction - required there - that ϕ (or $\phi(\sigma) - \nu\sigma$) be bounded; then Theorems 4 and 5 from the same paper become particular cases of the (thus generalized) Theorem 6. To treat this case (or more general ones) one uses the same approach as in this section, with obvious modifications of the index function J.

8. AN EXAMPLE WITH TWO NONLINEARITIES

Instead of 1.2 we consider now the equation

8.1 $$\sigma_k + \sum_{j=0}^{k} g_{k-j} (\phi_1(\sigma_j) - \phi_2(\sigma_{j-1})) = f_k, \quad k = 0,1,2,\ldots;$$

$$f, g \in l_1, \quad (\sigma_{-1}=0).$$

We assume that g_0 is nonnegative. We look for the simplest frequency-domain condition which implies stability. Let us assume that ϕ_1 and ϕ_2 are nondecreasing, continuous at the origin and have the properties

<u>8.2</u> $0 \leqq (\phi_1(\sigma) - \phi_2(\sigma))\sigma, \ \forall\sigma \in R$

<u>8.3</u> $\phi_1(0) = \phi_2(0) = 0.$

We proceed as in section 7. The definitions of F, T, H, Λ^2 and Λ^∞ remain the same. Instead of 7.3, we use a new definition: S_f is now the set of all sequences $\{u_k\}$ with the property that there exist two functions, ϕ_1 and ϕ_2 - which satisfy the above conditions - such that, if one denotes by $\{\sigma_k\}$ the corresponding solution of 8.1, one has the equalities $u_k = -\phi_1(\sigma_k) + \phi_2(\sigma_{k-1})$, $k = 0,1,2,\ldots.$ Now $\{\sigma_k\}$ satisfies again 7.4 and this shows that $\{\sigma_k\}$ is uniquely determined by $\{u_k\}$ (although ϕ_1 and ϕ_2 are not). As in section 7, one introduces the functions $\Psi_j(\sigma) = \int_0^\sigma \phi_j(\rho)d\rho$, $j = 1,2$ and one obtains the inequalities

$$- \phi_1(\sigma_k)(\sigma_k-\sigma_{k-1}) \leqq - \Psi_1(\sigma_k) + \Psi_1(\sigma_{k-1})$$

$$\phi_2(\sigma_{k-1})(\sigma_k-\sigma_{k-1}) \leqq \Psi_2(\sigma_k) - \Psi_2(\sigma_{k-1})$$

$$(-\phi_1(\sigma_k) + \phi_2(\sigma_{k-1}))\sigma_k \leqq -\phi_1(\sigma_k)\sigma_k + \phi_2(\sigma_{k-1})\sigma_{k-1} + \Psi_2(\sigma_k) - \Psi_2(\sigma_{k-1})$$

Choosing two nonnegative constants, α and β, we define the index function

$$J(f,u,n) = \sum_{k=0}^{n} u_k(\alpha\sigma_k+\beta(\sigma_k-\sigma_{k-1})).$$

From the above inequalities, it follows (as in section 7) that J is bounded in the sense 5.1. To secure the positivity of the index function J it suffices to introduce the frequency-domain condition

$$\text{Re}\ [\hat{g}(z)(\alpha+ \frac{\beta}{z}(z-1))] > 0, \ |z| = 1.$$

In conclusion, if there exist two nonnegative constants, α and β such that the above condition is satisfied, then (under our general assumptions) the stability of the system is secured.

9. FINAL COMMENTS

Having a systematic procedure for obtaining criteria of stability, the principal creative step in applications consists in finding suitable index functions, satisfying the conditions of boundedness and positivity. The positivity can often be obtained by using a Fourier transform and Parseval formula (and then it is secured by a criterion of the frequency-domain form). The boundedness is usually established by a direct investigation of the system. In our approach, in order to facilitate this step, we aimed at introducing the mildest requirements of boundedness which still secure the property of stability.

A specific application will normally necessitate the use of various auxiliary results from System Theory, the Theory of the Integral, Harmonic Analysis, etc. The approach we gave serves to show what kinds of auxiliary results are needed and how to combine them to answer the question. The applications we presented serve only to illustrate the process, keeping the technical details to a minimum. However, we hope that these simple examples help also indicate the way of extending such applications to more complicated and more unusual systems.

REFERENCES

1. Popov, V. M., Stability-spaces and frequency-domain conditions, in *Calculus of Variations and Control Theory*, edited by David L. Russell, Academic Press, Inc., 371-390.

2. Corduneau, C., *Integral Equations and Stability of Feedback Systems*, Academic Press, New York, 1973.

3. Miller, R. K., *Nonlinear Volterra Integral Equations*, W. A. Benjamin, Inc., 1971.

4. Nohel, J. A., and Shea, D. F., Stability of a nonlinear Volterra equation, (to appear).

5. Staffans, O. J., *Systems of Nonlinear Volterra Equations with Positive Definite Kernels*, (Helsinki University of Technology, Institute of Mathematics Report-HTKK-MAT-A72) 1975.

6. Narendra, K. S., and Cho, Y. S., Stability analysis of nonlinear systems, *SIAM J. Control 6*(No. 4, 1968).

7. Yakubovich, V. A., Absolute stability of pulse systems with several nonlinear or linear nonstationary blocks, II *Automation and Remote Control* (No. 2, 1968).

8. Halanay, A., private communication, 1962.

9. Datko, R., private communication, 1970.

10. Freedman, M. I., Falb, P. L., Zames, G., A Hilbert space stability theory over locally compact abelian groups, *SIAM J. Control 7*(No. 3, 1969).

11. Barbu, V., Nonlinear Volterra equations in Hilbert space, *SIAM J. Math. Anal. 6*(No.4, 1975).

12. Răsvan, V., *Stabilitatea Absoluta a Sistemelor Automate cu Intirziere*, Editura Academiei, Bucuresti, 1975.

13. Dunford, N., and Schwartz, J. T., *Linear Operators, Part I*, Interscience Publishers, Inc., New York.

14. Hewitt, E., and Stromberg, K., *Real and Abstract Analysis*, Springer Verlag, Berlin, 1969.

*This research was partially supported by the National Science Foundation, Grant No. MPS74-08184 A01.

OPTIMIZATION OF STRUCTURAL GEOMETRY

W. Prager
Division of Engineering and
Division of Applied Mathematics
Brown University

and

G. I. N. Rozvany
Department of Civil Engineering
Monash University

1. INTRODUCTION

Much of the literature on structural optimization is concerned with the optimal choice of cross-sectional dimensions in a structure whose layout has already been determined by other considerations. When the layout as well as the cross-sectional dimensions are at the choice of the designer, structural optimization becomes a more challenging problem.

The present paper reviews the state of the art in the optimization of the layout of trusses and grillages. Although there exist many analogies between the two problems, the theory of the optimal layout of trusses goes back to Maxwell, 1890, and Michell, 1904, whereas the theory of the optimal layout of grillages was only developed in the last decade. In veiw of this, the emphasis of the present paper is on grillages, and trusses are primarily discussed on account of the simplicity of their structural analysis, which will enable us to present the theory of the optimal layout of grillages by convenient analogy.

2. OPTIMALITY CONDITION FOR TRUSSES

To indicate typical arguments used in establishing a necessary and sufficient condition for the optimality of a truss layout, let us condider a truss that is to transmit a given load $\underline{P}$ from its point of application A to a rigid vertical wall by means of bars joining A to some or all of n given points W_1, W_2,...,W_n on the wall (Fig. 1).

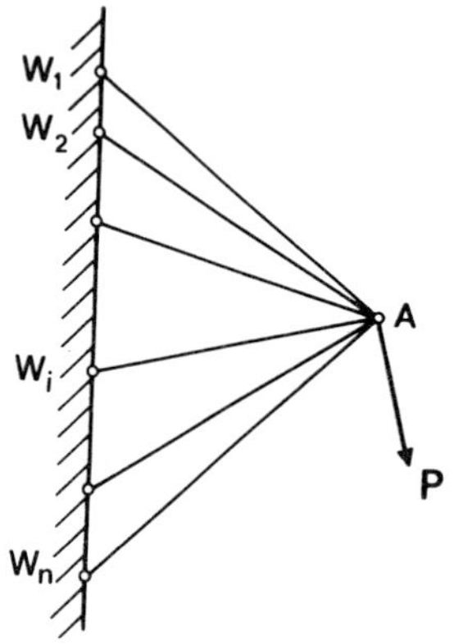

Figure 1

All bars of the truss are to consist of the same elastic material with Young's modulus E. The truss is to be designed for minimum total volume of its bars subject to the constraint that its compliance is to have the value C_0. (Compliance is a convenient measure of the deformability of a structure; it is defined as the work done by the loads in deforming the structure or, what amounts to the same, as the strain energy stored in the deformed structure.)

The bar joining A to W_i will be referred to as bar i, and its length and volume will be denoted by L_i and V_i. The elastic deformation of the truss displaces the loaded joint from its original position A to a neighboring position A'.

If the vector of this displacement is denoted by $\underline{\delta}$, and the unit vector of the direction W_iA by $\underline{u}_i$, the axial strain in bar i of the given layout is

$$\varepsilon_i = (\underline{u}_i \cdot \underline{\delta})/L_i. \tag{1}$$

With these notations, the compliance constraint takes the form

$$E \sum_i \varepsilon_i^2 V_i = 2C_0. \tag{2}$$

By its mechanical meaning, the sum in (2) includes only the bars of the optimal truss. Since, however, $V_i = 0$ for any other bar of the given layout, the sum may be extended over all bars of this layout. Note that the left side of (2) is twice the potential energy of the loaded truss.

We now consider a second design with the bar volumes V_i^* that also has the complicance C_0. For brevity, we shall refer to the two designs as design V_i and design V_i^*. According to the principle of minimum potential energy, we have $E \sum_i \varepsilon_i^2 V_i^* - 2C_0 \geq 0$ or, in view of (2),

$$\sum_i (V_i^* - V_i)\varepsilon_i^2 \geq 0. \tag{3}$$

Moreover, since the bar volumes are non-negative,

$$V_i^* - V_i \geq 0 \text{ if } V_i = 0, \tag{4}$$

that is, if bar i of the given layout is not used in the design V_i.

The difference between the bar volumes V_i^* and V_i of any two designs with the prescribed compliance are subject to the inequalities (3) and (4). We want the inequality

$$\sum_i (V_i^* - V_i) \geq 0, \tag{5}$$

which establishes the optimality of the design V_i, to follow from (3) and (4). According to a theorem of Farkas, 1902, a necessary and sufficient condition for this is that (5) is a non-negative linear combination of (3) and (4). If the coefficients of this combination are denoted by $1/\varepsilon_0^2$ and μ_i^2/ε_0^2, with $\mu_i^2 = 0$ if $V_i > 0$, this theorem furnishes the optimality condition $\varepsilon_i^2 + \mu_i^2 \geq \varepsilon_0^2$ or

$$|\varepsilon_i| \leq \varepsilon_0 \text{ with equality for } V_i > 0. \tag{6}$$

In view of (1), the optimality condition (6), which may be written as

$$|\underline{u}_i \cdot \underline{\delta}| \leq \varepsilon_0 L_i \text{ with equality for } V_i > 0, \tag{7}$$

restricts the position A' of the loaded joint in the deformed state to a strip whose boundaries are normal to $\underline{u}_i$ and have the distance $\varepsilon_0 L_i$ from A. This strip will be called the <u>admissible strip</u> for bar i. The intersection of the admissible strips for all bars of the given layout is a convex polygonal domain, which will be called the <u>admissible domain</u>.

When the point A' is on a side but not at a vertex of the admissible domain, only one bar of the given layout experiences a strain of the absolute value ε_0. The optimal truss then contains only this bar, and the design is only feasible if the given layout has a bar along the line of action of the load.

When the point A' is at a vertex of the admissible domain, there are, in general, only two bars of the given layout with strains of the absolute value ε_0, and the optimal truss consists of these two bars. Even in the exceptional case that A' is at a vertex in which more than two boundaries of admissible strips intersect, there exist optimal trusses that contain only two of the bars with strains of the absolute value ε_0. Indeed, whereas a positive V_i requires the equality sign to hold in (7), it does not follow from the validity of this sign that V_i must be positive.

As the preceding discussion shows, there always exists a statically determinate optimal truss. It can be shown that this truss is also an optimal design for (see Hegemier and Prager, 1969):

(i) given allowable magnitude of axial stress;

(ii) prescribed fundamental frequency if the masses of the bars are treated as negligible in comparison to a given point mass at the joint A; and

(iii) prescribed rate of compliance in stationary creep if the creep behaviour of the bars is governed by the law of Norton, 1929.

3. TRUSS-LIKE CONTINUA

In the problem considered above, the designer could only use joints provided by the layout, and the deformation of the structure was specified by the displacements of those joints, that is, by the displacement of the joint A and the fact that the joints on the rigid wall cannot move. The optimality condition (6), however, remains valid when joints may be placed anywhere. The deformation is then specified by a displacement field, and the optimality condition (6) stipulates that this field should have strains of the same absolute value ε_0 along the bars of the optimal truss and strains of smaller absolute value in any other direction. Accordingly, the bars of the optimal truss must follow lines of principal strain along which the strain has the absolute value ε_0.

Michell, 1904, primarily used plane displacement fields that have constant principal strains of the same magnitude and opposite signs, and hence vanishing mean normal strain.

If the displacement components along the rectangular axes x, y are denoted by u_x, u_y, the vanishing of the mean normal strain is expressed by the relation

$$\partial_x u_x + \partial_y u_y = 0, \tag{8}$$

where the symbols ∂_x and ∂_y indicate differentiation with respect to the coordinates x, y. It follows from (8) that u_x and u_y can be derived from a function $\psi(x,y)$ in accordance with

$$u_x = \partial_y \psi, \quad u_y = -\partial_x \psi. \tag{9}$$

The fact that the principal strains have the constant values $\pm \varepsilon_0$ then furnishes the hyperbolic differential equation

$$(\partial_{xx}\psi - \partial_{yy}\psi)^2 + 4(\partial_{xy}\psi)^2 - 4\varepsilon_0^2 = 0, \tag{10}$$

whose characteristics are the lines of principal strain.

In general, these characteristics form an orthogonal curvilinear net, and the optimal structure is a truss-like continuum rather than a truss in the usual sense of this term. While this kind of structure is not practical, it furnishes a useful lower bound on the volume of material needed for conventional designs that are subject to the same constraint. The efficiency of such a design will be defined as the ratio between the volumes of material used in the optimal structure and the considered conventional structure.

The following relation between Michell's theory of optimal truss layout and the theory of plane plastic flow is worth noting. If the function ψ in equation (10) is interpreted as the Airy function of the stress field in plane plastic flow, this equation requires the maximum shearing stress to be constant; its characteristics are the slip lines, which form a Hencky-Prandtl net. The geometry of these nets has been thoroughly investigated (see, for instance, Prager and Hodge, 1951), and this information is useful in the study of Michell's truss-like continua. Johnson, 1961, has used this analogy in the inverse sense.

If a truss-like continuum is to be attached to a rigid foundation arc along which the displacements u_x and u_y vanish, the function ψ has vanishing normal derivative along this arc and may, without loss of generality, be assigned the value zero along it. The layout of the optimal truss-like continuum may then be found by solving a Cauchy problem for the hyperbolic differential equation (10).

For a circular foundation arc of centre O and radius a, for instance, the characteristics are logarithmic spirals that form angles of $\pm 45°$ with the rays from O (Fig. 2). If O

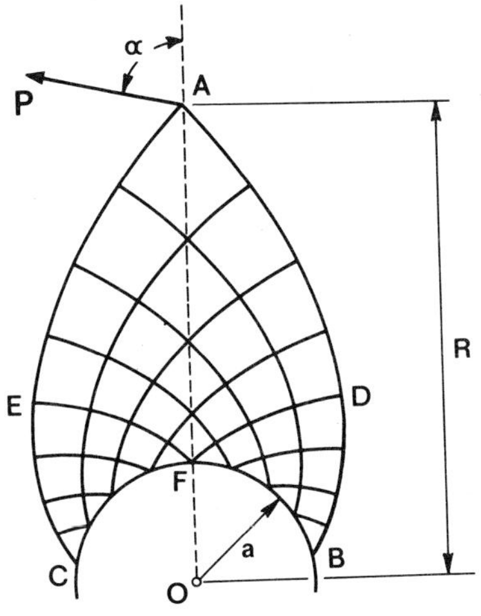

Figure 2

is taken as the origin of the polar coordinates r, θ, and the line joining O to the point of application A of the given load as the ray θ = 0, the positive sense of θ being counterclockwise, displacement components that vanish on the foundation arc and furnish the principal strains $\pm \varepsilon_0$ are given by

$$u_r = 0, \quad u_\theta = \pm 2\varepsilon_0 r \ln(r/a). \tag{11}$$

If the plus sign is taken in (11), the bars along the characteristic AB are stressed in tension and those along AC in compression. Equilibrium at the joint A then requires the angle α between OA and the direction of the load to $45° \le \alpha \le 135°$. Similarly, if the minus sign is taken in (11), $225° \le \alpha \le 315°$. The optimal design for other values of α will be discussed later.

To complete the present analysis, however, we must still determine the value of the constant ε_0 in (11) from the fact that the compliance, that is the work $\{P\, u_\theta(A) \sin\alpha\}/2$ done by the load in producing the displacements (11), should have the value C_0. If the distance OA is denoted by R, this condition yields

$$\varepsilon_0 = C_0/\{PR \ln(R/a) \sin\alpha\}. \tag{12}$$

Since the compliance may also be written as $(E/2)\sum_i \varepsilon_i^{\,2} V_i$ with $\varepsilon_i^2 = \varepsilon_0^2$ for $V_i > 0$, the volume of material needed for the optimal design is

$$V = 2\,\{PR \ln(R/a) \sin\alpha\}^2/EC_0). \tag{13}$$

For $\alpha = 90°$ and $R = 4a$, for instance, $V = 61.50\, P^2a^2/(EC_0)$. If, instead of using the layout in Fig. 2, we join A by two bars directly in B and C, the resulting truss is found to have an efficiency of only 0.493. The truss that uses only the joints A, B, C, D, E and F already has an efficiency of 0.849, but is cannot be asserted that this is the optimal symmetric truss with the joints A and F, two further joints on the founcation arc, and two additional joints. As a matter of fact, the optimization of a truss with a prescribed number of joints, some of which must be at the points of application of the loads and others on the foundation arc, while the rest may be placed anywhere, is a problem that, at present, can only be solved by tedious search procedures.

To enforce a finite number of bars in the optimal truss, one may, for instance, minimize the objective function

$$\Omega = \sum_i V_i (1+L_o/L_i) \tag{14}$$

instead of the total volume of bars. Here, L_o is a constant length, and $V_i L_o/L_i$, which is proportional to the cross-sectional area of bar i, may be interpreted as the contribution of bar i to the volume of material needed to connect the bars to each other. The reasoning that led to (6) now yields the optimality condition

$$\varepsilon^2 \le \varepsilon_0^2 \; (1+L_o/L_i) \text{ with equality for } V_i > 0, \tag{15}$$

which shows that long bars of the optimal truss should experience axial strains of smaller absolute value than short bars. Except, however, for trusses with very few joints, this optimality condition is not easy to apply. (For examples, see Prager, 1974, and Parkes, 1975; for another approach, see Prager, 1976).

Following Hu and Shield, 1961, we now discuss the optimal layout if the angle α in Fig. 2 is outside the range considered above. If

$$R\sqrt{2} \sin \alpha > a, \tag{16}$$

the optimal layout is of the kind shown in Fig. 3. The bar

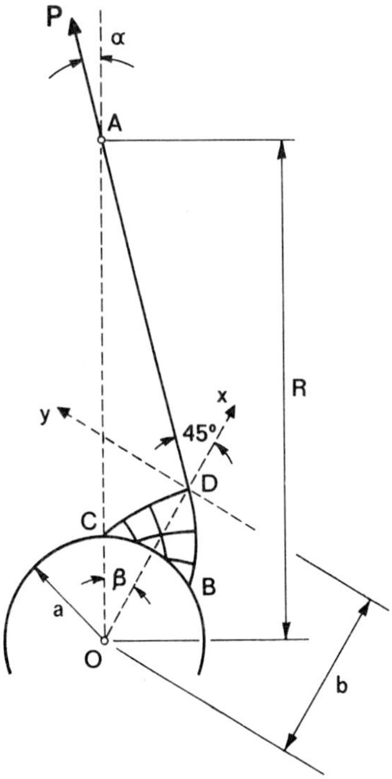

Figure 3

AD along the line of action of the load transmits this to the joint D of the truss BDC, which is formed by logarithmic spirals that intersect the rays from O under $\pm$ 45°. The bar AD is tangent to the spiral DB at D, and hence forms an angle of 45° with the ray OD. The distance b of D from O, and the angle AOD thus are

$$b = R\sqrt{2}\,\sin\alpha, \quad \beta = (\pi/4) - \alpha. \tag{17}$$

Within the annulus $a \le r \le b$, the displacement components are given by (11) with the plus sign. The displacement of D, which is purely circumferential, has the magnitude

$$U = 2\varepsilon_0 b \ln(b/a). \tag{18}$$

Between the circle of radius b and its tangent at D, the displacement components are

$$u_r = 0, \quad u_\theta = Ur/b \tag{19}$$

and correspond to a rigid body rotation. Finally, to the right of this tangent, the displacement results from the superposition of the rotation (19) and the simple shear specified by the components

$$u_x = 0, \quad u_y = 2\varepsilon_0 x \tag{20}$$

with respect to the rectangular axes x, y indicated in Fig. 3.

If the expression on the left of (16) has a value smaller than or equal to a, the optimal layout is as shown in Fig. 4: a single bar AB transmits the load directly to the foundation arc.

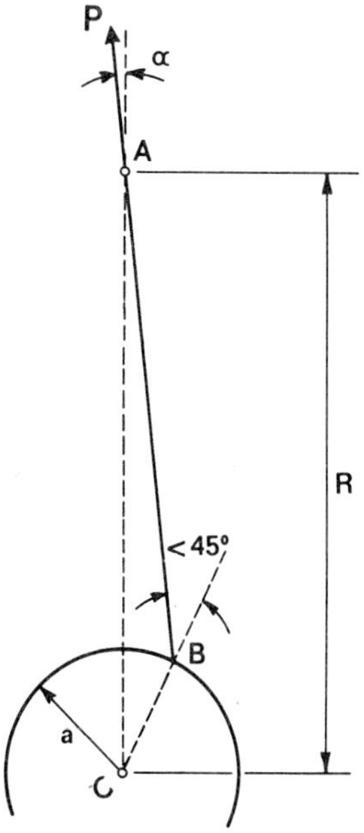

Figure 4

The extensive literature on Michell continua (see, for instance, Hemp, 1973) may create the impression that the subject has been nearly exhausted. Most examples found in this literature, however, are based on displacement fields with principal strains of the same magnitude ε_0 but opposite signs. The possibilities of combining regions of this kind with regions in which both principal strains have the values ε_0 or $-\varepsilon_0$, regions in which only one principal strain has the absolute value ε_0 while the other has a smaller absolute value, and regions of rigid body displacement, as in the example of Fig. 3, have not as yet been systematically exploited. In the theory of optimal grillages, the analogous combination of deflection fields of various types has received much more attention with the result that this theory has advanced further than the considerably older theory of optimal trusses.

4. OPTIMALITY CONDITION FOR GRILLAGES

The grillages considered in the following have a horizontal median plane and are subject to vertical downward loads. Their elastic beams have rectangular sections of uniform height and continuously varying width.

Considering first a single beam of this kind, we specify a cross-section by its distance s from a reference section, and denote the height and breadth of the section s by H and B(s). The product

$$V(s) = HB(s) \tag{21}$$

will be called the specific volume at the section s. The beam is to be supported in a given manner and to carry a given distributed load p(s). Subject to the condition that the compliance of the beam to this load has the given value C_0, we wish to minimize its volume

$$V = \int V(s)\, ds, \tag{22}$$

where the integration is extended over the entire length of

the beam. If the given loading causes the deflection v(s) and the curvature $\kappa(s) = -v''(s)$, where the prime indicates differentiation with respect to s, the compliance constraint is expressed by

$$c \int \kappa^2(s)\ V(s)\ ds - 2\ C_0 = 0, \qquad (23)$$

where $c = EH^2/12$.

If the beams with the specific volumes V(s) and V*(s) are supported in the same way and have the same compliance to the given loading, use of the principle of minimum potential energy and non-negativity of V* as in Sect. 2 yields the following necessary and sufficient condition for the optimality of the design V(s):

$$|\kappa(s)| \leq \kappa_0 \text{ with equality for } V(s) > 0. \qquad (24)$$

Here, κ_0 is a constant whose value must be determined from the compliance constraint in a manner similar to that yielding the value of ε_0 in (12).

The optimality condition (24) must be satisfied along each beam of an optimal grillage consisting of a finite number of beams. Unless, however, we explicitly stipulate that the grillage to be designed must have a finite number of beams, the optimal structure is not a grillage in the customary sense of this term but a grillage-like continuum, that is, a dense arrangement of beams with the prescribed height and infinitesimal widths. As Michell's truss-like continua, grillage-like continua are not practical structures, but furnish useful lower bounds for the volume of material needed in more realistic designs.

We shall take the median plane of a grillage-like continuum as the plane z = 0 of the rectangular coordinates x, y, z, and denote the deflection by v(x,y). Note that the deflection is supposed to be small in comparison to the dimensions of the grillage in the x, y plane. In analogy to what was found in Sect. 3, for a truss-like continuum, the surface z = v(x,y) for an optimal grillage-like continuum must have a principal curvature κ_1 of the constant absolute value κ_0, while the absolute value of the other principal curvature κ_2 must not exceed κ_0, and the beams of the structure must follow "lines of principal curvature" along which the curvature has the absolute value κ_0. For brevity, we have here used and shall continue to use the term "lines of principal curvature" for the projections of these lines of the surface z = v(x,y) onto the plane z = 0.

Depending on the values of κ_1 and κ_2, we distinguish five types of regions:

$$\begin{aligned}
&\text{Type } R^+: && \kappa_1 = \kappa_0,\ -\kappa_0 < \kappa_2 < \kappa_0,\\
&\text{Type } R^-: && \kappa_1 = -\kappa_0,\ -\kappa_0 < \kappa_2 < \kappa_0,\\
&\text{Type } S^+: && \kappa_1 = \kappa_2 = \kappa_0,\\
&\text{Type } S^-: && \kappa_1 = \kappa_2 = -\kappa_0,\\
&\text{Type } T\ : && \kappa_1 = -\kappa_2 = \kappa_0.
\end{aligned} \tag{25}$$

With the customary sign conventions, the curvatures κ_x, κ_y, and the twist τ for the coordinate directions are defined by

$$\kappa_x = -\partial_{xx}v,\ \kappa_y = -\partial_{yy}v,\ \tau = -\partial_{xy}v, \tag{26}$$

and therefore subject to the compatibility conditions

$$\partial_y\kappa_x - \partial_x\tau = 0,\ \partial_x\kappa_y - \partial_y\tau = 0. \tag{27}$$

With respect to the lines of principal curvature, these compatibility conditions take the form (Shield, 1960):

$$d_2\kappa_1 = (\kappa_2-\kappa_1)/\rho_1,\quad d_1\kappa_2 = (\kappa_1-\kappa_2)/\rho_2, \tag{28}$$

where d_1 and d_2 denote differentiation in the directions of the lines in which the principal curvatures have the values κ_1 and κ_2, and ρ_1 and ρ_2 are the radii of curvature of these lines, with the sign conventions indicated in Fig. 5. Since

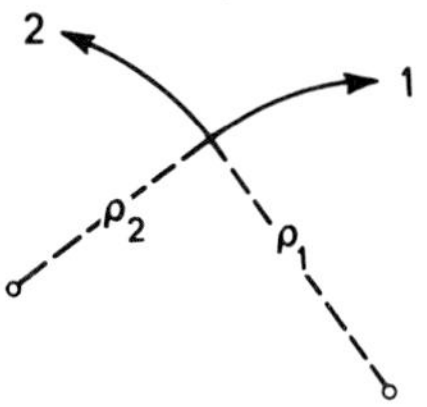

Figure 5

κ_1 has the constant magnitude κ_0, it follows from the first equation (28) that ρ_1 must be infinite if $\kappa_1 \neq \kappa_2$.

In regions of the types R^+, R^-, and T, lines along which the principal curvature has the absolute value κ_0 are therefore straight. This means that in regions of types R^+ and R^- one family of lines of principal curvature consists of straight lines, while the lines of principal curvature in a region of type T form an orthogonal net of straight lines. In regions of types S^+ and S^-, any direction is a principal direction. While curved beams could thus be used in regions of these types, straight beams are just as economic in structural material. The grillage-like continua discussed in the following will therefore use only straight beams.

5. MATCHING OF REGIONS

In view of the limited presentation time, the matching of regions of different types will only be discussed for straight-edged grillage-like continua.

The deflection v and its first derivatives $\partial_x v$ and $\partial_y v$ must be continuous across the common boundary of two regions of different types. In discussing a convenient way of satisfying these conditions of continuity and the boundary conditions along two edges forming a grillage corner, we shall use Mohr's circle to represent the variation of curvature and twist at a point O* of the deformed median plane. We take the projection O of O* onto the undeformed median plane $z = 0$ as the origin of the coordinates x, y, z, where the z-axis points downward.

The definitions of the curvature $\kappa(\phi)$ and the twist $\tau(\phi)$ for a direction through O that is normal to the z-axis and forms the angle ϕ with the positive x-axis involves differentiation of the infinitesimal deflection $v(x,y)$ in this direction and the direction obtained from it by a right-angle rotation about the z-axis in the sense that transforms the x- into the y-direction. Denoting differentiation in these directions by d_s and d_n, we adopt the definitions

$$\kappa(\phi) = - d_{ss}v, \quad \tau(\phi) = - d_{sn}v. \tag{29}$$

For the x- and y-directions, we thus have

$$\begin{aligned} \kappa_x &= \kappa(0) = - \partial^2 v/\partial x^2, & \tau_x &= \tau(0) = - \partial^2 v/\partial x\, \partial y, \\ \kappa_y &= \kappa(\pi/2) = -\partial^2 v/\partial y^2, & \tau_y &= \tau(\pi/2) = \partial^2 v/\partial x\, \partial y = -\tau_x. \end{aligned} \tag{30}$$

By plotting the points with the rectangular coordinates $\kappa(\phi)$, $\tau(\phi)$ for varying ϕ, we obtain a geometrical representation of the variation of curvature and twist at the considered points. It can be readily shown that the locus of these points is a circle (Mohr's circle). Figure 6 shows this circle and its pole P, which has the coordinates κ_y, τ_x. To find the values of curvature and twist for the direction ϕ, we draw a line of this direction through the pole. The

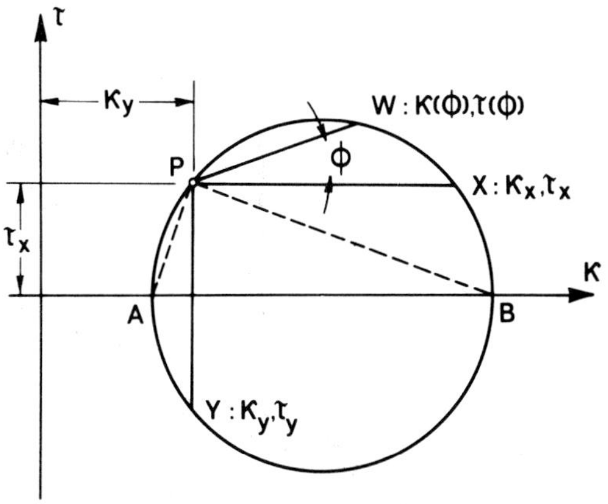

Figure 6

second point of intersection W of this line with the circle has the coordinates $\kappa(\phi)$, $\tau(\phi)$. For $\phi = 0$ and $\phi = \pi/2$, we thus obtain the points X and Y with the coordinates κ_x, τ_x and κ_y, τ_y. The segment XY is a diameter of the circle, whose center lies on the κ-axis. The intersections A and B of the circle with this axis have the principal curvatures as abscissas, and the lines PA and PB indicate the principal directions of curvature.

With the use of Mohr's circle, the optimal layout of beams near a corner formed by two straight edges is readily found. We illustrate this by discussing the corner formed by a built-in edge along the y-axis and a simply supported edge along a ray in the first quadrant.

Since $v = \partial_x v = 0$ along the built-in edge, we have $\kappa_y = \tau_y = 0$ on this edge. This indicates that the y-direction is a principal direction and the corresponding principal curvature vanishes. For positive v, optimality requires that the other principal curvature must have the value $\kappa_x = -\kappa_0$.

The left-hand circle in Fig. 7 is the Mohr circle for

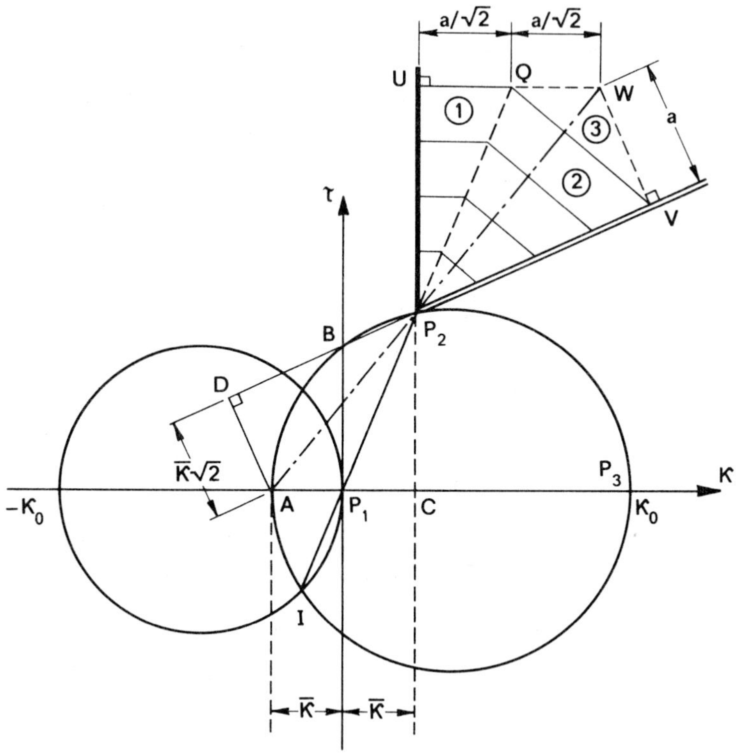

Figure 7

the region of the grillage that is adjacent to the y-axis. We shall refer to this circle and region as Circle 1 and Region 1. The pole P_1 of Circle 1 is at the origin.

Instead of giving the direction of the simply supported edge and determining the Mohr circle (Circle 2) for the region adjacent to this edge (Region 2), we choose this circle to correspond to the principal curvatures $-\bar{\kappa}$ and κ_0, where $0 < \bar{\kappa} < \kappa_0$, and ask for the direction of the simply supported edge.

The common boundary of Regions 1 and 2 is a ray across which the deflection and its first derivatives are continuous, which implies that curvature in the direction of the ray and twist are continuous across this ray. Conversely, since the deflection and its first derivatives vanish at the corner, the continuity of curvature and twist assures that of deflection and its first derivatives. It follows from these remarks that curvature and twist along this ray are given by the coordinates of the point of intersection I of the two circles in Fig. 7, and that the pole of Circle 2 is its point of intersection P_2 with the line IP_1.

The considered corner of the grillage is shown above the pole P_2 in Fig. 7. The common boundary of Regions 1 and 2 has the direction of IP_2, and the direction with the principal curvature $-\bar{\kappa}$ in Region 2 is given by AP_2. Note that the beams in Region 2 form right angles with this direction. Finally, since the curvature in the direction of the simply supported edge vanishes, this edge has the direction of the line joining P_2 to the intersection B of Circle 2 with the τ-axis.

If C is the projection of P_2 onto the κ-axis, and D the projection of A onto the line P_2B, it can be readily shown that the segments P_1C and AD have the lengths $\bar{\kappa}$ and $\bar{\kappa}\sqrt{2}$. This remark yields the construction of the common boundary of Regions 1 and 2, and the principal direction of smaller curvature in Region 2. This construction is indicated above P_2 in Fig. 7.

For the matching of the optimal beam patterns near the considered corner and other corners of the grillage, it is important to note that the surface $z = v(x,y)$ for Region 2 can be smoothly continued beyond the segment QV into the triangle QWV (Region 3) by a spherical surface of curvature κ_0. The Mohr circle for Region 3 shrinks to the point P_3 with the coordinates κ_0, 0, and P_2P_3 is parallel to QV. The point W is the projection of the center of the sphere onto the x,y-plane, and will therefore be called central point of Region 3. The derivatives of the deflection v in the directions normal to the segments QW and WV vanish along these segments.

Figure 8, which shows the optimal pattern of beams near

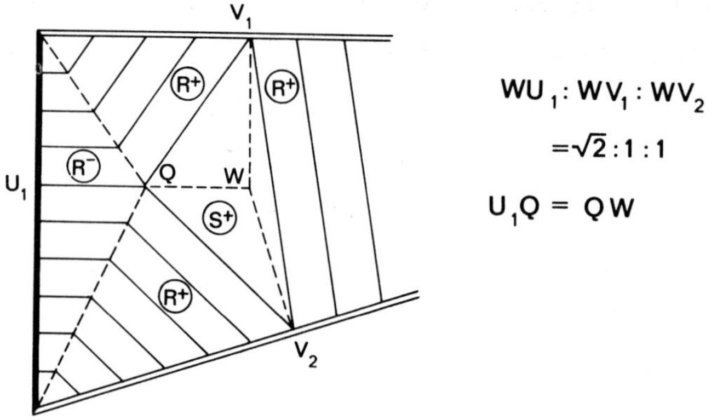

Figure 8

a short built-in edge and two longer simply supported edges, illustrates the use of this continuation. The immediate neighborhood of each corner corresponds to Fig. 7. The distances WU_1, WV_1, and WV_2 of the central point W from the built-in edge and the simply supported edges have the ratios $\sqrt{2}$:1:1. In the S^+ region $QV_1 V_2$, the optimal beam direction is not uniquely determined. We may, for instance, use beams of the direction V_1V_2 that are simply supported on the beams

QV_1 and QV_2, which are in turn simply supported on the cantilever U_1Q at Q and on the edges at V_1 and V_2. The beams to the right of the line V_1V_2 are simply supported on the edges.

The optimal beam patterns near corners formed by edges with other support conditions can be investigated in a similar fashion with the use of Mohr's circle. It is found that the concepts of a <u>central point</u> and its <u>associated edges</u> and <u>base points</u> (Rozvany, 1974) play an important role in most cases. The base points are the feet of the perpendiculars dropped from the central point W onto the associated edges. Base points on simply supported edges will be denoted by V_1, V_2,..., and base points on built-in edges by U_1, U_2,... and the centers of the segments WU_1, WU_2,..., by Q_1, Q_2,... The position of the central point is such that the segments WV_1, WV_2,... have the same length, say a, while the segments WU_1, WU_2, ... have the common length $a\sqrt{2}$. The points Q_1, Q_2, ..., V_1, V_2,... are the vertices of a polygonal S^+ region. The beams in an R^+ region adjacent to the S^+ region are parallel to the common boundary of the two regions. The beams in an R^- region adjacent to a built-in edge are normal to this edge.

Figures 9a and b illustrate these rules. Whereas the square grillage in Fig. 9a has only one central point, the rectangular grillage in Fig. 9b has two central points and hence two S^+ regions. In the T regions near the lower right corners of Fig. 9a and b, only one beam family is shown for simplicity. A single load acting at a point C of this

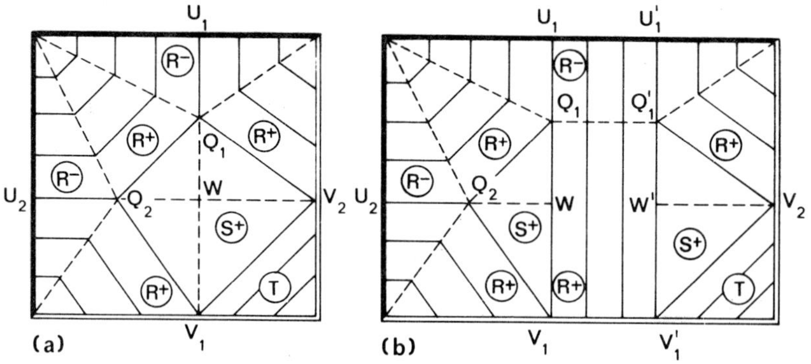

Figure 9

region can, however, be transmitted to the simply supported edges either by a beam AB of this family (Fig. 10a), or by a beam CD of the other family that is supported by an edge and a beam A'B' of the first family (Fig. 10b). The two ways

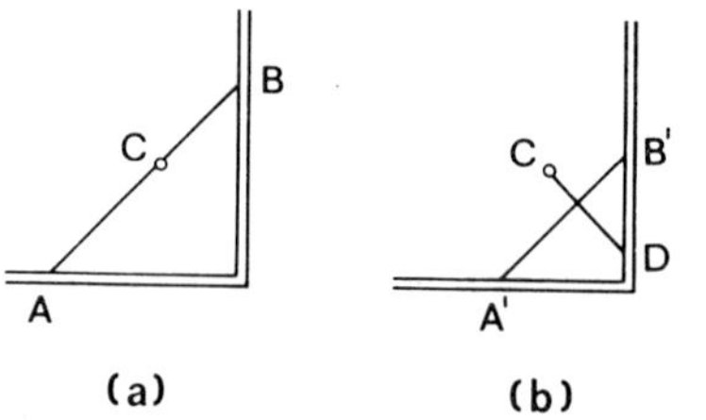

Figure 10

of transmitting the load require the same amount of material if the compliances are to have the same value C_0.

In determining optimal beam patterns near corners, we must keep in mind that the principal curvatures of the surface $z = v(x,y)$ must not exceed κ_0 in absolute value. For the pattern of Fig. 7, a limiting case is therefore reached when $\bar{\kappa} = \kappa_0$. The corresponding modification of Fig. 7 is shown in Fig. 11a, where the angle α formed by the two edges equals $3\pi/4$. Note that Region 2 is now a T region. For greater values of α, the optimal beam pattern is of the kind shown in Fig. 11b. Adjacent to the built-in edge, we have

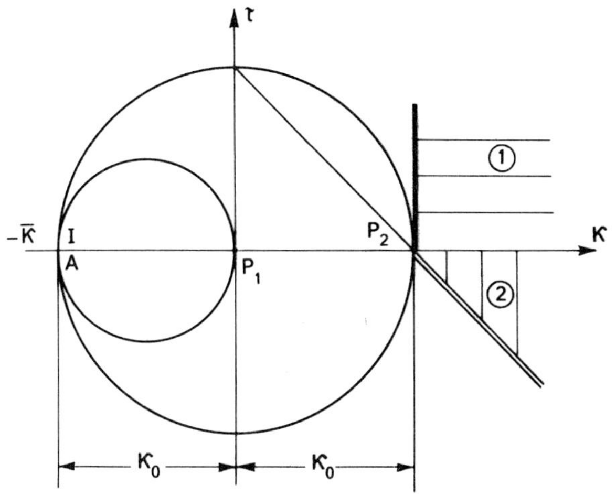

Figure 11a

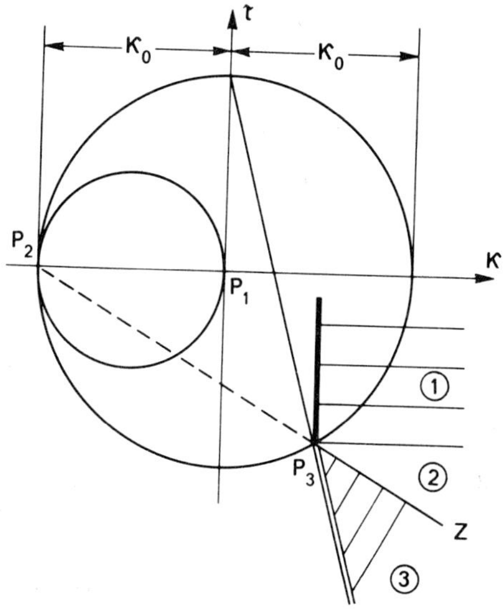

Figure 11b

an R^- region (Region 1) as in Fig. 7, with principal curvatures $-\kappa_0$ and 0 normal and parallel to the edge. Between the ray P_3Z that forms an angle of $\pi/4$ with the simply supported edge and this edge, we have a T region (Region 3) with principal curvatures $-\kappa_0$ and κ_0 parallel and normal to P_3Z. Finally, between Regions 1 and 3 we have an S^+ region (Region 2).

At a corner formed by two simply supported edges, the surface $z = v(x,y)$ has a horizontal tangent plane. Accordingly, any beam emanating from this corner should be regarded as being clamped there, and hence as being "normal" to the contour of the grillage. Figure 12, which shows half of a

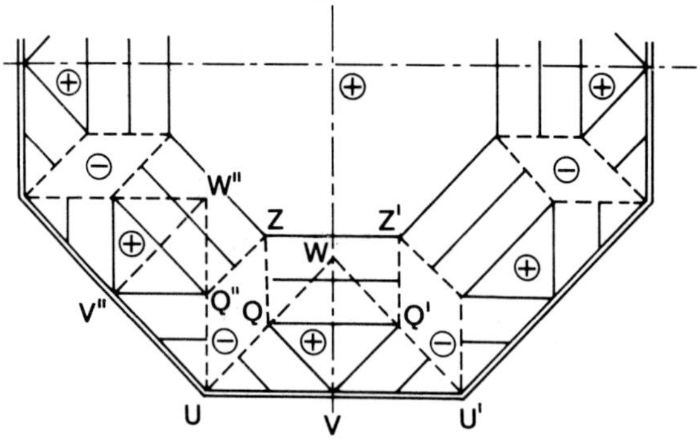

Figure 12

simply supported octagonal grillage, illustrates these concepts. The distances WV and WU of the central point W from the simply supported edge UU' and the clamped corner U have the ration $1:\sqrt{2}$, and the line joining V to the center Q of WU gives the direction of the beams of positive curvature in the T region QVU. The T regions Q'VU' and Q"V"U are constructed in similar ways. The two T regions near U are separated by an S^- region, and the two T regions near V by an S^+ region. The other boundaries of the S^- region near U are the parallel QZ to VW and the parallel Q"Z to UW. The region QQ'Z'Z is a T region whose beams of positive curvature are parallel to QQ' and supported by cantilevers from U and U'. Finally, the central octagon is an S^+ region.

The beams of each R^+ region used so far were parallel to each other. We now discuss R^+ regions of a more general kind. Figure 13, which shows the neighbourhood of two

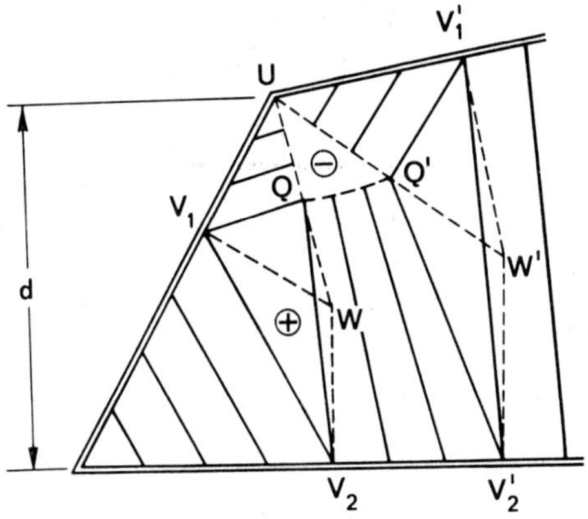

Figure 13

corners formed by simply supported edges, indicates the need for this general kind of R^+ region. The distances WU, WV_1 and WV_2 of the central point W from the clamped corner U and the left and bottom edges have the ratios $\sqrt{2}:1:1$, and the same ratios apply to the distances W'U, $W'V'_1$ and $W'V'_2$ of the central point W' from the clamped corner U and the top and botton edges. If the centers of the segments WU and W'U are denoted by Q and Q', we have the S^+ regions QV_1V_2 and $Q'V'_1V'_2$, between which we should have an R^+ region. The boundaries QV_2 and $Q'V'_2$ of this region, however, are not parallel to each other. Figure 14a shows how the directions of beams in this general R^+ region and the common boundary of it and the S^- region at U are found. With respect to the coordinates x, y shown in the figure, the equation of this boundary is

$$y^2 = x^2 + d^2/2 \qquad (31)$$

where d is the distance of U from the edge $V_2V'_2$. Figure 14b

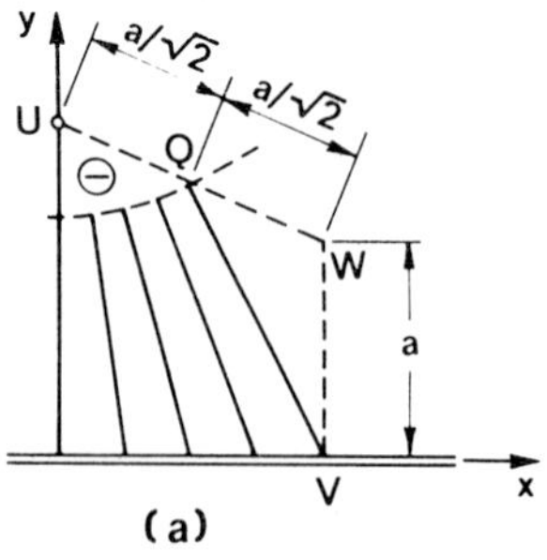

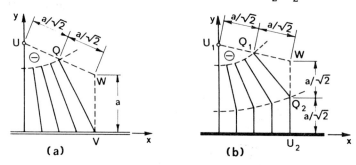

Figure 14

shows the analogous construction for the case when the edge $V_2V'_2$ is built in instead of being simply supported. The equations of the two curved boundaries of the R^+ region are

$$y = (x^2/4d) + d/4, \quad y = (x^2/d) + 3d/4. \qquad (32)$$

Note that in Figs. 14a and b the beams of the R^+ region are normal to the common boundary of this region with the S^- region near U.

Fig. 15 shows how the optimal beam pattern of a simply

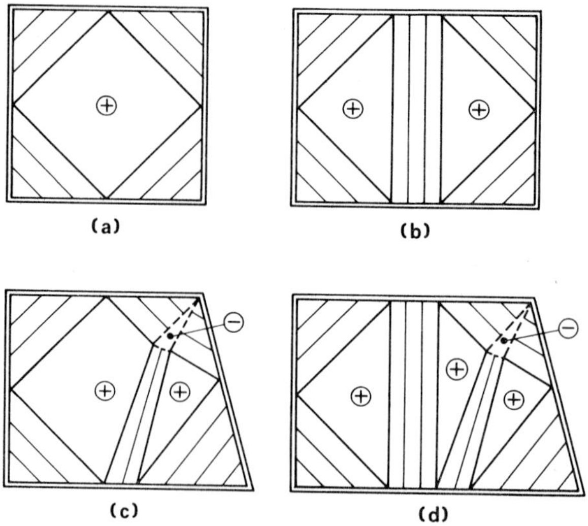

Figure 15

supported square grillage (Fig. 15a) changes when the right-hand side undergoes a translation (Fig. 15b), or a rotation (Fig. 15c) or both (Fig. 15d). Note that in the neighbourhood of the obtuse corners in Figs. 15c and d the arrangement of beams is of the kind discussed in connection with Fig. 13.

Fig. 16, in which the optimal beam pattern for a simply

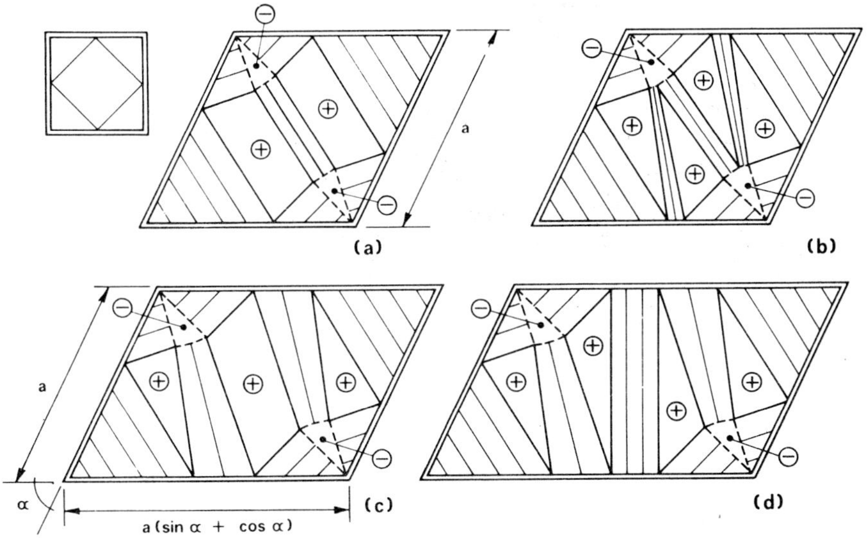

Figure 16

supported square grillage is recalled by the inset at the upper left, shows how this pattern changes when the square is transformed into a rhombus (Fig. 16a) or into parallelograms having the short sides a and the other sides shorter than, equal to and longer than [a (sin α + cos α)], respectively (Figs. 16b - d).

Finally, Fig. 17 shows how the pattern of Fig. 16a, which is recalled by the inset at the upper left, changes for rhomboids. Denoting the angles of the corners along the axis of symmetry by 2β and 2α (with $\beta > \alpha$), the range of

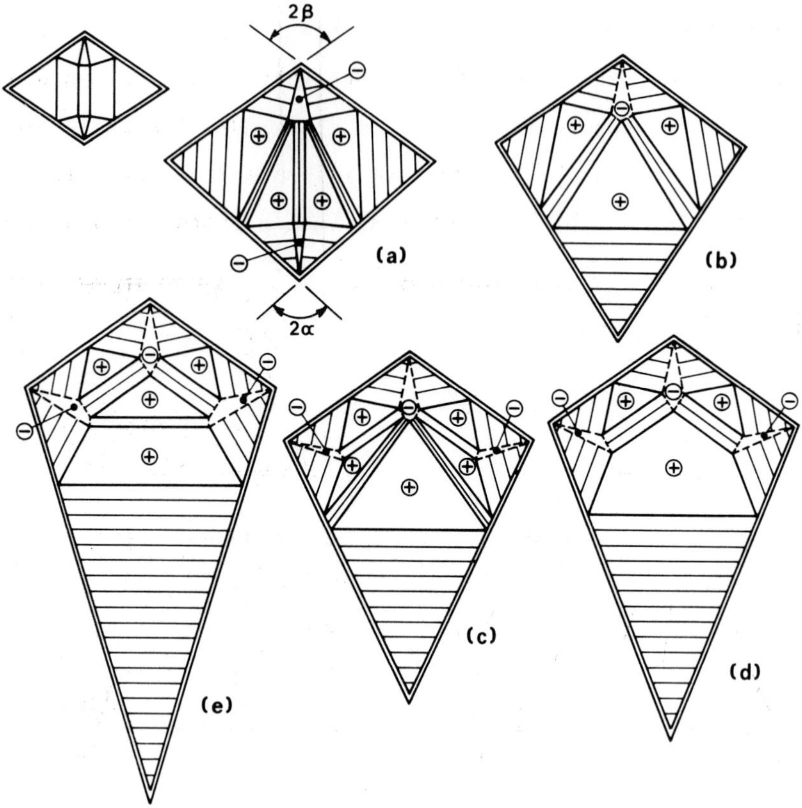

Figure 17

validity of various optimal patterns is the following:

Fig. 17a: $\alpha > 45°$

Fig. 17b: $45° \geq \alpha \geq 90° - \beta$

Fig. 17c: $90° - \beta > \alpha > 45° - \text{arc cos } (\sin 2\beta)$

Fig. 17d: $\alpha = 45° - \text{arc cos } (\sin 2\beta)$

Fig. 17e: $\alpha < 45° - \text{arc cos } (\sin 2\beta)$

In Fig. 17, the angle 2β at the top of the rhomboids is kept constant.

Optimal beam layouts for other quadrilateral grillages and for grillages with curved boundaries are discussed elsewhere (e.g. Rozvany and Hill, 1976).

REFERENCES

1. Farkas, J., Über die Theorie der einfachen Ungleichungen, *J. Reine & Angew. Math. 124*(1902), 1-24.

2. Hegemier, G. A., and Prager, W., On Michell trusses, *Internat. J. Mech. Sci. 11*(1969), 209-215.

3. Hemp, W. S., *Optimum Structures*, Clarendon Press, Oxford, Chap. 4, 1973.

4. Hu, T.-C., and Shield, R. T., Minimum volume design of discs, *J. Appl. Math. & Phys. (ZAMP) 12*(1961), 414-433.

5. Johnson, W., An analogy between upper bound solutions for plane strain metal working and minimum weight two-dimensional frames, *Internat. J. Mech. Sci. 3*(1961), 239-246.

6. Maxwell, J. C., On reciprocal figures, frames and diagrams of force, *Scientific Papers*, Vol. 2, University Press, Cambridge, 161-207.

7. Michell, A. G. M., The limits of economy in frame-structures, *Phil. Mag. 8*(6)(1904), 589-597.

8. Norton, F. H., *Creep of Steel at High Temperatures*, McGraw-Hill, New York, 1929.

9. Parkes, E. W., Joints in optimum frameworks, *Internat. J. Solids & Structures 11*(1975), 1017-1022.

10. Prager, W., and Hodge, P. G., *Theory of Perfectly Plastic Solids*, John Wiley & Sons, New York, Chapter 5, 1951.

11. Prager, W., A note on discretized Michell structures, *Computer Methods in Appl. Mech. & Engg. 3*(1974), 349-355.

12. Prager, W., Optimal layout of cantilever trusses, *J. Optimization Theory & Appls. 22*(1977), to appear.

13. Rozvany, G. I. N., A unified theory of optimal moment fields, *J. Struct. Mech. 3*(1974), 179-195.

14. Rozvany, G. I. N., and Hill, R., General theory of optimal load transmission by flexure, *Advances in Appl. Mech. 16*(1976), to appear.

15. Shield, R. T., Plate design for minimum weight, *Quarterly Appl. Math. 18*(1960), 131-144.

ON THE CESARI INDEX
AND THE BROWDER-PETRYSHIN DEGREE

E. H. Rothe
Department of Mathematics
University of Michigan

1. INTRODUCTION

Let E and F be real Banach spaces, let Ω be a bounded open subset of E, $\overline{\Omega}$ its closure, and let L and N be maps of $\overline{\Omega}$ into F where L is linear densely defined and N is continuous and bounded. If

$$f = L - N \tag{1.1}$$

we are concerned with solutions $x \in \Omega$ of the equation

$$f(x) = \Theta \tag{1.2}$$

where Θ denotes the zero element of F (the zero element of E will also be denoted by Θ).

It is well known (see e.g. [2]) that with the help of two projections $P : E \to E$ and $Q : F \to F$ with ranges of finite demensions p and q respectively problem (1.2) may be split into a system of two equations as follows (under assumptions to be specified later): define

$$T(x) = P(x) + H(I-Q)N(x), \quad x \in \overline{\Omega} \tag{1.3}$$

where H is a "partial inverse" of L (see section 2). For fixed $x^* \in P(\overline{\Omega})$ let T_{x^*} be the restriction of T to $p^{-1}(x^*) \cap \overline{\Omega}$. Then x is a solution of (1.2) if and only if for $x^* = P(x)$

$$x = T_{x^*}(x) \tag{1.4}$$

and

(1.5 $$Qf(x) = \Theta.$$

In many cases equation (1.4) has for given $x^* \in P(\Omega)$ one and only one solution $x = \tau(x^*)$ (see e.g. [2], section 3). Then (1.2) has a solution in Ω if and only if x^* satisfies

$$(1.6) \qquad Qf(\tau(x^*)) = \Theta.$$

Now $P\Omega$ is a bounded open set in the p-dimensional space PE, and the range of $Qf\tau$ is in the q-dimensional space QF. Therefore, if $p = q$ the Brouwer degree $d_B(f\tau, P\Omega, \Theta)$ is defined provided that $Qf\tau$ is continuous and that (1.6) has no solution x^* on the boundary of $P\Omega$. In this case the condition

$$(1.7) \qquad d_B(Qf\tau, P\Omega, \Theta) \neq 0$$

is sufficient for (1.6) to have a solution in $P\Omega$, and therefore for (1.2), to have a solution in Ω.

Following S. A. Williams ([10]) we call the degree (1.7) the Cesari index. Williams constructed a map $\bar{f}$ closely related to the map f such that the Leray-Schauder degree $d_{LS}(\bar{f}, \Omega, \Theta)$ exists and equals the Cesari index.

While the computation of the Cesari index seems to require the knowledge of the solution $x = \tau(x^*)$ of (1.4) it is one of the objects of the present paper to show that with proper choice of P and Q this index in independent of τ. Indeed in section 2, projections P^ν, Q^ν on E and F respectively will be constructed for $\nu = 0, 1, 2, \ldots$ such that (as shown in section 3) the decomposition described in (1.3) - (1.6) holds if P and Q are replaced by P^ν and Q^ν respectively. (In the Hilbert space case such P^ν, Q^ν were defined by P. J. McKenna in [5]; cf. also the different approach of D. Sweet in [9]). This is done without requiring the equality of the dimensions p_ν, q_ν of $P^\nu E$ and $Q^\nu F$.

Under certain assumptions it will be shown in section 4 that from a certain ν on

(1.8) $$d_B(Q^\nu f\tau_\nu,\ P^\nu\Omega,\ \Theta) = d_B(Q^\nu f,\ P^\nu\Omega,\ \Theta)$$

where τ_ν is defined with respect to P^ν, Q^ν as τ was defined with respect to P, Q.

Under further assumptions it will be shown (Theorem 4.2) that for large ν the right member of (1.8) equals the generalized topological degree $d_{BP}(f,\Omega,\Theta)$ introduced by F. E. Browder and W. V. Petryshin in [1].

While of course for these results concerning degrees the equality $p_\nu = q_\nu$ is necessary, the basic Lemma 4.2 establishing a homotopy between $Q^\nu f$ and $Q^\nu f\tau$ holds also if $p_\nu \neq q_\nu$. This makes it possible to assert the existence of a solution of (1.6) (with Q replaced by Q^ν) if f has a certain homotopy property (Theorem 5.1).

Definitions and results from the theory of the Browder-Petryshin degree needed in the present paper are recalled in section 4 (Definitions 4.1, 4.2, and Lemma 4.1).

2. THE CONSTRUCTION OF SUITABLE PROJECTIONS P^ν, Q^ν

Let E and F be real Banach spaces. The norm on either space will be denoted by $||\cdot||$. Let L be a linear map of $\mathcal{D}(L)$ into F where $\mathcal{D}(L)$ is a dense linear subset of E. The graph and range of L are supposed to be closed and the dimension p_0 of the kernel E_0 of L is supposed to be finite. Then there exists a projection P^0 on E such that $P^0E = E_0$ (A projection on E is a continuous linear map $E \to E$ with $P^2 = P$). We then have

(2.1) $$E = E_0 \dotplus E_1,\ E_0 = P^0E,\ E_1 = (I-P^0)E$$

where the symbol $\dotplus$ denotes direct sum.

We further assume the existence of a projection Q^0 on F such that

(2.2) $$F_1 = (I-Q^0)F = \text{range of } L$$

and such that with $F_0 = Q^0F$

(2.3) $$q_0 = \dim F_0$$

is finite.

Then

(2.4) $$F = F_0 \dotplus F_1, \; F_0 = Q^0F, \; F_1 = (I-Q^0)F.$$

It is clear from the foregoing that L restricted to $E_1 \cap \mathcal{D}(L)$ is a bijection on F_1. The inverse H of this restricted map is linear and, by the closed graph theorem (see e.g. [3] p. 57), bounded. We obviously have

(2.5) $$HL(x) = (I-P^0)x \text{ for all } x \in \mathcal{D}(L)$$

and

(2.6) $$LH(y_1) = y_1 \text{ for } y_1 \in F_1.$$

We subject the space E to

Assumption 2.1. There exists a sequence $P^1, P^2, \ldots$ of projections on E having the following properties:

i) the range $E_0^\nu = P^\nu E$ of P^ν is of finite dimension p_ν ($\nu = 1, 2, \ldots$)

ii) $E_0 \subset E_0^1 \subset E_0^2 \subset$ with the inclusions being proper

iii) $\lim P^\nu(x) = x$ for all $x \in E$ and $\lim LP^\nu(x) = L(x)$ for all $x \in \mathcal{D}(L)$ (obviously the second part of iii follows from the first one if L is bounded)

iv) $P^0P^\nu = P^\nu P^0$, $\nu = 0, 1, 2, \ldots$

v) $P^\nu\mathcal{D}(L) \subset \mathcal{D}(L)$, $\nu = 0, 1, 2, \ldots$

We note that

(2.7) $$E = E_0^\nu \dotplus E_1^\nu \text{ with } E_0 = P^\nu E, \; E_1^\nu = (I-P^\nu)E$$

for $\nu = 1, 2, \ldots$, and that by (2.1) the relation (2.7) holds also for $\nu = 0$ if we set

$$E_0^0 = E_0, \quad E_1^0 = E_1. \tag{2.8}$$

We verify the following two relations which will be needed later on:

$$P^\nu E_1 \subset E_1 \tag{2.9}$$

$$(I-P^\nu)E \subset E_1. \tag{2.10}$$

Proof of (2.9). Because of the direct decomposition (2.1) it will be sufficient to prove that $P^0 P^\nu E_1 = \Theta$. But by (2.1) and by iv) of Assumption 2.1 $P^0 P^\nu E_1 = P^0 P^\nu (I-P^0)E = P^\nu P^0 (I-P^0)E = \Theta$.

Proof of (2.10). Let $x \in (I-P^\nu)E$. Then $x = (I-P^\nu)x$ and by (2.1)

$$x = P^0(I-P^\nu)x + (I-P^0)(I-P^\nu)x. \tag{2.11}$$

Using Assumption 2.1 we see that the first summand at the right of (2.11) equals $(I-P^\nu)P^0 x$ and is contained in the set $(I-P^\nu)P^\nu E$ which has Θ as the only element. Thus by (2.11), $x = (I-P^0)(I-P^\nu)x \subset (I-P^0)E = E_1$. This obviously proves (2.10).

We note that by the principle of uniform boundedness (see e.g. [3] p. 60), the first part of iii) implies the existence of a $k > 0$ such that for all ν

$$||\tau^\nu x|| \leqq k||x||. \tag{2.12}$$

Remark 2.1. Since L is densely defined, it follows easily from part v of Assumption 2.1 that the linear map L restricted to $P^\nu \mathcal{D}(L)$ can be uniquely extended to a continuous linear map $\overline{L}_\nu : E_0^\nu = P^\nu E \to F$. In what follows we write $L_\nu(x)$ instead of $\overline{L}_\nu(x)$ for $x \in E_0^\nu$. With this convention LP^ν is a continuous linear map with domain E.

Definition 2.1. Let Q^0 be as above (see (2.2) and (2.3)). For $\nu \geqq 1$ we define a linear map $Q^\nu : F \to F$ as follows: by (2.4) every $y \in F$ admits the unique representation

(2.13) $$y = y_0 + y_1,\ y_0 \in F_0,\ y_1 \in F_1.$$

It will be sufficient to define $Q^\nu(y_0)$ and $Q^\nu(y_1)$. We set

(2.14) $$Q^\nu(y_0) = y_0,\ Q^\nu(y_1) = LP^\nu H(y_1),\ \nu \geqq 1$$

where H is as above (see (2.5) and 2.6)).

Lemma 2.1. The Q^ν have the following properties:

(2.15) i) $Q^\nu L(x) = LP^\nu(x)$ for all $x \in \mathcal{D}(L)$.

ii) Q^ν is a projection

iii) for all $y \in F$

(2.16) $$\lim Q^\nu(y) = y \text{ and } ||Q^\nu(y)|| \leqq k||y||$$

for some k independent of ν.

iv) let $F_0^\nu = Q^\nu F$ (and therefore $F_0^0 = F_0$).

Then,

(2.17) a) $F_0^\nu \subset F_0^{\nu+1}$, b) $Q^\nu(F_1) \subset F_1$, c) $(I-Q^\nu)F \subset F_1$

v) if p_ν and q_ν are the dimensions of E_0^ν and F_0^ν respectively, then

(2.18) $$q_\nu - p_\nu = q_0 - p_0.$$

Proof of i). It follows directly from the definition of P^0 and Q^0 and from (2.1) and (2.4) that both members of (2.15) equal Θ for $\nu = 0$, let then $\nu > 0$. For $x \in \mathcal{D}(L)$, let

(2.19) $$x = x_0 + x_1,\ x_0 \in E_0,\ x_1 \in E_1$$

(c.f. (2.1)). Then $L(x) = L(x_1) = y_1 \in F_1$, and $x_1 = H(y_1)$. From this and (2.14) we see that

(2.20) $$Q^\nu L(x) = Q^\nu(y_1) = LP^\nu H(y_1) = LP^\nu(x_1) = LP^\nu(x) - LP^\nu(x_0).$$

But $x_0 \in P^0E \in P^\nu E$. Thus $x_0 = P^\nu(x_0)$ and $LP^\nu(x_0) = L(x_0) = \Theta$. This together with (2.20) proves i).

Proof of ii). It is clear from Definition 2.1 that for each $\nu \geqq 0$, the linear map Q^ν is bounded since P^0, Q^ν, P^ν and H are bounded and since the restriction of L to the finite dimensional space $P^\nu E$ is also bounded (cf. Remark 2.1). It remains to show that $(Q_\nu(y))^2 = Q_\nu(y)$ for every $y \in F$. Since Q^0 is a projection by assumption we assume $\nu \geqq 1$. Then we obtain from (2.14), (2.15) and the fact that P^ν is a projection

$$\begin{aligned}(Q^\nu(y))^2 &= Q^\nu(y_0+Q^\nu(y_1)) = y_0 + Q^\nu Q^\nu(y_1)\\ &= y_0 + Q^\nu(LP^\nu H(y_1) = y_0 + L(P^\nu)^2 Hy_1\\ &= y_0 + LP^\nu Hy_1 = Q^\nu y.\end{aligned}$$

Proof of iii). Since $Q^\nu(y) = y_0 + Q^\nu(y_1)$ it will be sufficient to show that $\lim Q_\nu(y_1) = y_1$ for $y_1 \in F_1$. Now with $x_1 = H(y_1)$, $y_1 = L(x_1)$ we see from (2.14) that $Q^\nu(y_1) = LP^\nu(x_1)$. Here the right member converges to $L(x_1) = y_1$ by Assumption 2.1 iii). This proves the first part of iii). The second part now follows by virtue of the principle of uniform boundedness.

Proof of iv a). We have to prove $Q^\nu(y) \in F_0^{\nu+1}$ for every $y \in F$. Now with the notations used in the proof of iii), $Q^\nu(y) = y_0 + LP^\nu(x_1)$, $x = H(y_1)$. Since, by (2.14), $y_0 \in Q^{\nu+1}F = F_0^{\nu+1}$ and since $F_0^{\nu+1}$ is linear it will be sufficient to show that

(2.21) $$LP^\nu(x_1) \in F_0^{\nu+1}.$$

Now the domain $\mathcal{D} = \mathcal{D}(L)$ of L is dense in E. Therefore we see from Assumption 2.1 ii) that $P^\nu(x_1) \in P^\nu(E) \subset P^{\nu+1}(E) = P^{\nu+1}(\bar{\mathcal{D}}) = \overline{P^{\nu+1}(\mathcal{D})}$. This relation together with (2.15) and Remark 2.1 implies the inclusion $LP^\nu(x_1) \subset \overline{LP^{\nu+1}(\mathcal{D})} = \overline{LP^{\nu+1}(\mathcal{D})} = \overline{Q^{\nu+1}L(\mathcal{D})} = \overline{Q^{\nu+1}(F)}$.

This proves (2.21) since $Q^{\nu+1}(F) = F_0^{\nu+1}$ is closed.

Proof of iv b). By (2.14), $Q^{\nu}(F_1) = LP^{\nu}(E_1 \cap \mathcal{D})$ and the right member is contained in the range F_1 of L.

Proof of iv c). We will show that

$$Q^0 Q^{\nu} = Q^{\nu} Q^0 \tag{2.22}$$

for (2.17)c follows from (2.22) and (2.17)a in the same way (2.10) followed from parts iv) and ii) of Assumption 2.1. For the proof of (2.22) we note first that its right member equals y_0 as is seen immediately from the decomposition (2.13) and from (2.14). The left member of (2.22) equals by (2.14), $Q^0(y_0+LP^{\nu}H(y_1)) = y_0 + Q_0 LP^{\nu}H(y_1)$. Here the second term at the right equals Θ since the range of L is $F_1 = (I-Q_0)F$. Thus the left member of (2.22) also equals y_0.

Proof of v. Let $b_1, b_2, \ldots b_{p_\nu}$ be a base for $E_0^{\nu} = P^{\nu}E$. Because of Assumption 2.1 ii) we may assume that for $0 \leqq \mu < \nu$, $b_1, b_2, \ldots b_{p_\mu}$ form a base for E_0^{μ}. In particular $b_1, b_2, \ldots b_{p_0}$ then form a base for $E_0 = P^0 E$. Because of (2.1) we may also assume that $b_i \in E_1$ for $p_0 < i \leqq p_\nu$. Our assertion will be proved once a base $\beta_1, \beta_2, \ldots \beta_{q_\nu}$ for $F_0^{\nu} = Q^{\nu}F$ is constructed with $q_\nu = p_\nu + q_0 - p_0$.

This will be done as follows: by (2.3) we may choose $\beta_1, \beta_2, \ldots \beta_{q_0}$ as a basis for $F_0 = Q^0 F$. For $q_0 < i \leqq q_\nu$ we define

$$\beta_i = L(b_{i+j_0}), \quad j_0 = p_0 - q_0. \tag{2.23}$$

We proceed to prove that the β_i thus defined for $i = 1, 2, \ldots q_\nu$ from a base for F_0^{ν}.

We first claim that $\beta_i \in F_0^{\nu}$. Indeed for $i = 1, 2, \ldots q_0$, $\beta_i < F_0 = F_0^0$ by definition, and $F_0^0 \subset F_0^{\nu}$ by (2.17). For $q_0 < i \leqq q_\nu$ we see from (2.23) and (2.15) that

(2.24) $$Q^\nu(\beta_i) = Q^\nu L(b_{i+j_0}) = LP^\nu(b_{i+j_0}).$$

But $b_{i+j_0} \in P^\nu E$. Therefore $P^\nu(b_{i+j_0}) = b_{i+j_0}$ and, by (2.24), (2.23), $Q^\nu(\beta_i) = L(b_{i+j_0}) = \beta_i$. Thus $\beta_i \in Q^\nu F = F_0^\nu$ as asserted.

Next we show that the β_i are linearly independent, i.e., that a relation

(2.25) $$\sum_{i=1}^{q_\nu} \lambda_i \beta_i = \Theta, \ \lambda_i \text{ real}$$

implies

(2.26) $$\lambda_i = 0 \qquad i = 1, 2, \ldots q_\nu.$$

Now by our definitions, (2.25) implies

(2.27) $$\sum_{i=1}^{q_0} \lambda_i \beta_i + L\Big(\sum_{q_0+1}^{q_\nu} \lambda_i \beta_{i+j_0}\Big) = \Theta.$$

Here, again by our definition, the first sum is contained in F_0 and the second one in the range F_1 of L. Therefore (cf. (2.4)) each of these sums equals Θ, and since $\beta_1, \beta_2, \ldots \beta_{q_0}$ are linearly independent by definition we conclude that

(2.28) $$\lambda_1 = \lambda_2 = \ldots \lambda_{q_0} = 0.$$

From this and (2.27) we see that $\sum_{q_0+1}^{q_\nu} \lambda_i \beta_{i+j_0}$ is contained in the kernel E_0 of L. On the other hand this sum is a linear combination of $b_{p_0+1}, \ldots b_{p_\nu}$ and therefore also contained in E_1. From (2.1) we now conclude that the sum equals Θ, and since $b_{p_0+1}, \ldots, b_{p_\nu}$ are linear independent we see that $\lambda_{q_0+1} = \ldots \lambda_{q_\nu} = 0$. This together with (2.28) is our assertion (2.26).

It remains to show that the β_i span F_0^ν, i.e., that every $y \in F_0^\nu = Q^\nu F$ is a linear combination of the β_i. Now by (2.4) we have the unique decomposition $y = y_0 + y_1$, $y_0 \in F_0$, $y_1 \in F_1$, and y_0 is a linear combination of $\beta_1, \ldots \beta_{q_0}$. It is therefore sufficient to prove that every $y_1 \in F_1 \cap F_0^\nu$ is a linear combination of $\beta_{q_0+1}, \ldots \beta_{q_\nu}$. Noting that $y_1 = Q^\nu(y_1)$ we see from (2.15) that with $x_1 = H(y_1)$, $y_1 = L(x_1)$:

$$y_1 = Q^\nu(y_1) = Q^\nu L(x_1) = LP^\nu(x_1). \tag{2.29}$$

Now $b_{p_0+1}, \ldots b_{p_\nu}$ form a base for $P^\nu(E_1)$. Therefore $p^\nu(x_1) = \sum_{j=p_0+1}^{p_\nu} \mu_j b_j$ for some real μ_j, and by (2.23) $LP^\nu(x_1) = \sum_{j=p_0+1}^{p_\nu} \mu_j L(b_j) = \sum_{i=q_0+1}^{q_\nu} \mu_{i+j_0}\beta_i$. This together with (2.29) yields our assertion.

3. RETURN TO PROBLEM (1.2).

Let Ω, $\overline{\Omega}$ and N be as in the introduction. Let L and its "partial inverse" H be as in section 2. We recall that the domain of H is $F_1 = (I-Q^0)F$ which by (2.17c) contains $(I-Q^\nu)F$ for $\nu = 0, 1, 2, \ldots.$

Therefore for each such ν we may define

$$T^\nu(x) = P^\nu(x) + H(I-Q^\nu)N(x), \quad x \in \overline{\Omega} . \tag{3.1}$$

For fixed ν and $x^* \in P^\nu(\Omega)$ we denote by $T^\nu_{x^*}$ the restriction of T^ν to $(P^\nu)^{-1}(x^*) \cap \overline{\Omega}$.

<u>Theorem 3.1</u>. x is a solution of (1.2) if and only if

$$x = T^\nu_{x^*}(x), \quad P^\nu(x) = x^* \tag{3.2}$$

and

$$Q^\nu f(x) = Q^\nu(L(x)-N(x)) = \Theta. \tag{3.3}$$

Proof. It is well known that the assertion of the theorem is true if the following three conditions are satisfied:

k_1. $H(I-Q^\nu)L(x) = (I-P^\nu)x$ for all $x \in \mathcal{D}(L) \cap \overline{\Omega}$

k_2. $Q^\nu L(x) = LP^\nu(x)$ for all $x \in \mathcal{D}(L) \cap \overline{\Omega}$

k_3. $LH(I-Q^\nu)N(x) = (I-Q^\nu)N(x)$ for all $x \in \overline{\Omega}$

(see e.g. [2], Theorem 2.i). It remains to verify these conditions.

k_2 was already proved (see 2.15). To verify k_1 we note that by (2.15), (2.5) and (2.10)

$H(I-Q^{\nu})L(x) = HL(I-P^{\nu})x = (I-P^0)(I-P^{\nu})x = (I-P^{\nu})x.$

Finally, to verify k_3 we recall that by (2.17c), $(I-Q^{\nu})y \in F_1$ for every $y \in F$. Therefore we may apply (2.6) with $y_1 = (I-Q^{\nu})y$ and see that $LH(I-Q^{\nu})y = (I-Q^{\nu})y$ for every $y \in F$.

Theorem 3.2. Let H be completely continuous. Then

$$\lim_{\nu\to\infty} ||H(I-Q^{\nu})y|| = 0 \tag{3.4}$$

uniformly for y in a bounded set, say for $||y|| \leqq M$. (A related result by J. K. Hale may be found in [4], pp. 13, 14; see also [2], 5iii).

Proof. If the assertion were not true there would exist an $\varepsilon > 0$, a sequence $\{\nu_i\}$ of positive integers with $\nu_i \to \infty$, and points y_{ν_i} with

$$||y_{\nu_i}|| \leqq M \tag{3.5}$$

such that

$$||H(I-Q^{\nu_i})y_{\nu_i}|| \geqq \varepsilon. \tag{3.6}$$

Now by (2.4) we may write $y_{\nu_i} = (y_{\nu_i})_0 + (y_{\nu_i})_1$, where $(y_{\nu_i})_0 \in F_0 = F_0^0$, $(y_{\nu_i})_1 \in F_1$, and by (2.17a), $(y_{\nu_i})_0 \in F_0^{\nu_i}$. Therefore $(y_{\nu_i})_0 = Q^{\nu_i}((y_{\nu_i})_0)$, and $(I-Q^{\nu_i})(y_{\nu_i})_0 = (I-Q^{\nu_i})Q^{\nu_i}((y_{\nu_i})_0) = \Theta$. This means we may assume $y_{\nu_i} \in F_1$. Then $x_{\nu_i} = H(y_{\nu_i})$ is defined, and $y_{\nu_i} = L(x_{\nu_i})$. Setting

$$z_{\nu_i} = H(i-Q^{\nu_i})y_{\nu_i} \tag{3.6}$$

we see that $z_{\nu_i} = H(I-Q^{\nu_i})Lx_{\nu_i} - HL(I-P^{\nu_i})x_{\nu_i} = (I-P^{\nu_i})x_{\nu_i}$. Multiplying this relation by $I - P^{\nu_i}$ we conclude that

$$z_{\nu_i} = (I-P^{\nu_i})z_{\nu_i} \tag{3.7}$$

since $(I-P^{\nu_i})^2 x_{\nu_i} = (I-P^{\nu_i})x_{\nu_i} = z_{\nu_i}$. Now by (3.5) and (2.16) the sequence $(I-Q^{\nu_i})y_{\nu_i}$ is bounded and H is assumed to be completely continuous. Therefore we see from (3.6) that for some subsequence ν_{j_i} of the ν_i, the $z_{\nu_{j_i}}$ converge. If z_0 is the limit we may write by (3.7)

$$z_{\nu_{j_i}} = (I-P^{\nu_{j_i}})z_{\nu_{j_i}} - (I-P^{\nu_{j_i}})z_0 + (I-P^{\nu_{j_i}})(z_{\nu_{j_i}}-z_0).$$

Here the first term of the right member converges to Θ by assumption 2.1 iii, and, noting (2.12), we see that the second one also converges to Θ. Thus $z_{\nu_{j_i}} \to \Theta$ which contradicts (3.6).

4. TWO THEOREMS ON THE CESARI INDEX.

In this section we keep the notations and assumptions of the preceding sections while adding assumptions 4.1 - 4.5 below.

Assumption 4.1. Ω is convex and

(4.1) $$P^{\nu}(\Omega) \subset \Omega_{\nu}$$

where $\Omega_{\nu} = E^{\nu} \cap \Omega$.

Since the inclusion inverse to (4.1) is obvious we see that

(4.2) $$P^{\nu}(\Omega) = \Omega_{\nu}.$$

Moreover under Assumption 4.1

(4.3) $(1-t)x + tP^{\nu}(x) \in \Omega$ for $x \in \Omega$, $0 \leq t \leq 1$.

Assumption 4.2. f is Lipschitzean, i.e., there exists an $L > 0$ such that

(4.4) $||f(x_1)-f(x_2)|| < L||x_1-x_2||$ for $x_1, x_2 \in \overline{\Omega}$.

Assumption 4.3. Equation (3.2) has one and only one solution $x = \tau_{\nu}(x^*) \in \Omega \cap (P^{\nu})^{-1}(x^*)$ and this solution is continuous.

Assumption 4.4. H is completely continuous (cf. Theorem 3.2).

Assumption 4.5. f is A-proper with respect to $\gamma = (E, F, P^\nu, Q^\nu)$.

The concept "A-proper" was introduced by Browder and Petryshin in [1]. We recall the definition for our situation:

Definition 4.1. Let f_ν be the restriction of $Q^\nu f$ to Ω_ν. Then f is called A-proper with respect to the approximation scheme $\gamma = (E, F, P^\nu, Q^\nu)$ if the following is true: if for some subsequence $\{\nu_i\}$ of positive numbers with $\nu_i \to \infty$ there exists $x_{\nu_i} \in \Omega_{\nu_i}$ and a $y \in F$ such that

$$\lim f_{\nu_i}(x_{\nu_i}) = y \tag{4.5}$$

then there exists a subsequence $x_{\nu_{j_i}}$ of the x_{ν_i} which converges to some $x \in E$, and

$$f(x) = y. \tag{4.6}$$

The following lemma will be basic.

Lemma 4.1. Let f be A-proper with respect to $\gamma = (E,F,P^\nu,Q^\nu)$. Let y_0 be a point of F for which

$$y_o \notin f(\partial\Omega) \tag{4.7}$$

(As usual ∂ denotes "boundary of").

Then there exists a $\rho > 0$ and an integer ν_0 such that

$$||Q^\nu(f(x)-Q^\nu(y_0)|| \geqq \rho \text{ for } x \in \partial\Omega_\nu \text{ and } \nu \geqq \nu_o. \tag{4.8}$$

This lemma is a special case of lemma 1 page 220 in [1] which is stated for the case that the dimensions p_ν q_ν and $P^\nu E$ and $Q^\nu F$ respectively are equal whose proof however does not depend on this equality.

Lemma 4.2. (Homotopy lemma). In addition to the assumptions described at the beginning of this section we suppose that

$$\Theta \notin f(\partial\Omega). \tag{4.9}$$

Then there is a ν_0 such that for $\nu \geqq \nu_0$ there exists a continuous $g_\nu(t,x^*) : [0,1] \times \overline{\Omega}_\nu \to F_0^\nu = Q^\nu F$ having the following properties:

(4.10) $g_\nu(0,x^*) = Q^\nu f\tau_\nu(x^*)$, $g_\nu(1,x^*) = Q^\nu f(x^*)$ (cf. Assumption 4.3)

(4.11) $$\Theta \notin g_\nu(t,\partial\Omega_\nu),\ t \in [0,1].$$

Proof. We set

(4.12) $g_\nu(t,x^*) = Q^\nu f((1-t)\tau_\nu(x^*)+tx^*)$, $t \in [0,1]$, $x^* \in \overline{\Omega}_\nu$.

Then (4.10) is obviously satisfied. For the proof of (4.11) we assert the existence of a $\rho > 0$ and an integer $\overline{\nu}$ such that

(4.13) $||g_\nu(t,x^*)|| \geqq \overline{\rho}$ for $x^* \in \partial\Omega_\nu$, $t \in [0,1]$, $\nu \geqq \overline{\nu}_o$.

If this assertion were not true there would exist a sequence $\{\nu_i\}$ of integers with $\nu_1 \to \infty$, points $x_{\nu_i} \in \partial\Omega_{\nu_i}$, and $t_{\nu_i} \in [0,1]$ such that

(4.14) $$\lim g_{\nu_i}(t\nu_i,x_{\nu_i}) = \Theta.$$

Obviously we may assume that the t_{ν_i} converge to some $t_o \in [0,1]$:

(4.15) $$\lim t_{\nu_i} = t_o$$

We claim first that then

(4.16) $$\lim g_{\nu_i}(t_o,x_{\nu_i}) = \Theta.$$

Because of (4.14) it will be sufficient to show that

(4.17) $$||g_{\nu_i}(t_o,x_{\nu_i})-g_{\nu_i}(t_{\nu_i},x_{\nu_i})|| \to 0.$$

Now from (4.12), (2.16) and (4.4) we see that the left member of (4.17) is majorized by

$$kL|t_o-t_{\nu_i}|\ ||x_{v_i}-\tau_{v_i}(x_{v_i})||.$$

By (4.15) this estimate proves (4.17) since x_{ν_i} and $\tau_{\nu_i}(x_{\nu_i})$ are in the bounded set $\overline{\Omega}$.

Next we claim that

(4.18) $$\lim Q^{\nu_i}f(x_{\nu_i}) = \Theta.$$

By (4.16) this assertion is equivalent to

(4.19) $$||Q^{\nu_i}f(x_{\nu_i})-g_{\nu_i}(t_o,x_{\nu_i})|| \to 0.$$

Now from (4.12), (2.16) and (4.4) the left member of (4.19) is majorized by

$$kL(1-t_o)||x_{\nu_i}-\tau_{\nu_i}(x_{\nu_i})||.$$

But $||x_{\nu_i}-\tau_{\nu_i}(x_{\nu_i})-\tau_{\nu_i}(x_{\nu_i}|| = ||H(I-Q^{\nu_i})N(x_{\nu_i})||$ by (3.1), (3.2) and the definition of τ_{ν_i} (see Assumption 4.3), and the right member of the last equality converges to 0 by Assumption 4.4 and Theorem 3.2.

Thus (4.18) is established. But by (4.9) and Assumption 4.5 relation (4.18) contradicts the conclusion (4.8) (with $y_o = \Theta$) of Lemma 4.1. Lemma 4.2 is proved.

Theorem 4.1. In addition to the assumptions of Lemma 4.2 we assume that

$$p_0 = q_0. \tag{4.20}$$

Then for large enough ν the Brouwer degrees $d_B(Q^\nu f,\Omega_\nu,\Theta)$ and $d_B(Q^\nu f\tau_\nu,\Omega_\nu,\Theta)$ are defined and equal.

Proof. The restriction of $Q^\nu f$ to $\overline{\Omega}_\nu$ maps the q_ν dimensional set $\overline{\Omega}_\nu$ into the q_ν dimensional space $Q^\nu F$. But by (2.18) the assumption (4.20) implies $p_\nu = q_\nu$. Therefore $d_B(Q^\nu f,\Omega_\nu,\Theta)$ is defined provided that $Q^\nu f(\partial\Omega_\nu) \neq \Theta$. But this condition is satisfied for large enough ν on account of (4.9) and Lemma 4.1. Thus the assertion concerning $d_B(Q^\nu f,\Omega_\nu,\Theta)$ is proved. The remaining part of the theorem now follows from the homotopy Lemma 4.2 in conjunction with the invariance of the Brouwer degree under a suitable homotopy.

Before stating Theorem 4.2 we recall for our situation the definition of the Browder-Petryshin generalized degree (see [1], p. 220). This definition is made possible by Lemma 4.1 according to which (4.9) implies that the Brouwer degrees $d_B(Q^\nu f,\Omega_\nu,\Theta)$ are defined from a certain ν on provided that (4.20) (and therefore $p_\nu = q_\nu$) is satisfied.

Definition 4.2. Let $f : \overline{\Omega} \to F$ be A-proper with respect to $\gamma = (E,F,P^\nu,Q^\nu)$. Suppose (4.9) and (4.20) to be satisfied. Let Z be the set of all integers augmented by $+\infty$ and $-\infty$. Then the generalized degree $d_{BP}(f,\Omega,\Theta)$ is the set $\{z\}$ of those elements of Z for which $\lim d_B(Q^{\nu_i}f,\Omega_{\nu_i},\Theta) = z$ for some infinite subsequence $\{\nu_i\}$ of the positive integers. If this set consists of only one element z_0 we say the degree is one valued and denote it by z_0. If in addition z_0 if finite we call the degree finitely one valued.

For examples of A-proper maps see [1], section 3.2, (cf. also [7]), and for an important sufficient condition for $d_{BP}(f,\Omega,\Theta)$ to be finitely one valued, in fact to equal 1, see [1] section 3.1, theorem 5. In particular it follows from these two sections of [1] that for a certain class of monotone mappings f of a Hilbert space E into itself $d_{BP}(f,\Omega,y) = 1$ for every $y \in E$. For the proof of this latter fact see also [8], section 5, in particular the proof of theorem 5.1 in conjunction with corollary 4 in section 4 of that paper.

Theorem 4.2. In addition to the assumptions of Theorem 4.1 we assume $d_{BP}(f,\Omega,\Theta)$ to be finitely one valued. Then this degree equals the Cesari index $d_B(Q^\nu f\tau_\nu,\Omega_\nu,\Theta)$ for ν large enough.

Proof. It follows directly from Definition 4.2 that, for ν large enough, $d_{BP}(f,\Omega,\Theta) = d_B(Q^\nu f,\Omega_\nu,\Theta)$ if $d_{BP}(f,\Omega,\Theta)$ is finitely one valued. Our assertion is now an immediate consequence of Theorem 4.1.

5. A REMARK NOT ASSUMING (4.20).

Let $\Omega \subset E$ be a ball with the origin as center such that Ω_ν is a ball $B^{p_\nu} \subset E^\nu$ and $\partial\Omega^\nu$ a sphere $S^{p_{\nu-1}} \subset E^\nu$. Otherwise we assume the assumptions of Lemma 4.2 to be satisfied. Now by (4.9) and Lemma 4.1 $Q^\nu f(S^{p_{\nu-1}}) \neq \Theta$ for ν large enough. Therefore $\psi_\nu = \frac{Q^\nu f}{||Q^0_\nu f||}$ maps $S^{p_{\nu-1}}$ continuously into the unit-sphere $S^{q_{\nu-1}} \subset F^\nu$. We recall that a map $S^{p_{\nu-1}} \to S^{q_{\nu-1}}$ is called homotopically non-trivial if it is not homotopic to a map mapping $S^{p_{\nu-1}}$ into a single point of $S^{q_{\nu-1}}$.

Theorem 5.1. If in addition to the assumptions stated above Ψ_ν is supposed to be homotopically non-trivial for a large enough ν then $Q^\nu f\tau_\nu(x^*) = \Theta$ has at least one root $x^* \in B^{p_{\nu-1}} = \Omega_\nu$.

Proof. Let ϕ^0_ν be the restriction of $Q^\nu f\tau_\nu$ to $S^{p_{\nu-1}}$. It follows from the homotopy Lemma 4.2 that $\phi^\nu_0 \neq \Theta$ and that $\Psi^0_\nu = \phi^\nu_0/||\phi^\nu_0||$ is homotopic to the homotopically non-trivial map Ψ_ν. Therefore Ψ^0_ν is homotopically non-trivial. But as is well known (see e.g. [6], theorem 1.1.1.) this implies that for every extension $\overline{\phi^0_\nu}$ of ϕ^0_ν to $B^{p_\nu} = \Omega_\nu$ the equation $\overline{\phi^0_\nu} = \Theta$ has a solution in B^{p_ν}. In particular this is true for the extension $\overline{\phi^0_\nu} = Q^\nu f\tau_\nu$.

REFERENCES

1. Browder, F. E. and Petryshin, W. F., Approximation methods and the generalized topological degree for nonlinear mappings in Banach spaces, *Journal for Functional Analysis 3*(1969), 217-245.

2. Cesari, L., *Nonlinear Oscillations in the Framework of Alternative Methods*, (International Symposium on Dynamical Systems, Providence, R. I.) August, 1974.

3. Dunford, N. and Schwartz, I. T., *Linear Operators Part I*, Interscience Publishers, New York, 1958.

4. Hale, I. K., Applications of alternative problems, *Lecture Notes*, Brown University, 1971.

5. McKenna, P. I., *Non-selfadjoint Semilinear Problems in the Alternative Method*, University of Michigan thesis, 1976.

6. Nirenberg, L., *Topics in Nonlinear Functional Analysis*, (Lecture notes 1973-1974), Courant Institute of Mathematical Sciences, New York University.

7. Petryshin, W. V., On linear P-compact operators in Banach spaces with applications to constructive fixed point theorems, *Journal of Mathematical Analysis and Applications 15*(1966), 228-242.

8. Rothe, E. H., *Expository Introduction to Some Aspects of Degree Theory*, (State University of Michigan conference, East Lansing), April 9-12, 1975.

9. Sweet, D., An alternative method based on approximate solutions, *Mathematical Systems Theory 4*(1970), 306-317.

10. Williams, S. A., A connection between the Cesari and Leray-Schauder methods, *Michigan Mathematical Journal 15*(1968), 441-448.

DISPERSAL MANIFOLDS IN PARTIAL DIFFERENTIAL GAMES

Emilio O. Roxin
Department of Mathematics
University of Rhode Island

ABSTRACT. In a differential game with the heat equation, Nash equilibrium solutions are characterized and constructed backwards from the end-conditions. The main point is to discuss how far backwards these solutions are extendable and when they cease to be optimal (i.e. when they hit a dispersal manifold).

1. INTRODUCTION

In the development of the classical zero-sum two player differential games with ordinary differential equations, as given in the book by Isaacs [4], "singular surfaces" appeared from the very beginning as loci of discontinuities of the optimal feedback controls. Isaacs attempted a classification and distinguished, among others, "dispersal" or "efferent" surfaces as loci of points in the state-time space from where more than one optimal trajectory starts. While in two-dimensional problems such singular curves can be traced graphically (even if sometimes only approximately), this is no longer possible in higher dimensions, which makes the problem less tractable. In the case of games with partial differential equations (equivalent to dimension infinity), the general problem is of course even much more difficult, while still basically similar. In order to find out how far this similarity goes and what new features, if any, appear, the solution of particular examples seems to be the right approach. With this in mind, a simple differential game with the one-dimensional heat equation and boundary controls is considered here. Optimal solutions are characterized and constructed backwards from the end-conditions. The problems which arise are discussed.

2. THE DIFFERENTIAL GAME

Consider the real function $u(t,x)$ defined for $0 \leq x \leq 1$ and some not necessarily fixed t-interval, and let $u(t,x)$ be determined as solution of the partial differential equation

$$u_t(t,x) = u_{xx}(t,x) \tag{1}$$

with the boundary conditions

$$\begin{aligned} u(t,0) &= f(t) \\ u(t,1) &= g(t) \end{aligned} \tag{2}$$

and some suitable initial conditions

$$u(t_0,x) = U(x). \tag{3}$$

According to the smoothness of the conditions (2) and (3), the solution $u(t,x)$ has to be interpreted in a strict classical or some generalized (for example "weak") sense. Here we do not intend to go into these rather technical details; the optimal controls which we expect to find are piecewise continuous with a finite (or at worst denumerable) number of jumps, and for these there is no problem in interpreting the solution in a suitable sense.

We assume that $f(t)$ is the control function controlled by one player, whom we shall call player "f", and similarly that player "g" controls $g(t)$. Both controls are assumed to be restricted in magnitude by

(4) $$|f(t)| \leq m_0, \quad |g(t)| \leq m_1.$$

The game is assumed to be of perfect information, meaning that at time t_1 both players know the complete past, i.e. $u(t,x)$ and hence $f(t)$ and $g(t)$ for all $t < t_1$, starting from the beginning of the game. With such information they must choose $f(t_1)$ respectively $g(t_1)$.

For both players we will consider strategies $\bar{f}$ and $\bar{g}$ in the sense of "closed-loop" or "feedback" controls, i.e. the choice of $f(t_1)$ respectively $g(t_1)$ has to be made as a function of $u(t_1,.)$. Nevertheless, in what follows the type of strategy adopted will not play an important role.

As payoff we take

(5) $$P = \int_0^1 |u(T,x)|dx$$

for some prescribed value of $T > t_0$. Player "f" should maximize P and player "g" should minimize it. Therefore the game is as follows: given t_0, T, $U(x)$, m_0 and m_1, find optimal strategies $\bar{f}^*$, $\bar{g}^*$ (assuming these exist), which satisfy (4) and such that P obtained from (1), (2), (3), (5) is optimal in the sense of a Nash equilibrium point, meaning that

(6) $$P[\bar{f},\bar{g}^*] \leq P[\bar{f}^*,\bar{g}^*] \leq P[\bar{f}^*,\bar{g}]$$

holds for the optimal $\bar{f}^*$, $\bar{g}^*$ and all other admissible strategies $\bar{f}$, $\bar{g}$.

These equations can be interpreted as describing the temperature of a bar of length one, whose ends are kept at variable temperatures $f(t)$ and $g(t)$. One player wants to maximize and the other to minimize the deviation of the temperature at the final time T from some reference temperature taken as zero, and this deviation is measured in the L^1-norm (5).

This problem may lead to optimal controls with a denumerable infinity of jumps (see [1], where this case is analyzed as an optimal control problem). In order to avoid convergence problems as well as for practical implementation, we may assume that both players must apply zero controls during the last time interval $(T-\varepsilon, T)$, where $\varepsilon > 0$ is given in advance. With this restriction, optimal controls will have only finitely many jumps. This restriction on the controls will not always be mentioned explicitly in the following, but may be assumed whenever it is necessary to avoid the above mentioned problems.

3. SOLUTION OF THE INITIAL-BOUNDARY VALUE PROBLEM

The initial-boundary value problem (1), (2), (3) has a well known solution (see, for example, [2] or [3]). It can be written in the following form

$$u(t,x) = u_0(t,x) + u_1(t,x), \tag{6}$$

where $u_0(t,x)$ corresponds to the initial condition $U(x)$ and boundary conditions $f(t) = g(t) = 0$, while $u_1(t,x)$ corresponds to initial condition zero and the actual boundary controls.

We then have

$$u_0(t,x) = 2\sum_1^\infty e^{-n^2\pi^2(t-t_0)} \sin n\pi x \int_0^1 U(\xi) \sin n\pi\xi \, d\xi \tag{7}$$

and

$$\begin{aligned} u_1(t,x) &= 2\pi \sum_1^\infty n \sin n\pi x \int_{t_0}^t e^{-n^2\pi^2(t-\tau)} f(\tau)\, d\tau \,+ \\ &+ 2\pi \sum_1^\infty (-1)^{n+1} n \sin n\pi x \int_{t_0}^t e^{-n^2\pi^2(t-\tau)} g(\tau)\, d\tau. \end{aligned} \tag{8}$$

These can be written as

$$u_0(t,x) = \int_0^1 k(t-t_0,x,\xi)\, U(\xi)\, d\xi \tag{9}$$

and

$$u_1(t,x) = \int_{t_0}^{t} K_0(t-\tau,x)\, f(\tau)\, d\tau + \int_{t_0}^{t} K_1(t-\tau,x)\, g(\tau)\, d\tau, \tag{10}$$

where

$$\text{(a)}\quad k(t-t_0,x,\xi) = 2\sum_1^\infty \sin n\pi x \sin n\pi\xi\, e^{-n^2\pi^2(t-t_0)},$$

$$\text{(11)}\quad \text{(b)}\quad K_0(t-\tau,x) = 2\pi\sum_1^\infty n \sin n\pi x\, e^{-n^2\pi^2(t-\tau)},$$

$$\text{(c)}\quad K_1(t-\tau,x) = 2\pi\sum_1^\infty (-1)^{n+1}\, n \sin n\pi x\, e^{-n^2\pi^2(t-\tau)}$$

are the Green's functions of the corresponding problems. We may notice that (11)(b) and (c) have singularities at $\tau = t$, hence the above mentioned convenience of having the controls $f(t)$ and $g(t)$ restricted to be zero in some neighborhood of the final time T.

4. THE OPTIMIZATION PROBLEM

Substituting $u(T,x)$ from (6), (9), (10) into (5), we obtain the expression of the payoff in terms of the controls

$$P = \int_0^1 \left| u_0(T,x) + \int_{t_0}^{T} K_0(T-\tau,x)\, f(\tau)\, d\tau + \int_{t_0}^{T} K_1(T-\tau,x)\, g(\tau)\, d\tau \right| dx. \tag{12}$$

Assuming, as usual, the existence of the optimal solution and regularity conditions such that the following steps are valid, we denote by $f^*(\tau)$ and $g^*(\tau)$ the controls corresponding to the optimal strategies. Let also $u^*(t,x)$ be the corresponding payoff (which then is the "Value" of the game).

We define the sets

$$\text{(13)}\quad \begin{aligned} P &= \{x \in [0,1] \mid u^*(T,x) > 0\} \\ N &= \{x \in [0,1] \mid u^*(T,x) < 0\}. \end{aligned}$$

Then (12) can be written as

$$(14)\qquad P^* = \{\int_P - \int_N\}[u_0(T,x) + \int_{t_0}^{T} K_0(T-\tau,x)\ f^*(\tau)\ d\tau + \\ + \int_{t_0}^{T} K_1(T-\tau,x)\ g^*(\tau)\ d\tau]\ dx,$$

where we use the shorthand notation

$$\{\int_P - \int_N\} F(x)\ dx = \int_P F(x)\ dx - \int_N F(x)\ dx.$$

We now can split the expression (14) and interchange the order of integration, obtaining

$$(15)\qquad P^* = \{\int_P - \int_N\} u_0(T,x)\ dx + \\ + \int_{t_0}^{T} f^*(\tau)\ d\tau\{\int_P - \int_N\} K_0(T-\tau,x)\ dx + \\ + \int_{t_0}^{T} g^*(\tau)\ d\tau\{\int_P - \int_N\} K_1(T-\tau,x)\ dx.$$

Now we have to compare P^* with other values of the pay-off, obtained by changing the controls. First we consider open loop variations of the controls. We also consider that these variations are small in L^1 measure. Hence $u(T,x)$ will be only slightly defferent from $u^*(T,x)$, and these being continuous (even analytic) functions, also the sets P and N will change very slightly. An easy estimate shows that in first (linear) approximation, the optimal controls f^*, g^* also optimize the expression (15) where P and N are taken as fixed (not dependent on the particular solution). But if P and N are fixed, then $\delta P = P^* - P$ can be written as

$$(16)\qquad \delta P = \int_{t_0}^{T} \delta f(\tau)\ d\tau\{\int_P - \int_N\} K_0(T-\tau,x)\ dx + \\ + \int_{t_0}^{T} \delta g(\tau)\ d\tau\{\int_P - \int_N\} K_1(T-\tau,x)\ dx.$$

This shows that

$$f^*(t) = m_0 \operatorname{sgn} \{\int_P - \int_N\} K_0(T-t,x)\, dx \tag{17}$$
$$g^*(t) = -m_1 \operatorname{sgn}\{\int_P - \int_N\} K_1(T-t,x)\, dx$$

are necessary conditions for an open-loop Nash equilibrium point in linear approximation.

From standard comparison of inequalities it follows that an open-loop Nash equilibrium point is also a closed-loop one. Hence (17) should be regarded as the solution of our problem.

The solution can also be given without the aid of the kernels K_0, K_1 by the following methods, more similar to the standard treatment of the ordinary differential equation games.

For this prupose we define the "adjoint" function $v(t,x)$ on $0 \le x \le 1$, $t < T$ by

$$v_t(t,x) = -v_{xx}(t,x), \tag{18}$$

$$v(t,x) = v(t,1) = 0, \tag{19}$$

$$v(T,x) = \operatorname{sgn} u^*(T,x). \tag{20}$$

As easy calculation shows that

$$\phi(t) = \langle u,v\rangle = \int_0^1 u(t,x)\ v(t,x)\ dx$$

has the derivative

$$\phi'(t) = f(t)\ v_x(t,0) - g(t)\ v_x(t,1). \tag{21}$$

The optimal payoff can therefore be written as

$$P^* = \int_0^1 u^*(T,x)\ v(T,x)\ dx \tag{22}$$

and again, by standard procedures, our problem is in linear approximation equivalent to optimize

$$\begin{aligned} P &= \int_0^1 u(T,x)\ v(T,x)\ dx = \\ &= \phi(T) = \phi(t_0) + \int_{t_0}^T \phi'(t)\ dt = \\ &= \int_0^1 U(x)\ v(t_0,x)\ dx + \int_{t_0}^T [f(t)v_x(t,0)-g(t)v_x(t,1)]dt, \end{aligned} \tag{23}$$

where now $v(t,x)$ is fixed, independent of the variations of $f(t)$ and $g(t)$.

Hence we obtain from the conditions of optimality

$$f^*(t) = m_0 \operatorname{sgn} v_x(t,0)$$

(24)

$$g^*(t) = m_1 \operatorname{sgn} v_x(t,1).$$

Comparison of δP from (23) and (17) gives, by the way, an interesting interpretation of $v_x(t,0)$ and $v_x(t,1)$.

5. RETROGRESSIVE CONSTRUCTION OF OPTIMAL SOLUTIONS

We can apply here the basic method given by Isaacs [4] of constructing optimal solutions starting from the end-condition at $t = T$. Assuming a given $u^*(T,x)$, we can find backwards the optimal controls $f^*(t)$ and $g^*(t)$ by either (17) or (24). Both ways (which are obviously equivalent) determine in fact the optimal controls uniquely for all $t < T$, possibly up to the set of t-values where the sgn function has argument zero (and, $v(t,x)$ being analytic, this happens only at a finite number of points).

Now two basic questions arise:

A) How far backwards in time does the solution $u^*(t,x)$ exist?

B) How far backwards in time is this solution really optimal?

The first question arises because the backwards problem for the heat equation is ill-posed. The classical Fourier series giving the solution, rapidly becomes divergent except for particularly nicely chosen data. In particular, divergence from the initial-value problem can under some circumstances be cancelled out by properly chosen boundary values.

A systematic treatment of this problem, which seems interesting, is not known to the author. This problem is also related to recent work by Kloeden [5] on abstract dynamical systems without backwards extensions.

The second question of optimality appears also in the case of ordinary differential equation games (as in the book by Isaacs [4]), and in problems of optimal control. The loci of such states, at which the backwards extension of an extremal trajectory is no longer optimal, are precisely the dispersal manifolds.

6. AN EXAMPLE

We can easily give an example showing that such dispersal manifolds must exist. Consider the one-player game (optimal control problem) obtained from the above by setting $g(t) = m_1 = 0$. Let $m_0 = 1$ and the initial $U(x) = u(0,x) = -0.2$, for example. Then, for T sufficiently large and $f(t) \equiv 1$, the end-state $u(T,x)$ looks approximately as in the figure 1. It is quite obvious and can easily be checked by formula (17) that, "f" being the maximizing player, his control $f(t) \equiv 1$

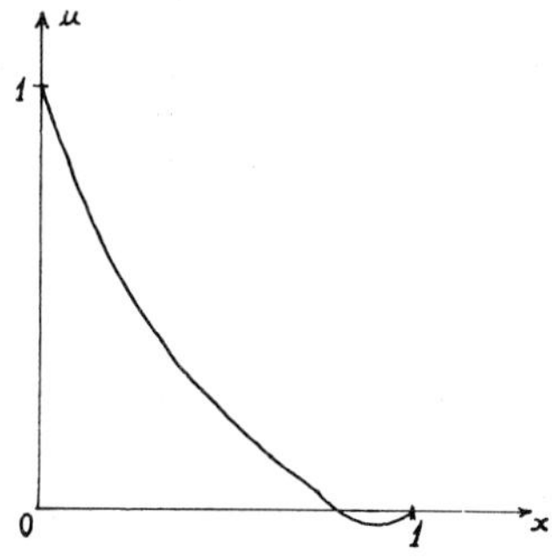

Figure 1

is really optimal shortly before $t = T$. On the other hand, it is also obvious that starting with $u(0,x)$ negative, he would do better by using $f(t) \equiv -1$, as he wants to maximize the absolute value in (5). Hence we see that $f(t) \equiv 1$ can only be optimal for some finite time-interval (t^*,T).

A method to find that minimal time t^* would be of importance. The loci of the corresponding states $u(t^*,x)$ would characterize the dispersal manifold in question. The uthor is not aware of any result along this line.

If, instead of the heat equation, we consider a similar game with the wave equation, even more explicit results can be obtained. The optimal solutions can be given constructively, as was shown in [6]. Due to the different expression of such optimal solutions, the characterization of states belonging to dispersal manifolds can acutally be done in that case.

REFERENCES

1. Butkovskiy, A. G., *Distributed Control Systems*, American Elsevier Publishing Company, New York, 1969 (Transl. from Russian, Nauka Press, Moscow, 1965).

2. Carslaw, H. S. and Jaeger, J. C., *Conduction of Heat in Solids*, 2nd ed., Clarendon Press, Oxford, 1959.

3. Hellwig, G., *Partial Differential Equations*, Blaisdell Publishing Co., New York, Toronto, London, 1964 (transl. from German, Teubner Verlag, Stattgart, 1960).

4. Isaacs, R., *Differential Games*, SIAM Series in Appl. Math., J. Wiley, New York, London, Sydney, 1965.

5. Kloeden, P., *On General Semi Dynamical Systems*, Thesis at the University of Queensland, Australia, 1974.

6. Roxin, E., Differential games with partial differential equations, in *Theory and Appl. of Differential Games*, Proc. NATO Adv. Study Inst. in Warwick, England, ed. J. D. Grote, D. Reidel Publ. Co., Dordrecht, Boston, 1975.

CONDITIONS FOR CONSTRAINED PARETO OPTIMA ON A BANACH SPACE WITH A FINITE NUMBER OF CRITERIA

Carl P. Simon*
Department of Mathematics
University of Michigan

1. INTRODUCTION

An important concept in mathematical models in economics, game theory, optimal control, and decision theory is that of a vector maximum or a Pareto optimum. Let $u_1,\ldots,u_p$ be a finite set of functions on a set X. (For example, X might be a space of commodity bundles or of sequences of commodity bundles and $u_j(x)$ may denote the jth consumer's level of satisfaction or utility with a given $x \in X$.) Let $P = \{\underline{x} \in \mathbb{R}^p | x_i \geq 0$ for all $i\}$, and write $\underline{x} \geq \underline{y}$ for $\underline{x},\underline{y} \in \mathbb{R}^p$ if $\underline{x} - \underline{y}$ is in P. A point $x^* \in X$ is called a vector maximizer or Pareto optimum (PO) of $u = (u_1,\ldots,u_p)\colon X \to \mathbb{R}^p$ if there is no y in X such that $u(y) \geq u(x^*)$ and $u(y) \neq u(x^*)$ in $\mathbb{R}^p$. If x^* has a neighborhood U in X such that x^* is a PO for $u|U$, then x^* is called a local Pareto optimum (LPO).

The purpose of this paper is to demonstrate how one can obtain rather strong necessary conditions and sufficient conditions for an element $x \in X$ to be a PO or LPO by first reducing a vector maximization problem to a system of scalar maximization problems and then using standard results in mathematical programming, as discussed in Luenberger [6] or Mangasarian [7], for example, This method has a number of advantages over the ad-hoc methods that are usually used in the recent literature on vector maximization. Not only does it give a unified and simple treatment for determining conditions for a PO put it also yields results which are at least as strong as and sometimes stronger than those obtained by the ad-hoc methods. (See especially section four.) In contrast to Simon [10], we will assume that X is a subset of a Banach space determined by a (possibly) infinite number of equality and inequality constraints.

In a later paper, we plan to develop the corresponding theory for u: X → Y where Y is a Banach space with a fixed positive cone. The technique will have to be modified to cover this more general setting; and judging by all the hypotheses used in recent papers in this setting, (e.g., Varsan [13]), one cannot expect results as strong as those described below.

The following result is our key to reducing vector maximization problems to systems of scalar maximization problems. Although I have heard that it is a folk theorem in optimization, I have not found any explicit statements of it in the literature. El-Hodiri [1] and Wan [14] use parts of this result in their discussions of vector maxima.

1. Proposition. Let X be any space and let $u: X \to \mathbb{R}^p$ be a mapping. Then u has a PO (LPO) at $x^\circ \in X$ if and only if x° (locally) maximizes each u_i on the constraint set

$$C_i \equiv \{x \in X | u_j(x) \geq u_j(x^\circ), j \neq i\}.$$

Proof: Suppose x° is a PO for u on X. If x° does not maximize u_k on C_k, then there exists $y \in X$ such that $u_j(y) \geq u_j(x^{\circ})$ for $j \neq k$, and $u_k(y) > u_k(x^{\circ})$, contradicting the Pareto optimatlity of x°.

Conversely, suppose x° maximizes each u_k on C_k. If x° is not a PO, there is a $y \in X$ such that $u_i(y) \geq u_i(x^{\circ})$ for all i and $u_h(y) > u_h(x^{\circ})$ for some h, contradicting the maximality of x° for u_h restricted to C_h.

Proposition 1 extends easily to the case where the range of u is a separable Hilbert space F with basis $\underline{e}_1, \underline{e}_2, \ldots$ and positive cone $P = \{\underline{w} \in F \mid \langle \underline{w}, \underline{e}_i \rangle \geq 0 \text{ for all } i\}$.

We now collect the assumptions that will be used in the rest of the propositions of this paper.

Assumption A: Let E,F,G, and H be Banach spaces. Suppose that P is a closed convex cone in G with its vertex at $\underline{0}$; and write $y \geq x$ for $x,y \in B$ if $y - x \in P$. Let $f: E \to \mathbb{R}$, $u: E \to \mathbb{R}^P$, $g: E \to G$, and $h: E \to H$ be continuous maps. Denote the constraint set by $C_{g,h} = \{x \in E \mid g(x) \geq 0 \text{ in } G \text{ and } h(x) = 0 \text{ in } H\}$.

2. FIRST ORDER NECESSARY CONDITIONS

To prove the basic first order necessary condition for an LPO, we will need the following Lemma which describes the weakest first order necessary condition for a constrained maximum. In the finite dimensional case, this Lemma was proved by F. John [4] for inequality constraints and by Mangasarian and Fromowitz [8] for equality-inequality constraints. In the infinite-dimensional case, Nagahisa-Sakawa [9] and Luenberger [6], for example, prove this Lemma for inequality constraints. One can add the equality constraints by using the technique described in Theorem 4.1 of Simon [10].

2. Lemma. Let $f:E \to \mathbb{R}$, $g:E \to G$, and $h:E \to H$ be C^1 maps as in Assumption A, where P has a non-empty interior. Suppose that: i) $x^\circ \in C_{g,h}$, ii) x° maximizes f on $C_{g,h}$, and iii) either the image of $Dh(x^\circ)$ is not dense in H or $Dh(x^\circ)$ is surjective and its kernel has a closed complement in E. Then, there exists a non-zero $(\alpha,\lambda,\mu) \in \mathbb{R} \times G^* \times H^*$ such that $\alpha \geq 0$ in $\mathbb{R}$, $\lambda \geq 0$ in G^*, $\langle\lambda,g(x^\circ)\rangle = 0$, and

$$\alpha Df(x^\circ) + \langle\lambda,Dg(x^\circ)\rangle + \langle\mu,Dh(x^\circ)\rangle = 0 \text{ in } E^*.$$

Remark: Of course, $\lambda \geq 0$ in G^* means that $\langle\lambda,y\rangle \geq 0$ for all $y \in P$ where $\langle\ ,\ \rangle$ denotes the usual pairing of G^* and G. Note that condition iii) on $Dh(x^\circ)$ in Lemma 2 holds automatically if H is finite dimensional.

To prove the corresponding first order necessary condition for a Pareto optimum of a mapping $u:E \to \mathbb{R}^P$, one notes that if x° is a PO for u on $C_{g,h}$, then by Proposition 1 x° maximizes u_1 on $\{x \in C_{g,h} \mid u_j(x) - u_j(x^\circ) \geq 0;\ j = 2,\ldots,p\}$. A straightforward application of Lemma 2 then yields the following Proposition.

3. Proposition. Let $u:E \to \mathbb{R}^p$, $g:E \to G$, and $h:E \to H$ be C^1 maps as in Assumption A. Assume that P has a non-empty interior. Suppose that i) $x^\circ \in C_{g,h}$ ii) x° is an LPO for u on $C_{g,h}$, and iii) either the image of $Dh(x^\circ)$ is not dense in H or $Dh(x^\circ)$ is surjective and its kernel has a closed complement in E. Then, there exists a non-zero $(\alpha,\lambda,\mu) \in \mathbb{R}^P \times G^* \times H^*$ such that

$$\alpha \geq 0 \text{ in } \mathbb{R}^P;\ \lambda \geq 0 \text{ in } G^*;\ \langle\lambda,g(x^\circ)\rangle = 0;$$

$$\text{and } \langle\alpha,Du(x^\circ)\rangle + \langle\lambda,Dg(x^\circ)\rangle + \langle\mu,Dh(x^\circ)\rangle = 0.$$

If some α_i is zero in Proposition 3, then the function u_i plays no role in the conclusion of Proposition 3 and the necessary condition has less impact. To guarantee that all the α_i's are strictly positive in Proposition 3, we must apply Proposition 1 to the situation where the constraint sets $C_1,\ldots,C_p$ satisfy the standard constraint qualifications of mathematical programming. We will only list a couple of these qualifications here. See Simon [10] for a more complete list in the finite-dimensional case.

4. Proposition. Let $u:E \to \mathbb{R}^p$, $g:E \to G$, and $h:E \to H$ be C^1 maps as in Assumption A. Suppose that $x^\circ \in C_{g,h}$ is an LPO for u on $C_{g,h}$. Let $u^{(i)} \equiv (u_1,\ldots,u_{i-1},u_{i+1},\ldots,u_p):E \to \mathbb{R}^{p-1}$ Suppose that one of the following hypotheses are satisfied for each $i \in \{1,\ldots,p\}$:

a) For each $v \in E$ such that $Dh(x^\circ)v = 0$, $g(x^\circ) + Dg(x^\circ)v \geq 0$, and $Du^{(i)}(x^\circ)v \geq 0$, there exists a C^1 $c:[0,\varepsilon) \to E$ with $c(0) = x^\circ$, $c'(0) = v$, $c(t) \in C_{g,h}$, and $u^{(i)}(c(t)) \geq u^{(i)}(x^\circ)$ for all $t \in [0,\varepsilon)$;

b) u and g are concave, h is linear, P has a non-empty interior $\mathring{P}$, and there exists $x^{(i)} \in E$ such that $h(x^{(i)}) = 0$, $g(x^{(i)}) \in \mathring{P}$, and $u_j(x^{(i)}) > u_j(x^\circ)$ for $j \neq i$;

c) $(Du^{(i)}(x^\circ),\ Dg(x^\circ),\ Dh(x^\circ)): E \to \mathbb{R}^{p-1} \times G \times H$ is onto, the kernel of $Dh(x^\circ)$ has a closed complement in E, and P has a non-empty interior. (If G is $\mathbb{R}^m$ with the usual positive cone, one can replace g by g_I where $I = \{j \mid g_j(x^\circ) = 0\}$.)

Then, there exists $(\alpha,\lambda,\mu) \in \mathbb{R}^p \times G^* \times H^*$ such that

$\alpha_i > 0$ for each i; $\lambda \geq 0$ in G^*; $<\lambda, g(x^\circ)> = 0$;

and $<\alpha, Du(x^\circ)> + <\lambda, Dg(x^\circ)> + <\mu, Dh(x^\circ) = 0$ in E^*.

Proof: By Proposition 1, x° maximizes each u_i on $u^{(i)} - u^{(i)}(x^\circ) \geq 0$, $g \geq 0$, $h = 0$. If a) or b) holds, Hurwicz [3] shows that one can find multipliers $(\alpha_1^i,\ldots,\alpha_p^i,\lambda^i,\mu^i) \in \mathbb{R}^p \times G^* \times H^*$ for each such scalar maximization problem so that the conclusion of Proposition 3 holds with $\alpha_i^i = 1$. Now, let $\alpha_k = \sum_{i=1}^{p} \alpha_k^i \geq 1$; let $\lambda = \sum_1 \lambda^i \geq 0$ in G*m and let $\mu = \sum_1^p \mu^i$.

If c) holds, choose (α,λ,μ) as in Proposition 3. If $\alpha_i = 0$, then the conclusion of Proposition 3 states that $(\alpha_1,\ldots,\alpha_{i-1},\alpha_{i+1},\ldots,\alpha_p\lambda,\mu)$ is a non-zero element in the kernel of the adjoint of $(Du^{(i)}(x^\circ), Dg(x^\circ), Dh(x^\circ)$, which contradicts the surjectivity in hypotheses c). See Chapter 6 of Luenberger [6].

3. FIRST ORDER SUFFICIENT CONDITIONS

Let E be a Banach space and let $f:E \to \mathbb{R}^1$. Then, f is quasi-concave if $f^{-1}[a,\infty)$ is convex for all $a \in \mathbb{R}^1$; and f is pseudoconcave if f is C^1 and $Df(x^\circ)(x'-x^\circ) \geq 0$ implies $f(x') \geq f(x^\circ)$. One checks that concavity ⇒ pseudoconcavity ⇒ quasiconcavity. (See Mangasarian [7], for example.) If G and P are as in Assumption A, then $g:E \to G$ is P-quasiconcave if $\{x|g(x) \in u + P\}$ is convex for all $y \in G$. Tagawa [12] proves the following result on concave programming.

5. Lemma. Let $f:E \to \mathbb{R}$, $g:E \to G$, and $h:E \to H$ be C^1 maps as in Assumption A with f pseudoconcave, g P-quasiconcave, and h linear. Let $x^\circ \in C_{g,h}$. Suppose there exists $\lambda \geq 0$ in G* and $\mu \in H^*$ such that $\langle\lambda,g(x^\circ)\rangle = 0$ and

$$Df(x^\circ) + \langle\lambda,Dg(x^\circ)\rangle + \langle\mu,Dh(x^\circ)\rangle = 0 \text{ in } E^*.$$

Then, x° maximizes f on $C_{g,h}$.

Actually, one need only find λ as in Lemma 5 so that $Df(x^\circ)v + \langle\lambda, Dg(x^\circ)v\rangle = 0$ for all v in the kernel of h. To prove the corresponding result for vector maxima, one simply uses Proposition 1 again and shows via Lemma 5 that x° maximizes <u>each</u> u_i subject to the constraints $u^{(i)} \geq u^{(i)}(x^\circ)$, $g \geq 0$, $h = 0$. Note that each $(u^{(i)}, g): E \to \mathbb{R}^{p-1} \times G$ is a quasiconcave with respect to the obvious product cone in $\mathbb{R}^{p-1} \times G$. The details of the proof of Proposition 6 are left to the reader. See Hurwicz [3] for a similar result where the range of u is infinite dimensional.

<u>6</u>. <u>Proposition</u>. Let $u: E \to \mathbb{R}^p$, $g: E \to G$, and $h: E \to H$ be C^1 maps as in Assumption A. Suppose that each u_i is pseudoconcave, g is P-quasiconcave, and h is linear. Suppose $x^\circ \in C_{g,h}$ and there exists $(\alpha_1, \ldots, \alpha_p; \lambda, \mu) \in \mathbb{R}^p \times G^* \times H^*$ such that

$$\text{each } \alpha_i > 0;\ \lambda \geq 0 \text{ in } G^*;\ \langle\lambda, g(x^\circ)\rangle = 0;$$

and $\langle\alpha, Du(x^\circ)\rangle + \langle\lambda, Dg(x^\circ)\rangle + \langle\mu, Dh(x^\circ)\rangle = 0$ in E^*

$$(\text{or } \langle\alpha, Du(x^\circ)\rangle + \langle\lambda, Dg(x^\circ)\rangle = 0 \text{ on kernel of } h).$$

Then, x° is a PO for u on $C_{g,h}$.

4. SECOND ORDER SUFFICIENT CONDITIONS

To prove second order sufficient conditions for an LPO, we will drop the condition that all the α_i's be non-zero but we will restrict ourselves to the case of finitely many inequality constraints. We will apply Theorem 1 to the following theorem on scalar maximization, which generalizes a result of Goldstine [2].

<u>7</u>. <u>Lemma</u>. Let $f: E \to \mathbb{R}$, $g: E \to \mathbb{R}^m$, and $h: E \to H$ be C^2 maps as in Assumption A with $P = \{\underline{x} \in \mathbb{R}^m | x_1 \geq 0, \ldots, x_m \geq 0\}$. Suppose that $x^\circ \in C_{g,h}$, that the kernel of $Dh(x^\circ)$ has a closed

complement in E, and that the image of $Dh(x^\circ)$ is closed in H. Suppose that there exist <u>non-negative</u> reals $\beta, \lambda_0, \lambda_1, \ldots, \lambda_m$ and that there exists $\mu \in H^*$ such that $\beta > 0$, $\lambda_i g_i(x^\circ) = 0$ for $i = 1,\ldots,m$, and such that if $L(x) \equiv \lambda_0 f(x) + \sum_1^m \lambda_i g_i(x) + \langle \mu, h(x)\rangle$, then $DL(x^\circ) = 0$ and $D^2L(x^\circ)(v,v) \le -\beta|v|^2$ for all $v \in E$ such that $Dh(x^\circ)\underline{v} = \underline{0}$ and $\lambda_i Dg_i(\underline{x}^\circ)v = 0$ for $i = 1,\ldots,m$. Then, f restricted to $C_{g,h}$ has a strict local maximum at x°.

<u>Proof</u>: (Note that the hypotheses on $Dh(x^\circ)$ hold automatically if H is finite dimensional.) Since Goldstine [2] assumes that $\lambda_0 > 0$, that there are no inequality constraints, and that $Dh(x^\circ)$ is onto (i.e., x° is "regular"), we will sketch a proof of Lemma 7. First, assume that $g \equiv 0$. Let K denote the kernel of $Dh(x^\circ)$ and let M be its closed complement in E. Let $T_1:E \to K$ and $T_2:E \to M$ be projection operators and assume without loss of generality that $||T_1|| \le 1$. Let N denote the image of $Dh(x^\circ)$ in H. Using Taylor's Theorem and the assumption that $DL(x^\circ) = 0$ and $h(x^\circ) = 0$, one can write

(1) $$L(x) - L(x^\circ) = D^2L(x^\circ)(x-x^\circ, x-x^\circ) + R_1(x)$$

where $\dfrac{R_1(x)}{|x-x^\circ|^2} \to 0$ as $x \to x^\circ$, and

(2) $$h(x) = Dh(x^\circ)(x-x^\circ) + R_2(x),$$

where $\dfrac{R_2(x)}{|x-x^\circ|} \to 0$ as $x \to x^\circ$. Let b denote the norm of $D^2L(x^\circ)$. By the Banach Inverse Theorem applied to $Dh(x^\circ):M \to N$ (e.g., see chapter six of Luenberger [6]), there is an $\varepsilon > 0$ such that if $|Dh(x^\circ)w| < \varepsilon$, then $|w| < \min\{\frac{\beta}{16b}, \sqrt{\frac{\beta}{8b}}\}$. By (1) and (2), there is a neighborhood U of x° in E such that if $x \in U$, then

$$(3) \qquad \frac{|R_1(x)|}{|x-x_0|^2} < \frac{\alpha}{4} \text{ and } \frac{|R_2(x)|}{|x-x_0|} < \varepsilon.$$

If $x \in U$ and $h(x) = 0$, let $v = \frac{x-x^\circ}{|x-x^\circ|}$. Then,

$$|Dh(x^\circ)v| = |Dh(x^\circ)(T_2v)| < \varepsilon \text{ and}$$

$$\begin{aligned} |D^2L(x^\circ)(v,v) - D^2L(x^\circ)(T_1v,T_1v)| \\ &\le 2|D^2L(x^\circ)(T_1v,T_2v)| + |D^2L(x^\circ)(T_2v,T_2v)| \\ &\le 2b||T_1|| \ |T_2v| + b|T_2v|^2 \\ &\le 2b\frac{\beta}{16b} + b\frac{\beta}{8b} = \frac{\beta}{4}. \end{aligned}$$

Then, since $D^2L(x^\circ)(T_1v,T_1v) < -\beta|T_1v|^2 < -\beta||T_1||^2 < -\beta$,

$$D^2L(x^\circ)(v,v) < -\frac{\beta}{2}.$$

Combining this with (1) and (3) yields, $\frac{L(x)-L(x^\circ)}{|x-x^\circ|^2} < -\frac{\beta}{4}$ if $x \in U$ and $h(x) = 0$. For such x, $L(x) = \lambda_0 f(x)$. Since $\lambda_0 \ge 0$, $f(x) < f(x^\circ)$ if $x \ne x^\circ$, $h(x) = 0$, and $x \in U$; and x° is a strict local maximizer of f on $h^{-1}(0)$. Notice that if $\lambda_0 = 0$, the above argument implies that $h^{-1}(0) \cap U = \{x^\circ\}$.

To include the inequality constraints $g \ge 0$, let $J = \{i \in \{1,\ldots,m\} | \lambda_i > 0\}$. If x° maximizes f subject to $g_J \ge 0$, $h = 0$, then x° maximizes f on $C_{g,h}$. Since both problems have the same Lagrangian, we can assume without loss of generality that $\lambda_i > 0$ (and thus $g_i(x^\circ) = 0$) for $i = 1,\ldots,m$. We now use a technique of Karush [5] to represent inequality constraints as equality constraints. Define $k: E \times \mathbb{R}^m \to \mathbb{R}^m$ by $k_i(x,z) = g_i(x) - z_i^2$. Define $\hat{f}: E \times \mathbb{R}^m \to \mathbb{R}$ and $\hat{h}: E \times \mathbb{R}^m \to H$ by $\hat{f}(x,z) = f(x)$ and $\hat{h}(x,z) = h(x)$. One checks easily that x° is a strict local max of f on $C_{g,h}$ if and only if there exists a $z^\circ \in \mathbb{R}^m$ such that (x°,z°) is a strict local max of $\hat{f}$ on $k^{-1}(0)$, $\hat{h}^{-1}(0)$. Since $g(x^\circ) = 0$, $z^\circ = 0$. If the weak regularity hypotheses hold for $Dh(x^\circ)$, they will hold for $D(k,\hat{h})(x^\circ)$. Furthermore, the $\lambda_0, \lambda_1, \ldots, \lambda_m, \mu$ of the hypothesis satisfy

$$D_{x,z}(\lambda_0 f + \langle\lambda,k\rangle + \langle\mu,\hat{h}\rangle)(x^\circ,0)$$
$$= (D_x(\lambda_0 f + \langle\lambda,g\rangle + \langle\mu,h\rangle)(x^\circ);0) = \underline{0} \text{ and}$$
$$D^2_{x,z}(\lambda_0 f + \langle\lambda,k\rangle + \langle\mu,h\rangle)(x^\circ,0)[(w_1,w_2),(w_1,w_2)]$$
$$= D^2_x(\lambda_0 f + \langle\lambda,g\rangle + \langle\mu,h\rangle)(x^\circ)(w_1,w_1) - 2\Sigma\, \lambda_i w_{2,i}^2$$
$$\le -\beta|w_1|^2 - 2\gamma|w_2|^2, \text{ where } \gamma = \min\{\lambda_1,\ldots,\lambda_m\} > 0,$$

for $(w_1,w_2) \in \ker D(k,\hat{h})(x^\circ,0)$, i.e., $w_1 \in \ker\,(Dg(x^\circ), Dh(x^\circ))$.

By the result of the previous paragraph, $\hat{f}$ restricted to $k = 0$, $\hat{h} = 0$ has a strict local maximum at $(x^\circ,0)$, i.e., f restricted to $C_{g,h}$ has a strict local maximum at x°. □

Combining Proposition 1 and Lemma 7, one obtains the following second order sufficient condition for x° to be a strict LPO which generalizes the corresponding result in Smale [11], Wan [14], and Varsan [13].

8. Proposition. Let $u:E \to \mathbb{R}^p$, $g:E \to \mathbb{R}^m$, and $h:E \to F$ be C^2 maps as in Assumption A with $P = \{x \in \mathbb{R}^m | x_1 \ge 0,\ldots,x_m \ge 0\}$. Suppose that $x^\circ \in C_{g,h}$, that the kernel of $Dh(x^\circ)$ has a closed complement in E, and that the image of $Dh(x^\circ)$ is closed in H. Suppose that there exist $\mu \in H^*$ and non-negative scalars $\beta, \alpha_1,\ldots,\alpha_p, \lambda_1,\ldots,\lambda_m$ such that $\beta > 0$, $\lambda_i g_i(x^\circ) = 0$ for $i = 1,\ldots,m$, and if

$$L(x) \equiv \sum_1^p \alpha_i u_i + \sum_1^m \lambda_j g_j + \langle\mu,h\rangle,$$

then $DL(x^\circ) = 0$ and $D^2L(x^\circ)(v,v) \le -\beta|v|^2$ for all $v \in E$ such that $\alpha_i Du_i(x^\circ)v = 0$, $i = 1,\ldots,p$; $\lambda_j Dg_j(x^\circ)v = 0$, $j = 1,\ldots,m$; and $Dh(x^\circ)v = 0$. Then, u restricted to $C_{g,h}$ has a strict LPO at x°.

Proof: By Proposition 1, we need only show that for each $i \in \{1,\ldots,p\}$ x° maximizes u_i locally on $\{x \in C_{g,h} | u_j(x) \geq u_j(x^\circ), j \neq i\}$. Work with $i = 1$ for simplicity of notation. Let $L' = \alpha_1 u_1 + \sum_2^p \alpha_i(u_i - u_i(x^\circ)) + \sum_1^m \lambda_i g_i + \langle\mu, h\rangle$. Then $DL'(x^\circ) = DL(x^\circ) = 0$. Choose non-zero $\underline{v}$ such that

$$\alpha_i D(u_i - u_i(x^\circ))(x^\circ)v = \alpha_i Du_i(x^\circ)v = 0, \ i = 2,\ldots,p,$$

$$\lambda_j Dg_j(x^\circ)v = 0, \ j = 1,\ldots,m, \text{ and } Dh(x^\circ)v = 0.$$

Since $DL(x^\circ)v = 0$, $\alpha_1 Du_1(x^\circ)v = 0$. By hypothesis, $D^2L'(x^\circ)(v,v) = D^2L(x^\circ)(v,v) < -\beta|v|^2$. By Lemma 7, u_1 has a strict local maximum at x° on $\{x \in C_{g,h} | u_j(x) \geq u_j(x^\circ), j > 1\}$. Since the same argument works for $i = 2,\ldots,p$, u restricted to $C_{g,h}$ has a strict LPO at x° by Proposition 1. □

It should be clear that if h is linear or affine in Lemma 7 or Proposition 8, then one can drop all the assumptions on the kernel and image of $Dh(x^\circ)$. One merely replaces E by the Banach space $h^{-1}(0)$ in the proofs.

REFERENCES

1. El-Hodiri, M., *Constrained Extrema: Introduction to the Differentiable Case with Economic Applications*, Springer, New York, 1971.

2. Goldstine, H., Minimum problems in functional calculus, *Bulletin of the American Math. Society 46*(1940), 142-149.

3. Hurwicz, L., Programming in linear spaces, in Arrow, Hurwicz, and Uzawa (eds.), *Studies in Linear and Non-linear Programming*, Stanford University Press, Stanford, California, 1958.

4. John, F., Extremum problems with inequalities as subsidiary conditions, in Friedrichs, Neugebauer, and Stoker (eds.), *Studies and Essays: Courant Anniversary Volume*, Interscience, New York, 1948, 187-204.

5. Karush, W., Minima of functions of several variables with inequalities as side conditions, Master's thesis, University of Chicago, 1939.

6. Luenberger, D., *Optimization by Vector Space Methods*, John Wiley, New York, 1969.

7. Mangasarian, O., *Nonlinear Programming*, McGraw-Hill, New York, 1969.

8. Mangasarian, O., and Fromowitz, S., The Fritz John necessary optimality conditions in the presence of equality and inequality constraints, *J. Math. Analysis and Applications 17*(1967), 37-47.

9. Nagahisa, Y., and Sakawa, Y., Nonlinear programming in Banach spaces, *Jour. Optimization Theory and Applications 4*(1969), 182-190.

10. Simon, C., Scalar and vector maximization: calculus techniques with economics applications, in S. Reiter (ed.), *MAA Studies in Mathematical Economics*, (1977), to appear.

11. Smale, S., Sufficient conditions for an optimum, in A. Manning (ed.), *Dynamical Systems - Warwick 1974*, Springer, New York, 1975, 287-292.

12. Tagawa, S., Generalized convexities of continuous functions and their applications to mathematical programming, *Bulletin Math. Statistics 16*(1974), 115-125.

13. Varsan, C., Sufficient conditions of optimality of second order in Banach spaces, *Rev. Roum. Math. Pures et Appl. 20*(1976), 239-251.

14. Wan, Y. H., On local Pareto optima, *Journal of Math. Economics 2*(1975), 35-42.

*Supported in part by NSF Grant MPS 75-08563

REMARKS ON EXISTENCE THEOREMS FOR PARETO OPTIMALITY

M. B. Suryanarayana
Department of Mathematics *
SUNY at Albany

1. INTRODUCTION

Optimization problems with vector valued cost criteria have been of considerable interest in the past, particularly in Economics. Existence theorems for such problems have been recently obtained by Cesari and Suryanarayana in [1], where the cost functional assumed values in a partially ordered Banach space. In [1], we have defined a weak extremum for a suitably bounded nonempty set and showed that under hypotheses, at least one extremum belongs to the attainable set. It is the purpose of this paper to show that a stronger conclusion can be drawn in the case where the Banach space is finite dimensional; namely, that every (weak) extremum of the set of values of the functional is "attained" (that is, belongs to the attainable set). More precisely, if Ω is the admissible class, then for any Pareto extremum i, $\{x,u)|I(x,u) = i\}$ is contained in Ω. This is in line with the notion of Pareto extremality found in Mathematical Economics.

We shall follow closely the notations and definitions of [1], and explicitly state here only those that are relevant. However, we restrict ourselves to the case where the functionals take values in a Euclidean Space. The proof of our result uses the multidimensional Fatou Lemma, (stated below in a revised form) proved by Schmeidler in [5] and also by Hildenbrandt and Mertens in [4].

2. A LOWER CLOSURE THEOREM FOR PARETO OPTIMALITY

Let (Y,d), (G,d_1) be two given metric spaces and let (G,α,μ) be also a measure space with a finite regular complete nonnegative measure μ. Let $(B,||\cdot||)$ be a Banach space and E^ρ be a Euclidean space. We shall consider the closed convex cone $\Lambda = \{(\eta,0) \in E^\rho \times B \mid \eta = (\eta^1,\ldots,\eta^\rho),\ \eta^i \geq 0,\ i = 1,\ldots,\rho\}$. Let $A \subset G\times y$ and $Q(t,y) \subset E^\rho \times B$, $(t,y) \in A$ be given. Let $\tilde{\tilde{Q}}(t,y) = Q(t,y) + \Lambda$ and $A(t) = \{y \mid (t,y) \in A\}$.

We say that the sets $\tilde{\tilde{Q}}(t,y)$ satisfy property (Q) with respect to y at (t_o,y_o) if

$$\tilde{\tilde{Q}}(t_o,y_o) = \cap_{\delta>0} \operatorname{clco} \cup\{\tilde{\tilde{Q}}(t_o,y) \mid d(y,y_o) \leq \delta,\ (t_o,y) \in A\}$$

The following version of multidimensional Fatou lemma is needed below and is an immediate consequence of that found in [5].

<u>2.1</u> <u>Lemma</u>. Let λ,λ_k,η_k, $k = 1,2,\ldots$ be integrable functions on a measurable space G (with finite measure), taking values in E^ρ such that $\eta_k - \lambda_k$ takes values in the positive orthant E^ρ_+ of E^ρ, that is $\eta^\ell_k \geq \lambda^\ell_k$, $\ell = 1,\ldots,\rho$. Let $\int_G \eta_k \to a \in E^\rho$ and $\lambda_k \to \lambda$ weakly in $L_1(G,E^\rho)$ as $k \to \infty$. Then there exists an integrable function η from G into E^ρ such that for almost all $t \in G$, $\eta(t)$ is a limit point of $\{\eta_k(t),\ k=1,2,\ldots\}$ and $\int_G \eta^\ell \leq a^\ell$, $\ell=1,\ldots,\rho$.

<u>2.1</u> <u>Remark</u>. If in the above lemma, $\{\int_G \eta_k,\ k=1,2,\ldots\}$ (instead of being convergent) is bounded in E^ρ, so that there are $a,b \in E^\rho$ with $a^\ell \leq \int_G \eta^\ell_k \leq b^\ell$, $= 1,\ldots,\rho$, then there is a convergent subsequence, which denote by [k] again, and an element $\Theta \in E^\rho$ with $\int_G \eta_k \to \Theta$ as $k \to \infty$. Using above lemma for

this subsequence, we obtain $\eta \in L_1(G,E^\rho)$ with $a^\ell \le \int_G \eta^\ell \le b^\ell$.

We now state and prove a lower closure theorem analogous to that found in [1].

<u>2.1</u> <u>Theorem</u>. Let $T_o \in \alpha$ with $\mu(T_o) = 0$ be such that $A(t)$ is closed for $t \in G-T_o$ and let $\tilde{\tilde{Q}}(t,y)$ satisfy property (Q) with respect to y in $A(t)$ for $t \in G-T_o$. Let $\xi(t), y(t), \lambda(t), \eta_k(t)$, $\xi_k(t), y_k(t), \lambda_k(t)$, $t \in G$, $k = 1,2,\ldots$, be a.e. finite measurable functions $\xi, \xi_k \in L_1(G,B), \lambda, \lambda_k, \eta_k \in L_1(G,E^\rho)$ such that $y_k(t) \in A(t)$, $(\eta_k(t), \xi_k(t)) \in \tilde{\tilde{Q}}(t, y_k(t))$, $t \in G$(a.e), $k = 1,2,\ldots$. Let $j_k = \int_G \eta_k(t)\,d\mu$, $k = 1,2,\ldots$ and let $i = \lim_{k\to\infty} j_k \in E^\rho$. Let $\xi_k \to \xi$ weakly in $L_1(G,B)$, $\lambda_k \to \lambda$ weakly in $L_1(G,E^\rho)$, $y_k \to y$ in measure in G, as $k \to \infty$. Let $\eta_k^\ell(t) \ge \lambda_k^\ell(t)$, $\ell = 1,\ldots,\rho$, $k = 1,2,\ldots$, $t \in G$ (so that $(\eta_k(t),0) \in (\lambda_k(t),o) + \Lambda$ for $t \in G$).

Then there exists $\eta(t)$, $t \in G$, $\eta \in L_1(G,E^\rho)$ such that $y(t) \in A(t)$, $(\eta(t),\xi(t)) \in \tilde{\tilde{Q}}(t,y(t))$ and $i^\ell \ge \int_G \eta^\ell(t)\,d\mu$, $\ell = 1,\ldots,\rho$.

<u>Proof</u>. The proof is same as that found in [1], except for appropriate modifications. By taking suitable subsequence we may take $y_k(t) \to y(t)$, $y(t)$ finite for all $t \in G-(T_o \cup T_1')$ with $\mu(T_o') = 0$. Since $j_k \to i$ as $k \to \infty$ and $i \in E^\rho$, we have $\mu_s = \max\{\|j_k - i\|, k \ge s+1\} \to 0$ as $s \to \infty$. For each $s = 1,2,\ldots$, using the weak convergence of λ_{s+k}, ξ_{s+k}, $k = 1,2,\ldots$, to λ and ξ and using Mazur Theorem, we obtain a set of real numbers $C_{sNk} \ge 0$, $k = 1,\ldots,N$, $N = 1,2,\ldots$, with $\sum_{k=1}^N C_{sNk} = 1$ such that if $\lambda_{sN}(t) = \sum_{k=1}^N C_{sNk}\,\lambda_{s+k}(t)$ and $\xi_{sN}(t), \eta_{sN}(t)$ are defined similarly for $t \in G$, $N = 1,2,\ldots$, then $\lambda_{sN} \to \lambda$ in $L_1(G,E^\rho)$, $\xi_{sN} \to \xi$ is $L_1(G,B)$ strongly and $\eta_{sN}^\ell(t) \ge \lambda_{sN}^\ell(t)$ for almost all t in G, as well as $i^\ell - \mu_s \le \int \eta_{sN}^\ell(t)\,dt \le i^\ell + \mu_s$, $\ell = 1,\ldots,\rho$.

This is true for each $s = 1,2,\ldots$. Thus, for each s, there is a subset $T_s \subset G, \mu(T_s) = 0$ and a subsequence N_h (depending possibly on s) such that $\int_G \eta_{sN_h} \to \Theta_s \in E^\rho$, with $i^\ell - \mu_s \le \Theta_s^\ell \le i^\ell + \mu_s$, $\ell = 1,\ldots,\rho$ and $\lambda(t), \xi(t)$ are finite and $\xi_{sN_h}(t) \to \xi(t), \lambda_{sN_h}(t) \to \lambda(t)$ as $h \to \infty$ for $t \in G-T_s$. By using Fatou Lemma 2.1 above, there is an integrable function $\eta_s \in L_1(G,E^\rho)$ with $\eta_s(t) \in \mathrm{cl}\{\eta_{sN_h}(t), h=1,2,\ldots\}$ and finite for all $t \in G-T_s'$ with $\mu(T_s') = 0$. Also, $\eta_s^\ell(t) \ge \lambda^\ell(t)$ and

$$\int_G \lambda^\ell(t)d\mu < \int_G \eta_s^\ell(t)d\mu \le \Theta_s^\ell \le i^\ell + \mu_s, \ell = 1,\ldots,\rho, s = 1,2,\ldots$$

Since $\int_G \eta_s$ is bounded in E^ρ, there is a convergent subsequence $\{s_\sigma\}$ so that $\int_G \eta_{s_\sigma} \to \xi = (\xi^1,\ldots,\xi^\rho)$ with $\xi^\ell \le i^\ell$, $\ell = 1,\ldots,\rho$.

Using Fatou lemma above again, we obtain an integrable function $\eta \in L_1(G,E^\rho)$ with $\eta(t) \in \mathrm{cl}\{\eta_{s_\sigma}(t), \sigma=1,2,\ldots\}$ and $\eta(t)$ is finite for all $t \in G-T_o''$ with $\mu(T_o'') = 0$. Also $\int_G \eta^\ell(t)d\mu \le \xi^\ell \le i^\ell$, $\ell = 1,\ldots,\rho$. Let T denote the union of all sets $T_o, T_o', T_o'', T_s, T_s'$, $s = 1,2,\ldots$, so that $\mu(T) = 0$. For $t_o \in G-T$, then $(t_o, y_k(t_o)) \to (t_o, y_o) \in A$ where $y_o = y(t_o)$. The sets $\tilde{\tilde{Q}}(t_o,y)$ have property (Q) with respect to y at y_o. Given $\varepsilon > 0$ there is an integer s_o such that $d(y_s(t_o), y_o) \le \varepsilon$ for $s \ge s_o$. Let σ_o be such that $\sigma \ge \sigma_o$ implies $s_\sigma \ge s_o$. Then, $(\eta'_{\sigma,k}(t_o), \xi'_{\sigma,k}(t_o)) \in \tilde{\tilde{Q}}(t_o, y'_{\sigma,k}(t_o))$ and $d(y'_{\sigma,k}(t_o), y_o) \le \varepsilon$ for all $k = 1,2,\ldots$ and $\sigma \ge \sigma_o$ where $\eta'_{\sigma,k} = \eta_{s_\sigma+k}$ and similarly ξ', y' are defined. Thus, $(\eta_{s_\sigma N}(t_o), \xi_{s_\sigma N}(t_o))$ belongs to co K_ε where K_ε stands for $\cup\{\tilde{\tilde{Q}}(t_o,y) \mid d(y,y_o) \le \varepsilon\}$. Hence $(\eta_{s_\sigma}(t_o), \xi_{s_\sigma}(t_o)) \in \mathrm{cl\ co}\ K_\varepsilon$. This is true for $\sigma \ge \sigma_o$. Since $\eta(t_o) \in \mathrm{cl}\{\eta_{s_\sigma}(t_o)\}$ we have $(\eta(t_o), \xi(t_o)) \in \mathrm{cl\ co}\ K_\varepsilon$. This is true for each ε. Taking intersection over ε and using property (Q) we get $(\eta(t_o),\xi(t_o)) \in \tilde{\tilde{Q}}(t_o,y_o)$. This concludes the proof.

2.2 Remark. In the above theorem we required property (Q) of $\tilde{\tilde{Q}}(t,y)$ while in the corresponding Theorem (Theorem 7.1) of [1] we could require property (Q) merely of the sets $a\cdot \tilde{\tilde{Q}}(t,y)$ (where a is the element appearing in the definition of Λ). The reason is that here in the conclusion, we obtain inequality in each coordinate. It is known that if $\tilde{\tilde{Q}}(t,y) \subset E^{\rho+r}$ then property (Q) of $a\cdot \tilde{\tilde{Q}}(t,y)$ for all a in some basis of $E^{\rho+r}$ implies property (Q) of $\tilde{\tilde{Q}}(t,y)$, (see [7]).

2.3 Remark. Several variations of above lower closure theorem are analogously proved by suitable modifications of the corresponding theorems in [3,6]. We state here some of these variations.

2.2 Theorem. (A lower closure theorem without property (Q)). Let Y,G,B, and A be as in theorem 2.1. Let $\Lambda = \{(\lambda,0) \mid \lambda = (\lambda^1,\ldots,\lambda^\rho),\ \lambda^\ell \geq 0, \ell = 1,\ldots,\rho\}$. Let $\tilde{\tilde{Q}}(t,y) \subset E^\rho \times B$ be given nonempty closed sets for almost all t and $y \in A(t)$. Let $y(t)$, $\xi_k(t)$, $\xi(t)$, $\eta_k(t)$, $\eta(t)$, $\overline{\eta}_k(t)$, $\overline{\xi}_k(t)$, $\lambda_k(t)$, $\lambda(t)$ be a.e. finite measurable functions on G. Let $\xi,\xi_k,\overline{\xi}_k \in L_1(G,B)$, $\lambda,\lambda_k \in L_1(G,E^\rho)$, $y(t) \in A(t)$, $(\overline{\eta}_k(t),\overline{\xi}_k(t)) \in \tilde{\tilde{Q}}(t,y(t))$, $t \in G$(a.e), $k = 1,2,\ldots.$ Let $\eta_k^\ell(t) \geq \lambda_k^\ell(t)$, $\ell = 1,\ldots,\rho$, $k = 1,2,\ldots$ and let $\xi_k \to \xi$ weakly in $L_1(G,B)$, $\lambda_k \to \lambda$ weakly in $L_1(G,E^\rho)$. Let $\gamma_k = \eta_k - \overline{\eta}_k$ and $\delta_k = \xi_k - \overline{\xi}_k$. Let $(\gamma_k,\delta_k) \to 0$ in measure in G as $k \to \infty$. Also, let $j_k = \int_G \eta_k(t)\,d\mu \to i \in E^\rho$ as $k \to \infty$. Finally, let $\tilde{\tilde{Q}}(t,y(t))+\Lambda$ be convex for almost all $t \in G$. Then, there exists a function $\eta(t)$, $t \in G$, $\eta \in L_1(G,E^\rho)$ such that $(\eta(t),\xi(t)) \in \tilde{\tilde{Q}}(t,y(t))+\Lambda$, $t \in G$ (a.e.), and such that $i^\ell \geq \int_G \eta^\ell(t)\,d\mu$, $\ell = 1,\ldots,\rho$.

With the use of McShane-Warfield implicit function lemma, we could obtain analogs of theorems 2.1, 2.2, for control problems. To this end we introduce the control space (U,d_2) with the metric d_2. With notations as before, let $U(t,y)$ be a closed subset of U for each $(t,y) \in A$. Let $S = \{(t,y,u) \mid (t,y) \in A,\ u \in U(t,y)\}$ and $f:S \to B$, $g:S \to E^\rho$ be two given functions. We say that (A,S,f,g) (or simply f,g) satisfy property (C) if for each $\varepsilon > 0$ there is compact subset K of G with $\mu(G-K) < \varepsilon$ such that the sets $A_K = \{(t,y) \mid t \in K\}$ and $S_K = \{(t,y,u) \in S \mid t \in K\}$ are closed and f,g restricted to S_K are continuous. We introduce the sets $\tilde{\tilde{Q}}(t,y) = \{(z^0,z) \in E^\rho \times B \mid z^0 = g(t,y,u),\ z = f(t,y,u)\ u \in U(t,y)\}$ and $\tilde{Q}(t,y) = \tilde{\tilde{Q}}(t,y) + \Lambda$ where $\Lambda = \{(\lambda^1,\dots,\lambda^\rho,0) \in E^\rho \times B \mid \lambda^i \ge 0,\ i=i,\dots,\rho\}$.

2.3 Theorem. Let G,A,S,U,B,f,g be as above satisfying condition (C). Let $T_o \in \alpha$, $\mu(T_o) = 0$ be such that for $t \in G-T_o$, $A(t)$ is closed, $\tilde{Q}(t,y)$ is closed and convex and satisfy property (Q) with respect to $y \in A(t)$. Let $\xi_k(t),\xi(t)$, $y_k(t),y(t),u_k(t),\eta_k(t),\phi_k(t),\phi(t)$, $t \in G$ be measurable functions on G such that

(i) $y_k(t) \in A(t)$, $u_k(t) \in U(t,y_k))$, $\xi_k(t) = f(t,y_k(t),u_k(t))$, $\eta_k(t) = g(t,y_k(t),u_k(t))$, μ a.e. in G, $k = 1,2,\dots$.

(ii) $\xi_k,\xi \in L_1(G,B)$, $\eta_k,\phi_k,\phi \in L_1(G,E^\rho)$

(iii) $-\infty < i^\ell = \lim_{k\to\infty} \int \eta_k^\ell(t)\,d\mu < \infty$, $\ell = 1,\dots,\rho$.

(iv) $\xi_k \to \xi$ weakly in $L_1(G,B)$, $\phi_k \to \phi$ weakly in $L_1(G,E^\rho)$, $y_k \to y$ in measure on G, as $k \to \infty$,

(v) $\eta_k^\ell(t) \ge \phi_k^\ell(t)$, $\ell = 1,\dots,\rho$.

Then there are μ-measurable functions $\bar{\eta}(t),u(t),t \in G$ such that $\bar{\eta} \in L_1(G,E^\rho)$, $y(t) \in A(t)$, $u(t) \in U(t,y(t))$, and if $\eta(t)$ denotes $g(t,y(t),u(t))$ then $\bar{\eta}^\ell(t) \ge \eta^\ell(t)$, $\ell = 1,\dots,\rho$,

$\xi(t) = f(t,y(t),u(t))$ μ-a.e. in G and

$$i^{\ell} \geq \int_G \eta^{\ell}(t)d\mu, \quad \ell = 1,\dots,\rho.$$

We can guarantee that $\eta \in L_1(G,E^{\rho})$ if there exists $\phi = (\phi^1,\dots,\phi^{\rho}) \in L_1(G,E^{\rho})$ such that $g(t,y,u) \geq \phi^{\ell}(t)$ for all $(t,y,u) \in S$.

2.4 Remark. In theorem 2.2 we avoided the requirement of property (Q) by assuming the existence of a particular sequence $(\overline{\xi}_k,\overline{\eta}_k)$. In the following theorem we reduce the requirement by a different observation, in the case where $B = E^r$.

2.4 Theorem. Let $Q(x) \subset E^{\rho+r}$ and $\tilde{\tilde{Q}}(x) = Q(x) + \Lambda$ where $\Lambda = \{(\lambda^1,\dots,\lambda^{\rho},0,\dots,0) \mid \lambda^i \geq 0,\ i=1,\dots,\rho\}$. Let R be so chosen that $\tilde{\tilde{Q}}(x) \cap \{E^{\rho}\times S_R\}$ is nonempty for all $x \in A$, where S_R is the sphere $\{z \mid |z| \leq R\}$ of radius R around origin. If the sets $\tilde{\tilde{Q}}(x)$ are closed and convex (that is Q(x) is Λ-convex) and if the sets $\tilde{\tilde{Q}}(x) \cap \{E^{\rho}\times S_R\}$ are uniformly bounded below and satisfy property (K) for all $x \in A$, then the sets $\tilde{\tilde{Q}}_R(x) = \tilde{\tilde{Q}}(x) \cap \{E^{\rho}\times S_R\}$ satisfy property (Q) for all $x \in A$. [Here, the term "uniformly bounded below" means that there is some $\lambda_0 \in \Lambda$ such that for all $(y,z) \in \tilde{\tilde{Q}}_R(x)$, we have $(y,o)+\lambda_0 \in \Lambda$]. We use the following lemma in the proof of the theorem.

2.2 Lemma. If x_k, y_k, z_k, $k=1,2,\dots$, are sequences in E^n such that $x_k = y_k + z_k$, $z_k^i \geq 0$, $y_k^i \geq m^i$, $i = 1,\dots,n, k = 1,2,\dots$, and $x_k \to x \in E^n$ as $k \to \infty$, then there is a $y \in E^n$ and a subsequence $\{k_h\}$ such that $y_{k_h} \to y$ as $h \to \infty$.

Proof. Since $x_k \to x$, $x_k^i \to x^i$, $i = 1,\dots,n$. Thus, given $\varepsilon > 0$ there is a k_i such that $k \geq k_i$ implies $x_k^i \leq x^i+\varepsilon$. Let $k_0 = \max(k_1,\dots,k_n)$. Then $x_k^i \leq x^i+\varepsilon$ for all $k \geq k_0$ and $i = 1,\dots,n$. Now, $m^i \leq y_k^i = x_k^i - z_k^i \leq x_k^i \leq x^i + \varepsilon$ for all

$k \geq k_o$ and $i = 1,\ldots,n$. Thus $\{y_k^i\}$ is bounded and hence so is $\{y_k\}$. Thus $\{y_k\}$ has a convergent subsequence $\{y_{kh}\}$ with limit y. Thus, $y_{k_h} \to y$ as $k \to \infty$.

<u>Proof of Theorem 2.4</u>. Let $\tilde{\tilde{Q}}_R(x)$ denote $\tilde{\tilde{Q}}(x) \cap E^\rho \times S_R$. Let $(y,z) \in \cap_{\delta>0}$ clco $\cup \{\tilde{\tilde{Q}}_R(x) | x \in N_\delta(x_o)\}$. Then there are real numbers $c_{ik} \geq 0$, $i = 1,\ldots,K = r + \rho + 1$, $k = 1,2,\ldots,$ $|x_{ik}-x_o| < k^{-1}$, such that $\sum_{i=1}^{K} c_{ik} = 1$, $k = 1,2,\ldots$ and $(y_k,z_k) \to (y,z)$ where $y_k = \sum_{i=1}^{K} c_{ik}y_{ik}$, $z_k = \sum_{i=1}^{K} c_{ik}z_{ik}$ and $(y_{ik},z_{ik}) \in \tilde{\tilde{Q}}_R(x_{ik})$. Since $|z_k| \leq R$, we may assume that $c_{ik} \to c_i \geq 0$ and $z_{ik} \to z_i \in E^r$ as $k \to \infty$, $i = 1, \ldots,K$. Since $y_{ik}^j + \lambda_o^j \geq 0$, $j = 1,\ldots,\rho$ and $\sum_{i=1}^{K} c_{ik}(y_{ik}^j+\lambda_o^j) \to y^j + \lambda_o^j$ as $k \to \infty$. Using the lemma above, we may also assume that $c_{ik}(y_{ik}^j+\lambda_o^j) \to w_i^j \geq 0$ as $k \to \infty$, $i = 1,\ldots,K$. If $c_i > 0$ then $y_{ik}^j \to y_i^j$ where $y_i^j = c_i^{-1}w_i^j - \lambda_0^j$. Thus, for all i with $c_i > 0$, $(y_i,z_i) \in \tilde{\tilde{Q}}_R(x_o)$ because the sets $\tilde{\tilde{Q}}(x)$ have property (K) at x_o. Thus $\sum_{c_i>0} c_i(y_i,z_i) \in \text{co}\, \tilde{\tilde{Q}}_R(x_o) = \tilde{\tilde{Q}}_R(x_o)$. Furthermore, $y^j+\lambda_o^j = \lim_{k\to\infty} \sum_{i=1}^{k} c_{ik}(y_{ik}^j+\lambda_o^j) = \sum_{i=1}^{k} w_i^j = \sum_{c_i>0} c_i(y_i^j+\lambda_o^j) + \sum_{c_i=0} w_i^j \geq \sum_{c_i>0} c_i y_i^j + \lambda_o^j$ and $z = \lim_{k\to\infty} \sum_{i=1}^{K} c_{ik}z_{ik} = \sum_{i=1}^{K} c_i z_i = \sum_{c_i>0} c_i z_i$. Thus, $y^j \geq \sum_{c_i>0} c_i y_i^j$ and $(y,z) \in \tilde{\tilde{Q}}_R(x_o)$ since $\tilde{\tilde{Q}}_R(x_o) + \Lambda \subseteq \tilde{\tilde{Q}}_R(x_o)$.

<u>2.5 Remark</u>. In view of the above theorem, in the case $B = E^r$, we may replace the requirement of property (Q) in Theorem 2.1, by the following milder assumption.

There exist measurable functions $\mu_k(t), \mu(t), \rho_k(t), \rho(t)$, $k = 1,2,\ldots,$ defined on G such that (i) $(\mu_k(t),\rho_k(t)) \in \tilde{\tilde{Q}}(t,y_k(t))$ a.e. in G (ii) $\rho,\rho_k \in L_1(G,E^r), \mu,\mu_k \in L_1(G,E^\rho)$ (iii) $\rho_k \to \rho$ strongly in $L_1(G,E^r)$ and $\mu_k \to \mu$ weakly in $L_1(G,E^\rho)$ (iv) $\lambda_k \to \lambda$ strongly in $L_1(G,E^\rho)$ (where λ_k, λ are as in Theorem 2.1). Finally, the sets $\tilde{\tilde{Q}}(t,y)$ satisfy Kuratowski property (K) with respect to $y \in A(t)$, $t \in G$(a.e), that is for almost all $t_o \in G$, and for all $y_o \in A(t_o)$ we have $\tilde{\tilde{Q}}(t_o,y_o) = \cap_{\delta>o} \text{cl} \cup \{\tilde{\tilde{Q}}(t_o,y) \mid |y-y_o| \leq \delta, y \in A(t_o)\}$ (It is to be noted that here $\tilde{\tilde{Q}}(t,y) = Q(t,y) + \Lambda$ and the so called "upper set property," used in case $\rho = 1$, is automatically verified).

3. MAYER PROBLEMS

Let G,A,S,U,B,f be as in Section 2, (see Theorem 2.4 in particular). Let $L_1(G,B)$ denote the set of all measurable functions $z(t)$ from G into B such that $\int_G ||z(t)||\, d\mu < \infty$. Let $\mu(G,U)$ denote the set of all μ-measurable functions on G with values in U. Let (X,τ) denote a topological space and let $X_o \subset X$. Let L and M be given (not necessarily linear) operators, $L:X_o \to L_1(G,B), M:X_o \to \mu(G,U)$. We say that a pair (x,u) is admissible (for a Mayer problem) if $x \in X_o$, $u \in \mu(G,U)$, $Mx(t) \in A(t)$, $u(t) \in U(t,Mx(t))$ and $Lx(t) = f(t,Mx(t),u(t))$, μ a.e. in G.

We shall consider a nonempty class Ω of admissible pairs. We assume that Ω is closed, that is whenever $(x_k,u_k) \in \Omega, x_k \to x$ in X, $x \in X_o$, and (x,u) is admissible for some u, then there exists $\bar{u}$ such that $(x,\bar{u}) \in \Omega$.

For $(t,y) \in A$, we define the sets $Q(t,y) = \{f(t,y,u) \mid u \in U(t,y)\}$ and say as before that the sets $Q(t,y)$ satisfy property (Q) with respect to $y \in A(t)$, if for any $t_o \in G$, and $y_o \in A(t_o)$, we have $Q(t_o,y_o) = \cap_{\delta>o} \text{cl co} \cup \{Q(t_o,y) \mid y \in A(t_o), d(y,y_o) \leq \delta\}$

The following Lemma needed below, is proved in [1].

3.1 Lemma. Let Z be any Banach space and let Λ be a closed convex cone in Z satisfying the following property (π): there is an element $a \in Z^*$, dual of Z, such that (i) $a\lambda \geq 0$ for all $\lambda \in \Lambda$ and (ii) for each $\delta > 0$ the set $\{\lambda \in \Lambda | a\lambda \leq \delta\}$ if nonempty is relatively weakly compact in Z. Then for any nonempty subset A of Z with $A \subset c + \Lambda$ for some $c \in Z$, the set of weak Λ-Pareto extrema is nonempty.

3.1 Theorem. Let A, S, f satisfy condition (C) and let Ω be a nonempty closed class. Let $\{x\}_\Omega = \{x | (x,u) \in \Omega$, for some $u\}$ be sequentially relatively compact in X. Let $I: \{x\}_\Omega \to E^\rho$, be a ρ-vector valued functional whose coordinates $I^1, \ldots, I^\rho$ satisfy the following two conditions: (i) there exists $c = (c^1, \ldots, c^\rho) \in E^\rho$ with $I^\ell(x) \geq c^\ell$, $\ell = 1, \ldots, \rho$ and (ii) whenever $x_k \in \{x\}_\Omega$, $x_k \to x$ in X as $k \to \infty$ and $x \in \{x\}_\Omega$ then $I^\ell(x) \leq \liminf_{k\to\infty} I^\ell(x_k)$, that is I^ℓ is lower semicontinuous for every ℓ. We assume that whenever $x_k \in X_0, x_k \to x_0$ in X as $k \to \infty$, we have a subsequence $[k_s]$ for which Lx_{k_s}, $s = 1,2,\ldots,$ converges to Lx_0 weakly in $L_1(G,B)$ and Mx_{k_s}, $s = 1,2,\ldots,$ converges in measure to Mx_0. Finally, let the sets $Q(t,y)$ satisfy property (Q) with respect to y in $A(t)$ for almost all $t \in G$.

When every Pareto minimum for I belongs to $\{x\}_\Omega$.

Proof. By hypothesis $I^j(x) \geq c^j$ for all $x \in \{x\}_\Omega$ so that the attainable set $A = I(\{x\}_\Omega)$ is nonempty and "Λ-bounded" in E^ρ with $\Lambda = \{\lambda \in E^\rho | \lambda^i \geq 0, i=1,\ldots,\rho\}$. Hence, by Lemma 3.1 above, Pareto extrema exist. Let z_0 be any Pareto extremum for A so that there exist $z_k = I(x_k)$, $k = 1,2,\ldots,$ converging (coordinate wise) to z_0 as $k \to \infty$. Since $\{x\}_\Omega$ is compact and $x_k \in \{x\}_\Omega$, we have (after extracting a subsequence if needed) $Lx_k \to Lx_0$ weakly $L_1(G,B)$ and $Mx_k \to Mx_0$ in measure as $k \to \infty$.

By the lower semicontinuity of I if $\bar{z}$ denotes $I[x_0]$ then $\bar{z} \le z_0$, that is $\bar{z}^j \le z_0^j$, $j = 1,\dots,\rho$. Using lower closure Theorem 2.3 with $\eta_k = 0$, $\phi_k = 0$, we observe that there is a function $u(t)$ measurable on G such that (x_0,u) is admissible. Since Ω is closed, there is a $\bar{u}$ with $(x_0,\bar{u}) \in \Omega$ or $x_0 \in \{x\}_\Omega$ and $\bar{z} \in A$. But since z_0 is Pareto optimal, $\bar{z} \in A$ and $\bar{z} \le z_0$ imply $\bar{z} = z_0$ and thus $z_0 = I[x_0] \in A$. This completes the proof.

4. LAGRANGE PROBLEMS

With G,A,S,U,B,f as in section 2, let $g(t,y,u)$ be a function defined on S with values in E^ρ. Let A,S,f,g satisfy condition (C). We consider operators L and M as in section 3, $L:X_0 \to L_1(G,B)$ $M:X_0 \to \mu(G,U)$ with $X_0 \subset X$. We consider the sets $\tilde{\tilde{Q}}(t,y) = \{(z^0,z) \in E^\rho \times B \mid z^{0j} \ge g_j(t,y,u),\ j=1,\dots\rho,$ $z=f(t,y,u),\ u \in U(t,y)\}$ as in Theorem 2.3.

We say that a pair (x,u), $x \in X_0$, $u \in \mu(G,U)$ is admissible provided $Mx(t) \in A(t)$, $u(t) \in U(t,Mx(t))$, $Lx(t) = f(t,Mx(t),u(t))$, μ.a.e. in G, and $||g(t,Mx(t),u(t))||$ is an L_1-function on G. Let Ω be a nonempty class of admissible pairs. Let $I:\Omega \to E^\rho$ be defined as $I[X,u] = \int_G g(t,Mx(t),u(t))d\mu$. We shall say that the class Ω is closed (with respect to Lagrange problem and functional I) provided whenever $(x_k,u_k) \in \Omega$, $k = 1,2,\dots$, $x_k \to x$ in (X,τ) as $k \to \infty$, $x \in X_0$ and there is some $u \in \mu(G,U)$ such that (x,u) is admissible then there is also some $\bar{u} \in \mu(G,U)$ such that $(x,\bar{u}) \in \Omega$ and $I^j(x,\bar{u}) \le I^j(x,u)$ for all $j = 1,\dots,\rho$.

We say that a condition (α_0) holds, if there are functions $\phi_j(t) \in L_1(G)$, such that $\phi_j(t) \le g_j(t,y,u)$ for all $(t,y,u) \in S$, $j = 1,\dots,\rho$.

4.1 Theorem. Let G,S,U,A,B,f and g be as above. Let condition (α_o) hold and let A,S,f,g satisfy condition (C). Let Ω be a nonempty class of admissible pairs, closed (with respect to I). We shall assume that whenever $x_k \in X_o$, and $x_k \to x$ in X as $k \to \infty$, we can extract a subsequence $[k_s]$, such that $Lx_{k_s} \to Lx$ weakly in $L_1(G,B)$ and $Mx_{k_s} \to Mx$ in measure. Further let $\{x\}_\Omega$ be sequentially relatively compact in X. Finally, let the sets $\tilde{\tilde{Q}}(t,y)$ have property (Q) with respect y in $A(t)$. Then, the E^ρ-valued functional $I[x,u]$ attains its Pareto minimum in Ω. That is, if $A = I(\Omega)$ and z_o is any Pareto extremum of A, then there exists an (x_o,u_o) in Ω such that $I[x_o,u_o] = z_o$.

Proof. As in the proof of Theorem 3.1, let us observe that the set $A = I(\Omega)$ is Λ-bounded with $\Lambda = \{\lambda \in E^\rho \mid \lambda^i \geq 0, i=1,\dots,\rho\}$ due to condition (α_o). Also A is nonempty. If z_o is any Pareto extremum of A (which exists due to Lemma 3.1 above), then $z_o \in w\text{-}cl(A) = cl(A)$ and there exists a sequence $z_k \in A$, $z_k \to z_o$ as $k \to \infty$. Let $z_k = I[x_k,u_k]$. Since $x_k \in \{x\}_\Omega$ which is compact, there is a $x_o \in X$ with $x_k \to x_o$. But then $Lx_k \to Lx_o$ weakly and $Mx_k \to Mx_o$ in measure. Now, by the lower closure Theorem 2.4, there exist measurable functions $\bar{\eta}(t), u(t), t \in G$ such that $y(t) = Mx_o(t) \in A(t)$, $u(t) \in U(t,y(t))$ and if $\eta(t)$ denotes $g(t,Mx(t),u(t))$ then $\bar{\eta}^\ell(t) \geq \eta^\ell(t)$, $\ell = 1,\dots,\rho$ and $\xi(t) = Lx_o(t) = f(t,Mx_o(t),u(t))$ μ a.e. in G and

$$\int_G \eta^\ell(t)\,d\mu \leq \lim \int_G g_\ell(t,Mx_o(t),u(t))\,d\mu, \quad \ell = 1,\dots,\rho.$$

Also, $\eta \in L_1(G,E^\rho)$. But then $\bar{z} = \int_G \eta \leq z_o$ and $\bar{z} \in A$. By Pareto optimality $\bar{z} = z_o$. Thus, $z_o \in A$. This concludes the proof.

4.1 Remark. The above results are valid in a finite dimensional Banach Space B and for any closed convex cone Λ which is polyhedral.

REFERENCES

1. Cesari, L. and Suryanarayana, M. B., Existence theorems for Pareto optimization, multivalued and Banach space valued functionals, to appear.

2. Cesari, L. and Suryanarayana, M. B., Existence theorems for Pareto optimization in Banach spaces, *Bulletin of the AMS 82* (No. 2, 1976), 306-308.

3. Cesari, L. and Suryanarayana, M. B., Closure theorems without seminormality conditions, *Journal of Optimization Theory and Applications 15* (No. 4, 1975), 441-465.

4. Hildenbrandt, W. and Mertens, J. F., On Fatou's lemma in several dimensions, *Zeitschrift für Wahrscheinlichkeitstheorie und verwandte Gebiete 17* (1971), 151-155.

5. Schmeidler, D., Fatou's lemma in several dimensions, *Proceedings of the AMS 24* (1970), 300-306.

6. Suryanarayana, M. B., Remarks on lower semicontinuity and lower closure, *Journal of Optimization Theory and Applications 19* (No. 4, 1976).

7. Suryanarayana, M. B., Upper semicontinuity of set valued functions, to appear.

---

*Current address: Eastern Michigan University, Ypsilanti, Michigan 48197.

THE STABILITY OF SOLUTIONS BIFURCATING FROM STEADY OR PERIODIC SOLUTIONS

H. F. Weinberger*
School of Mathematics
University of Minnesota

1. INTRODUCTION

The method of Lyapounov [11] and Schmidt [15] allows one to obtain branches of solutions of the functional equation

$$F(x,\lambda) = 0 \tag{1.1}$$

in a neighborhood of a point, say (0,0), where the Fréchet derivative $F_x(0,0)$ is a Fredholm operator. Here x lies in a Banach space X, and the parameter λ lies in a Banach space Λ. If, in particular, $F_x(0,0)$ has a one-dimensional null space, the solutions of (1.1) near (0,0) can be expressed in the form $\{x(\alpha,\lambda),\lambda\}$ where α is a real parameter defined in terms of a suitable linear functional d* as

$$\alpha = d^*x$$

and one has a solution if and only if the bifurcation equation

$$\phi(\alpha,\lambda) = 0$$

is satisfied.

The linearized stability of a solution $\{x(\alpha,\lambda),\lambda\}$ of (1.1) with respect to the equation of motion

$$F(x,\lambda) - K\frac{dx}{dt} = 0$$

is expressed in terms of the spectrum of the family of linear transformations

$$F_x(x(\alpha,\lambda),\lambda) - \sigma K. \tag{1.2}$$

If this spectrum lies in a half-plane Re $\sigma \le \eta < 0$, the solution is linearly stable, while if the spectrum contains a point with Re $\sigma > 0$, the solution is linearly unstable.

By hypothesis the family (1.2) has the simple eigenvalue $\sigma = 0$ when $\alpha = \lambda = 0$. If, in addition, this eigenvalue is K-simple in the language of Crandall and Rabinowitz [3] (that is, if for a the null vector Ka does not lie in the range of $F_x(0,0)$) then there is a continuous family $\sigma(\alpha,\lambda)$ of simple eigenvalues of (1.2) near (0,0).

If one assumes that the remainder of the spectrum of (1.2) lies in a half-plane Re $\alpha \le \eta < 0$, then the stability of a solution $\{x(\alpha,\lambda),\lambda\}$ is determined by the sign of $\sigma(\alpha,\lambda)$.

We have shown in [16] that near (0,0) this sign is equal to that of the partial derivative $\phi_\alpha(\alpha,\lambda)$:

$$\operatorname{sgn} \sigma(\alpha,\lambda) = \operatorname{sgn} \phi_\alpha(\alpha,\lambda). \tag{1.3}$$

Moreover the analogous result was established for the Hopf bifurcation of periodic solutions from an equilibrium solution. This means that the stability of the bifurcating branches can be determined from the same bifurcation function which gives the branches themselves.

In addition one also obtains qualitative information from (1.3). For example, if for $\lambda \ne 0$ the function $\phi(\alpha,\lambda)$ has only simple zeros and if the bifurcating branches of solutions are ordered by the value of α, then every other branch is stable and every other one is unstable.

On the other hand, if λ is a real parameter and if the zero at $\alpha = 0$ of $\phi(\alpha,0)$ is isolated, then the maximal (or minimal) branches on the two sides $\lambda < 0$ and $\lambda < 0$ of the bifurcation have the same stability properties.

In the present work we shall repeat the proof of (1.3) for the case of the Lyapounov-Schmidt bifurcation in Section 2. We shall then show how to obtain the analogous result for the bifurcations from a nonconstant periodic solution originally treated by Poincaré [12, §§37, 38]. Section 3 deals with the case where the system is time dependent and periodic, while in Section 4 we discuss an autonomous system.

In Section 5 we sketch the proof for the bifurcation of periodic solutions from a steady state solution which was obtained by E. Hopf [8].

The qualitative results deduced from (1.3) and its analogues generalize the results on the stability of subcritical bifurcating solutions first found by E. Hopf [8], which have been extended to various situations in infinite dimensions by Sattinger [13,14], Iooss [17], Crandall and Rabinowitz [3,4], Joseph and Sattinger [10], and Joseph and Nield [9].

We have only considered some particular classical bifurcations here. It remains to be seen whether our method can be applied to more general bifurcations such as those discussed by Cesari [1,2] and Hale [5,6,7]. However, the simplicity of the zero eigenvalue of $F_x(0,0)$ seems to be needed in order to obtain a total ordering of the bifurcating branches. Examples in [16] show that the K-simplicity of the zero eigenvalue is also essential.

2. THE LYAPOUNOV-SCHMIDT BIFURCATION

We begin with the class of problems treated by the method of Lyapounov [11] and Schmidt [15].

Let X_1, X_2, and Λ be real Banach spaces, let $F(x,\lambda)$ be a continuously Fréchet differentiable map from a neighborhood

N_1 of $(0,0)$ in $X_1 \times \Lambda$ into X_2 with the property that

$$F(0,0) = 0. \tag{2.1}$$

We wish to investigate the linearized stability of the solutions near $(0,0)$ of the equilibrium equation

$$F(x,\lambda) = 0. \tag{2.2}$$

Stability is defined with respect to the equation of motion

$$F(x,\lambda) - K\frac{dx}{dt} = 0 \tag{2.3}$$

where K is a bounded linear transformation from X_1 to X_2.

We assume that the Fréchet derivative $F_x(0,0)$ has the K-simple eigenvalue zero. By this we mean that the null-space of $F_x(0,0)$ is spanned by $a \in X_1$, that the range of $F_x(0,0)$ is the null space of $b^* \in X_1^*$, and that $b^*Ka \neq 0$.

The method of Lyapounov and Schmidt can be formulated as follows: Choose an element w of X_2 such that

$$b^*w \neq 0 \tag{2.4}$$

and a linear functional d^* in X_1^* such that

$$d^*a \neq 0. \tag{2.5}$$

When $\alpha = \lambda = 0$ the problem

$$\begin{aligned} F(x,\lambda) - \phi w &= 0 \\ d^*x &= \alpha \end{aligned} \tag{2.6}$$

has the solution $x = 0$, $\phi = 0$. Moreover, the Fréchet derivative of the left-hand side with respect to (x,ϕ) at these values is

$$\begin{pmatrix} F_x(0,0) & -w \\ d^* & 0 \end{pmatrix},$$

which is an isomorphism from $X_1 \times R$ to $X_2 \times R$ because of (2.4) and (2.5). Hence by the implicit function theorem the problem (2.4) has a unique C^1 solution

$$x = x(\alpha,\lambda)$$

$$\phi = \phi(\alpha,\lambda)$$

in a neighborhood $\hat{N}_1$ of $(0,0)$ in $R \times \Lambda$. Moreover, $x(0,0) = \phi(0,0) = 0$.

We note that when

$$\phi(\alpha,\lambda) = 0, \tag{2.7}$$

$(x(\alpha,\lambda),\lambda)$ is a solution of the equilibrium equation (2.2). Moreover, if (x,λ) is a solution of (2.2) in the neighborhood

$$N_2 = \{(x,\lambda) \in N_1 : (d^*x,\lambda) \in \hat{N}_1\},$$

then x must be equal to $x(\alpha,\lambda)$ where $\alpha = d^*x$ and $\phi(\alpha,\lambda) = 0$.

Thus the bifurcation equation (2.7) gives all the solutions of (2.2) in N_2.

The linearized stability of a solution (x,λ) of (2.2) with respect to the equation of motion (2.3) is to be found from the spectrum of the family of linear transformations

$$F_x(x,\lambda) - \sigma K \tag{2.8}$$

from X_1 to X_2.

We have assumed that $F_x(0,0)$ has a one-dimensional null space spanned by a and that its range is the null space of b^*. Hence $\sigma = 0$ is an isolated K-simple eigenvalue of (2.8) when $x = \lambda = 0$. To find the eigenvalues σ of (2.8) near $\sigma = 0$ when $x = x(\alpha,\lambda)$ and (α,λ) is near $(0,0)$ and to find an associated eigenfunctional c^*, we consider the problem

$$\begin{aligned} c^*(F_x(x(\alpha,\lambda),\lambda) - \sigma K) &= 0 \\ c^*w &= 1. \end{aligned} \tag{2.9}$$

When $\alpha = \lambda = 0$ this problem has the solution

$$\begin{aligned} c^*(0,0) &= \frac{1}{b^*w} b^* \\ \sigma(0,0) &= 0. \end{aligned} \tag{2.10}$$

The Fréchet derivative with respect to (c^*,σ) of the left-hand side at these values is

$$\begin{pmatrix} F_x(0,0) & -\frac{1}{b^*w}\, b^*K \\ w & 0 \end{pmatrix},$$

which is an isomorphism from $X_2^* \times R$ to $X_1^* \times R$ because of (2.4) and (2.5). Thus by the implicit function theorem a C^1 solution $\{c^*(\alpha,\lambda), \sigma(\alpha,\lambda)\}$ exists in a neighborhood $\hat{N}_2$ of $(0,0)$.

We define the standard function

$$\operatorname{sgn} \xi = \begin{cases} 1 \text{ if } \xi > 0 \\ 0 \text{ if } \xi = 0 \\ -1 \text{ if } \xi < 0 . \end{cases}$$

We shall now state and prove our principal result.

1. Theorem. Let $F(x,\lambda) \in C^1(N_1, X_2)$ where N_1 is a neighborhood of $(0,0)$ in $X_1 \times \Lambda$, and let $K \in L(X_1, X_2)$. Suppose that $F(0,0) = 0$, that $F_x(0,0)$ has the K-simple eigenvalue 0 with the eigenvector a, and that the range of $F_x(0,0)$ is the null space of b^*.

If

$$b^*w\ b^*Ka\ d^*a > 0, \tag{2.11}$$

if $x(\alpha,\lambda)$ and the bifurcation function $\phi(\alpha,\lambda)$ are defined by (2.6), and if the eigenvalue $\sigma(\alpha,\lambda)$ is defined by (2.9), then there is a neighborhood $\hat{N}$ of $(0,0)$ in $R \times \Lambda$ such that for (α,λ) in $\hat{N}$

$$\operatorname{sgn} \sigma(\alpha,\lambda) = \operatorname{sgn} \phi_\alpha(\alpha,\lambda). \tag{2.12}$$

Proof. We differentiate the first equation in (2.6) with respect to α and apply $c^*(\alpha,\lambda)$. By using (2.9), we see that

$$\sigma c^*Kx_\alpha = \phi_\alpha . \tag{2.13}$$

Since $\sigma(0,0) = 0$, we conclude that $\phi_\alpha(0,0) = 0$.

We differentiate both equations (2.6) with respect to α and set $\alpha = \lambda = 0$ to conclude that

$$x_\alpha(0,0) = \frac{1}{d^*a}\, a.$$

Therefore by (2.10)

$$c^*(0,0)Kx_\alpha(0,0) = \frac{b^*Ka}{b^*wd^*a} .$$

We see from the condition (2.11) that this is positive. By continuity there is a neighborhood $\hat{N}$ of (0,0) where $c^*(\alpha,\lambda)Kx_\alpha(\alpha,\lambda) > 0$. In this neighborhood (2.12) follows from (2.13).

3. THE POINCARÉ BIFURCATION FOR A TIME PERIODIC SYSTEM

In this section and the next one we shall show how our stability result can be extended to the bifurcation of periodic solutions from a periodic solution, which was discussed by Poincaré [12, §§37, 38].

Let Π_1 be a Banach space of 2π-periodic functions $x(\tau)$ from R to X_1, and let Π_2 be a Banach space of 2π-periodic functions $y(\tau)$ from R to X_2. Let $F(x,\lambda)$ be a C^1 mapping of an open subset N_1^0 of $\Pi_1 \times \Lambda$ to Π_2, and let $K \in L(X_1,X_2)$ and $Kd/d\tau \in L(\Pi_1,\Pi_2)$.

If the problem

$$F(x,\lambda) - K\frac{dx}{d\tau} = 0 \tag{3.1}$$

has a solution $(x_0,0)$ in N_1^0 and if the linear transformation

$$F_x(x_0,0) - K\frac{d}{d\tau} \tag{3.2}$$

from Π_1 to Π_2 is invertible, then there are neighborhoods U of x_0 in Π_1 and V of $\lambda = 0$ in Λ such that for each $\lambda \in V$ the problem (3.1) has a unique solution $x = x(\lambda)$ in U. Moreover, $x(\lambda) \in C^1$.

If, on the other hand, the transformation (3.2) has a one-dimensional null space spanned by $u \in \Pi_1$, if its range is the null space of $p^* \in \Pi_2^*$, and if $p^*Ku \neq 0$, we are in the case treated in Section 2. That is, we can choose $w \in \Pi_2$ and $n^* \in \Pi_1^*$ with $p^*w\, n^*u\, p^*Ku > 0$ and obtain a bifurcation function $\phi(\alpha,\lambda)$ by solving the problem

$$F(x,\lambda) - K\frac{dx}{d\tau} - \phi w = 0$$
$$n^*x = \alpha. \tag{3.3}$$

The Floquet theory relates the stability of a solution of (3.1) to the real parts of the spectrum of the family

$$F_x(x,\lambda) - K\frac{d}{d\tau} - \sigma K \tag{3.4}$$

of transformations from Π_1 to Π_2.

Theorem 1 applied to the transformation $F - K\frac{d}{d\tau}$ gives the following result.

2. Theorem. Let $F(x,\lambda) \in C^1(N_1^0,\Pi_2)$ where N_1^0 is a neighborhood of $(x_0,0)$ in $\Pi_1 \times \Lambda$ and let $K \in L(X_1,X_2)$ and $K\frac{d}{d\tau} \in L(\Pi_1,\Pi_2)$. Let

$$F(x_0,0) - K\frac{dx_0}{d\tau} = 0,$$

let the range of the transformation (3.2) be the null space of p^* in Π_2^*, and let the null space of (3.2) be spanned by u in Π_1. If

$$p^*w\; n^*u\; p^*Ku > 0, \tag{3.5}$$

if the bifurcation problem (3.3) is solved, and if the branch of eigenvalues $\sigma(\alpha,\lambda)$ of (3.4) is constructed as in Section 2, then in a neighborhood of $(0,0)$

$$\operatorname{sgn}\sigma(\alpha,\lambda) = \operatorname{sgn}\phi_\alpha(\alpha,\lambda). \tag{3.6}$$

4. THE POINCARÉ BIFURCATION IN AN AUTONOMOUS SYSTEM

From now on we shall assume that the spaces Π_1 and Π_2 of 2π-periodic functions satisfy the following hypotheses:

(i) The family of τ-translations is defined and continuous on Π_1 and on Π_2.

(ii) The injections $a \to a\cos n\tau$, $n = 0,1,2,\ldots$ and $a \to a\sin n\tau$, $n = 1,2,\ldots$ of X_1 into Π_1 and of X_2 into Π_2 are continuous and have continuous inverses.

(iii) If $F(x,\lambda)$ is a map in $C^{\ell}(N_1,X_2)$ where N_1 in an open set in $X_1 \times \Lambda$, then the induced map $(x(\tau),\lambda) \to F(x(\tau),\lambda)$ is in $C^{\ell}(N_1^0,\Pi_2)$, where

$$N_1^0 = \{(x,\lambda) \in \Pi_1 \times \Lambda : (x(\tau),\lambda) \in N_1 \ \forall \ \tau \in R\}.$$

(iv) If $K \in L(X_1,X_2)$, then $K\, d/d\tau \in L(\Pi_1,\Pi_2)$.

All of the standard spaces satisfy these hypotheses, as long as Π_1 has one more derivative than Π_2. For example, we may take Π_1 to be the set of 2π-periodic functions in $C^{k_1,\alpha}(R,X_1)$ and Π_2 to be those of $C^{k_2,\alpha}(R,X_2)$, if $k_1 \geq k_2 + 1$. Similarly, we may let Π_1 be the space of 2π-periodic functions with the norm of the Sobolev space $H^{k_1,p}([0,2\pi],X_1)$ and Π_2 to be the 2π-periodic functions with the norm of $H^{k_2,p}([0,2\pi],X_2)$, provided $k_1 \geq k_2 + 1$.

We shall assume from now on that $F \in C^1(N_1,X_2)$ where N_1 is an open set in $X_1 \times \Lambda$, that $K \in L(X_1,X_2)$, and that the maps F and K in (3.1) are the induced maps of N_1^0 into Π_2. That is, we assume that the system (3.1) is autonomous.

We note that F and $K\, d/d\tau$ commute with τ-translations. Hence if the autonomous system (3.1) has a solution $\{x(\tau),\lambda\}$, it has the whole one-parameter family of solutions $\{x(\tau-\beta),\lambda\}$. Consequently, if $\{x_0,0\}$ with x_0 not constant is a solution of (3.1), the transformation (3.2) cannot be invertible. In fact, if $dx_0/d\tau \in \Pi_1$ we see by differentiating (3.1) that $dx_0/d\tau$ lies in the null space of the transformation (3.2).

In order to remove the ambiguity introduced by the trivial process of τ-translation of the solution, we introduce a phase normalization. We suppose that $dx_0/d\tau$ lies in Π_1 and is not the zero function. We choose $m^* \in \Pi_1^*$ such that

$$m^* \frac{dx_0}{d\tau} \neq 0. \tag{4.1}$$

We agree to look only at those solutions of (3.1) for which

$$m^*x = m^*x_0. \tag{4.2}$$

In an autonomous system the period is not predetermined. For this reason we replace the equation (3.1) by

$$F(x,\lambda) - \omega K \frac{dx}{d\tau} = 0 \tag{4.3}$$

If $\{x(\tau),\lambda\} \in N_1^0$ is a solution of this equation, then $\{x(\omega t),\lambda\}$ is a solution of the autonomous equation

$$F(x,\lambda) - K \frac{dx}{dt} = 0. \tag{4.4}$$

The period of $x(\omega t)$ is, of course, $2\pi/\omega$.

Suppose now that we have a solution $\{x_0, 0, \omega_0\} \in N_1^0 \times R$ of (4.3) with $dx_0/d\tau \neq 0$ and $dx_0/d\tau \in \Pi_1$.

We also suppose that the transformation

$$F_x(x_0,0) - \omega_0 K \frac{d}{d\tau}$$

is a Fredholm operator with Fredholm index zero whose range is orthogonal to $p^* \in \Pi_2^*$, and that

$$p^*K \frac{dx_0}{d\tau} \neq 0 \tag{4.5}$$

If we define the transformation

$$\mathcal{F}: N_1^0 \times R \to \Pi_2 \times R$$

by

$$\mathcal{F}(x,\lambda,\omega) = \{F(x,\lambda) - \omega K \frac{dx}{d\tau},\ m^*x - m^*x_0\} \tag{4.6}$$

the equation (4.3) and the phase normalization (4.1) can be written together as

$$\mathcal{F}(x,\lambda,\omega) = 0. \tag{4.7}$$

$\{x_0, 0, \omega_0\}$ is a solution of this equation, and the Fréchet derivative of $\mathcal{F}$ with respect to $\{x,\omega\}$ at $\{x_0,0,\omega_0\}$ is

$$\begin{pmatrix} F_x(x_0,0) - \omega_0 K \frac{d}{d\tau} & -K\frac{dx_0}{d\tau} \\ m^* & 0 \end{pmatrix}. \tag{4.8}$$

If the null-space of $F_x(x_0,0) - \omega_0 K d/d\tau$ is one-dimensional, this transformation is an isomorphism, so that there is a unique C^1 branch $(x(\lambda),\lambda,\omega(\lambda))$ of solutions of (4.7) for λ near 0.

If, on the other hand, the null-space of $F_x(x_0,0)-\omega_0 Kd/d\tau$ is two-dimensional, then that of the Fréchet derivative (4.8) is one-dimensional. Therefore we can apply the proof of Theorem 1 to F.

More specifically, we assume that the range $F_x(x_0,0) - \omega_0 Kd/d\tau$ is the common null space of p^* and q^* in Π_2^* and that there is a solution $v \in \Pi_1$ of

$$F_x(x_0,0)v - \omega_0 K \frac{dv}{d\tau} = 0, \tag{4.9}$$

such that

$$\begin{vmatrix} p^*K \dfrac{dx_0}{d\tau} & p^*Kv \\ q^*K \dfrac{dx_0}{d\tau} & q^*Kv \end{vmatrix} \neq 0.$$

We choose $w \in \Pi_2$ and $n^* \in \Pi_1^*$ such that

$$m^* \frac{dx_0}{d\tau} \begin{vmatrix} p^*K \dfrac{dx_0}{d\tau} & p^*w \\ q^*K \dfrac{dx_0}{d\tau} & q^*w \end{vmatrix} \begin{vmatrix} p^*K \dfrac{dx_0}{d\tau} & p^*Kv \\ q^*K \dfrac{dx_0}{d\tau} & q^*Kv \end{vmatrix} \begin{vmatrix} m^*\dfrac{dx_0}{d\tau} & m^*v \\ n^*\dfrac{dx_0}{d\tau} & n^*v \end{vmatrix} > 0. \tag{4.10}$$

The problem

$$\begin{aligned} F(x,\lambda) - \omega K \frac{dx}{d\tau} - \phi w &= 0 \\ m^*x &= m^*x_0 \\ n^*x &= n^*x_0 + \alpha \end{aligned} \tag{4.11}$$

has the solution $x = x_0$, $\omega = \omega_0$, $\phi = 0$ when $\alpha = \lambda = 0$. The Frećhet derivative of the left-hand side with respect to $\{x,\omega,\phi\}$ at these values is

$$\begin{pmatrix} F_x(x_0,0) - \omega_0 K \dfrac{d}{d\tau} & -K \dfrac{dx_0}{d\tau} & -w \\ m^* & 0 & 0 \\ n^* & 0 & 0 \end{pmatrix}, \tag{4.12}$$

which is an isomorphism by (4.10). Hence there is a neighborhood of $(0,0)$ in $R \times \Lambda$ in which (4.11) has a unique C^1

solution $(x(\alpha,\lambda),\omega(\alpha,\lambda),\phi(\alpha,\lambda))$. All the solutions of (4.3) near $(x_0,\omega_0,0)$ with the normalization (4.2) are obtained by setting $\phi(\alpha,\lambda) = 0$.

According to the Floquet theory the linear stability of a solution of (4.3) is obtained from the spectrum of the family of linear transformations

$$F_x(x,\lambda) - \omega K \frac{d}{d\tau} - \sigma K \tag{4.13}$$

with the trivial eigenvalue $\sigma = 0$ corresponding to the eigenvector $dx/d\tau$ removed. Because (4.3) is autonomous and because of hypothesis (ii) above, we see that the spectrum of (4.13) is periodic of period $i\omega$. Therefore one only needs to examine the spectrum for $|\text{Im}\ \sigma| \le \frac{1}{2}|\omega|$.

The implicit function theorem shows that if $r^*(\alpha,\lambda) \in \Pi_2^*$ is chosen so that

$$r^*Kv = 0,$$
$$r^*K \frac{dx(\alpha,\lambda)}{d\tau} \ne 0,$$

then for (α,λ) in a neighborhood of $(0,0)$ one can find $\{c^*,\sigma,\eta\} \in \Pi_2^* \times R^2$ such that

$$\begin{aligned} c^*[F_x(x(\alpha,\lambda),\lambda) - \omega(\alpha,\lambda)K \tfrac{d}{d\tau}] - \sigma c^*K - \eta r^*K &= 0 \\ c^*K \frac{dx(\alpha,\lambda)}{d\tau} &= 0 \\ c^*Ka &= 1 \end{aligned} \tag{4.14}$$

and $\sigma(0,0) = \eta(0,0) = 0$. If, moreover, $\phi(\alpha,\lambda) = 0$ so that $\{x(\alpha,\lambda),\omega(\alpha,\lambda),\lambda)\}$ is a solution of (4.3), we see by applying the first equation (4.14) to $dx/d\tau$ that $\eta = 0$. Thus for $\phi(\alpha,\lambda) = 0$, $\sigma(\alpha,\lambda)$ is a nontrivial eigenvalue of (4.13) and $c^*(\alpha,\lambda)$ is the corresponding eigenfunctional.

An argument like that which was used to prove Theorem 1 now gives the following result.

3. Theorem. Let $F(x,\lambda) \in C^1(N_1,X_2)$ where N_1 is an open set in $X_1 \times \Lambda$ and let $K \in L(X_1,X_2)$.

If $\phi(\alpha,\lambda)$ is the bifurcation function which is obtained by solving (4.11) and if (4.10) holds, then there is a neighborhood $\hat{N}$ of $(0,0)$ in $R \times \Lambda$ such that at each point (α,λ) of $\hat{N}$ where $\phi(\alpha,\lambda) = 0$ the relation

$$\operatorname{sgn} \sigma(\alpha,\lambda) = \operatorname{sgn} \phi_\alpha(\alpha,\lambda) \tag{4.15}$$

is satisfied.

5. THE HOPF BIFURCATION

We now consider the case which was treated by E. Hopf [8] and, for the infinite-dimensional case, by Iudovich [18], Joseph and Sattinger [10], and Sattinger [19]. We again consider a real autonomous system

$$F(x,\lambda) - \omega K \frac{dx}{d\tau} = 0 \tag{5.1}$$

but we now assume that we are given a solution $\{x_0,\omega_0,0\}$ with x_0 independent of τ.

We suppose that the intersection of the spectrum of the family $F_x(0,0) - \sigma K\colon X_1 \to X_2$ with the imaginary axis contains the two K-simple eigenvalues $\pm i\omega_0$, $\omega_0 > 0$ and no other integral multiples of $i\omega_0$. Let a be the (complex) eigenvector corresponding to $i\omega$. Then $\bar{a}$ corresponds to $-i\omega_0$.

Since 0 does not lie in the spectrum, there is a branch $x = x_0(\lambda)$ of constant solutions of (5.1) for λ near 0 with $x_0(0) = x_0$. By replacing x by $x - x_0(\lambda)$ we assume without loss of generality that

$$F(0,\lambda) = 0 \tag{5.2}$$

for all λ near 0.

It is easily seen that $e^{i\tau}a$ and $e^{-i\tau}\bar{a}$ lie in the null space of

$$F_x(0,0) - \omega_0 K \frac{d}{d\tau} : \Pi_1 \to \Pi_2. \tag{5.3}$$

We make the stronger hypotheses that (5.3) has a two-dimensional null space, and that its range is the common null space of the real and imaginary parts of $e^{-i\tau}b^*$, where b^* lies in the complexification of X_2^* and satisfies $b^*F_x(0,0) = i\omega_0 b^*K$. We make the real and imaginary parts of $e^{-i\tau}b^*$ into members of Π_2^* by defining

$$(e^{-i\tau}b^*, w) = \frac{1}{2\pi}\int_0^{2\pi} e^{-i\tau}b^*w\,d\tau.$$

Since the eigenvalues $\pm i\omega_0$ of $F_x(0,0)$ are K-simple, $b^*Ka \neq 0$.

We shall imitate the development of Section 4. Since x_0 is constant, $dx_0/d\tau$ is not one of the null functions. We replace $dx_0/d\tau$ and v by independent linear combinations u and v of the real and imaginary parts of $e^{i\tau}a$. We take p^* and q^* to be the real and imaginary parts of $e^{-i\tau}b^*$, and choose w in Π_2 and m^* and n^* in Π_1^* in such a way that (4.10) holds. If u is chosen so that m^*u is real and positive, the condition (4.10) takes the form

$$\text{Im}[m^*e^{i\tau}a\ \overline{b^*Ka}\ (e^{-i\tau}b^*,w)]\,\text{Im}[m^*\ e^{i\tau}a\ \overline{n^*e^{i\tau}a}] > 0. \tag{5.4}$$

The analogue of the system (4.11) is

$$\begin{aligned} &F(x,\lambda) - \omega K\frac{dx}{d\tau} - \phi w = 0 \\ &m^*x = 0 \\ &n^*x = \alpha. \end{aligned} \tag{5.5}$$

When $\alpha = \lambda = 0$ this system has the solution $x = 0$, $\omega = \omega_0$, $\phi = 0$. However, the Fréchet derivative, which is given by (4.12), is singular because $dx_0/d\tau = 0$.

This difficulty is overcome by the following trick. For $\alpha \neq 0$ we introduce the new variables

$$y = x/\alpha$$

$$\Psi = \phi/\alpha$$

and divide the equations in (5.5) by α. In this way we obtain

$$\begin{aligned} &\frac{1}{\alpha} F(\alpha y,\lambda) - \omega K \frac{dy}{d\tau} - \Psi w = 0 \\ &m^* y = 0 \\ &n^* y = 1. \end{aligned} \tag{5.6}$$

We define the transformation $F(\alpha y,\lambda)/\alpha$ at $\alpha = 0$ to be the limiting value $F_x(0,\lambda)y$. Then if $F \in C^2(N_1,X_2)$, $F(\alpha y,\lambda)/\alpha \in C^1(N_1,X_2)$.

With this extension the system (5.6) with $\alpha = \lambda = 0$ has the solution $y = \beta_1 e^{i\tau}a + \beta_2 e^{-i\tau}\bar{a}$, $\omega = \omega_0$, $\Psi = 0$ where

$$\beta_1 m^*(e^{i\tau}a) + \beta_2 m^*(e^{-i\tau}\bar{a}) = 0$$

$$\beta_1 n^*(e^{i\tau}a) + \beta_2 n^*(e^{-i\tau}\bar{a}) = 1.$$

The Fréchet derivative of the left-hand side of (5.6) at these values is

$$\begin{pmatrix} F_x(0,\lambda) - \omega_0 K \frac{d}{d\tau} & -K \frac{d}{d\tau}(\beta_1 e^{i\tau}a + \beta_2 e^{-i\tau}\bar{a}) & -w \\ m^* & 0 & 0 \\ n^* & 0 & 0 \end{pmatrix},$$

which is an isomorphism because of (5.4). Thus we obtain a unique C^1 solution $\{y(\alpha,\lambda),\omega(\alpha,\lambda),\Psi(\alpha,\lambda)\}$ of (5.6) for (α,λ) in a neighborhood of $(0,0)$. We then set

$$\begin{aligned} x(\alpha,\lambda) &= \alpha y(\alpha,\lambda), \\ \phi(\alpha,\lambda) &= \alpha \Psi(\alpha,\lambda), \end{aligned} \tag{5.7}$$

to obtain a unique C^1 solution of (5.5). Note that $x(0,\lambda) = 0$ and $\phi(0,\lambda) = 0$, so that (5.5) is satisfied. All the solutions near $\{0,\omega_0,0\}$ of (5.1) are obtained in the form $(x(\alpha,\lambda),\omega(\alpha,\lambda),\lambda)$ where $\phi(\alpha,\lambda) = 0$.

We construct the eigenvalue $\sigma(\alpha,\lambda)$ and the eigenfunctional $c^*(\alpha,\lambda)$ as in Section 4 by solving

$$c^*[F_x(x(\alpha,\lambda),\lambda) - K\frac{d}{d\tau} - \sigma K] - \eta r^*K = 0$$

$$c^*K\,\frac{dy(\alpha,\lambda)}{d\tau} = 0$$

$$c^*Ky = 1$$

where $y(\alpha,\lambda) = x(\alpha,\lambda)/\alpha$ is obtained from (5.6) and $r^* \in \Pi_2^*$ is chosen so that $r^*Kdy/d\tau \neq 0$. Proceeding as in Section 4 we find that (4.15) holds near (0,0) when $\phi(\alpha,\lambda) = 0$ and $\alpha \neq 0$.

The linearized stability of the steady state solution $x = 0$ which occurs for $\alpha = 0$ is determined by the real parts of the spectrum of the family $F_x(0,\lambda) - \sigma K$: $X_1 \to X_2$. An argument which is carried out in [16] shows that if $\sigma(\lambda)$ is the branch of simple eigenvalues of this family with $\sigma(0) = i\omega_0$, then (4.15) remains valid if σ is replaced by Re σ.

We summarize these results in the following theorem, which is contained in [16].

4. <u>Theorem</u>. Let the system (5.1) be autonomous, let $F \in C^2(N_1,X_2)$ where N_1 is a neighborhood of (0,0) in $X_1 \times \Lambda$, and let $K \in L(X_1,X_2)$.

Suppose that

$$F(0,\lambda) = 0.$$

Let the null space of $F_x(0,0) - Kd/d\tau$ be spanned by the real and imaginary parts of $e^{i\tau}a$ where a lies in the complexification of X_1, and let its range be the common null space of the real and imaginary parts of $e^{-i\tau}b^*$.

Let the inequality (5.4) be satisfied, and let $x(\alpha,\lambda)$, $\omega(\alpha,\lambda)$, and $\phi(\alpha,\lambda)$ be defined by (5.6) and (5.7). Then there is a neighborhood $\hat{N}$ of (0,0) in $R \times \Lambda$ such that for $(\alpha,\lambda) \in \hat{N}$ and $\phi(\alpha,\lambda) = 0$

$$\begin{aligned} \operatorname{sgn} \ \sigma(\alpha,\lambda) &= \operatorname{sgn} \ \phi_\alpha(\alpha,\lambda) \quad \text{if } \alpha \neq 0, \\ \operatorname{sgn} \operatorname{Re} \sigma(\tau) &= \operatorname{sgn} \ \phi_\alpha(0,\lambda). \end{aligned} \tag{5.8}$$

We remark that if, as in [16], we choose m* and n* to be the real and imaginary parts of $e^{-i\tau}d^*$ with d* in the complexification of X_1^*, then the second factor in (5.4) is equal to $\frac{1}{4}|d^*a|^2$, which is automatically positive if the first factor is not zero.

REFERENCES

1. Cesari, L., Existence theorems for periodic solutions of nonlinear Lipschitzian differential systems and fixed point theorems, Contributions to the Theory of Nonlinear Oscillations 5, *Annals of Math. Studies 45*(1960), Princeton, 115-172.

2. Cesari, L., Alternative methods in nonlinear analysis, *International Conference on Differential Equations*, Academic Press 1975, 95-148.

3. Crandall, M. G. and Rabinowitz, P. H., Bifurcation, perturbation of simple eigenvalues, and linearized stability, *Arch. for Rat. Mech. and Anal. 52*(1973), 161-180.

4. Crandall, M. G., and Rabinowitz, P. H., The principle of exchange of stability, *Proceedings of the International Symposium on Dynamical Systems*, Academic Press, Gainesville, 1976.

5. Hale, J. K., Periodic solutions of nonlinear systems of differential equations, *Riv. Mat. Univ. Parma 5*(1954), 281-311.

6. Hale, J. K., On the stability of periodic solutions of weakly nonlinar periodic and autonomous differential systems, Contributions to the Theory of Nonlinear Oscillations 5, *Annals of Math. Studies 45*(1960), Princeton, 91-113.

7. Hale, J. K. *Oscillations in Nonlinear Systems*. McGraw-Hill, New York, 1963.

8. Hopf, E., Abzweigung einer periodischen Lösung eines Differential-systems, *Akad. d. Wiss., Leipzig, Berichte, Math.-Phys. Kl. 94*(1942), 1-22.

9, Joseph, D. D. and Nield, D. A., Stability of bifurcating time-periodic and steady state solutions of arbitrary amplitude, *Arch. for Rat. Mech. and Anal.* *58*(1975), 369-380.

10. Joesph, D. D. and Sattinger, D. H., Bifurcating time-periodic solutions and their stability, *Arch. for Rat. Mech. and Anal.* *45*(1972), 79-109.

11. Lyapounov, A. M., Sur les figures d'équilibre peu différentes des ellipsoides d'une masse liquide homogène donnée d'un mouvement de rotation, *Zap. Akad. Nauk*, St. Petersburg (1906), 1-225. See also idid., (1908), 1-175, (1912), 1-228, (1914), 1-112.

12. Poincaré, H., *Les Méthodes Nouvelles de la Mécanique Céleste*, t. I. Gautheir-Villars, Paris, 1892.

13. Sattinger, D. H., Stability of bifurcating solutions by Leray-Schauder degree, *Arch. for Rat. Mech. and Anal.* *43*(1971), 154-166.

14. Sattinger, D. H., Stability of solutions of nonlinear equations, *J. Math. Anal. and Appl.* *39*(1972), 1-12.

15. Schmidt, E., Zur theorie der linearen und nichtlinearen integralgleichungen. 3. Teil: Über die Auflösung der nichtlinearen intergralgleichungen und die Verzweigung ihrer Lösungen, *Math. Ann.* *69*(1908), 370-399.

16. Weinberger, H. F., On the stability of bifurcating solutions, *Nonlinear Analysis: A Collection of Papers in Honor of Erich Rothe*, Academic Press (in print).

17. Iooss, G., Existence et stabilité de la solution périodique secondaire intervenant dans les problèmes d'évolution du type Navier-Stokes, *Arch. for Rat. Mech. and Anal.* *47*(1972), 301-329.

18. Iudovic, V. I., Appearance of auto oscillations in a fluid, *Prikl. Math. Meckh.* *35*(1971), 638-655, *J. of Appl. Math. and Mech.* *35*(1971), 587-603.

19. Sattinger, D. H., Bifurcation of periodic solutions of the Navier-Stokes equations, *Arch. for Rat. Mech. and Anal.* *41*(1971), 66-80.

---

*This work was supported by the National Science Foundation through Grant GP 37660 X.

MODELING GONORRHEA

James A. Yorke*
and
Annett Nold
*Institute for Fluid Dynamics
and Applied Mathematics
University of Maryland*

INTRODUCTION

Gonorrhea is clearly the most frequently reported communicable disease. In calendar year 1973, it accounted for over 60% of the total cases of specified notifiable diseases reported to the Public Health Service. Gonorrhea continues to increase and during Fiscal 1974 reached a total of reported cases more than double that of six years before [12]. The total of 874,161 cases reported during Fiscal year 1974 was the greatest number ever reported since the Public Health Service started keeping records 55 years ago. Figure 1 displays this recent increase.

Since the incidence of asymptomatic male gonorrhea is low and males usually develop symptoms of urethral gonorrhea, the majority of them seek medical care; consequently the trend of male cases may be considered as one indicator of the trend of gonorrhea in the population. Because clinical symptoms are lacking in the majority of women with gonorrhea, many cases are undetected and untreated. The increase in reported cases of females with gonorrhea is partly a result of increased efforts in detection and treatment of females

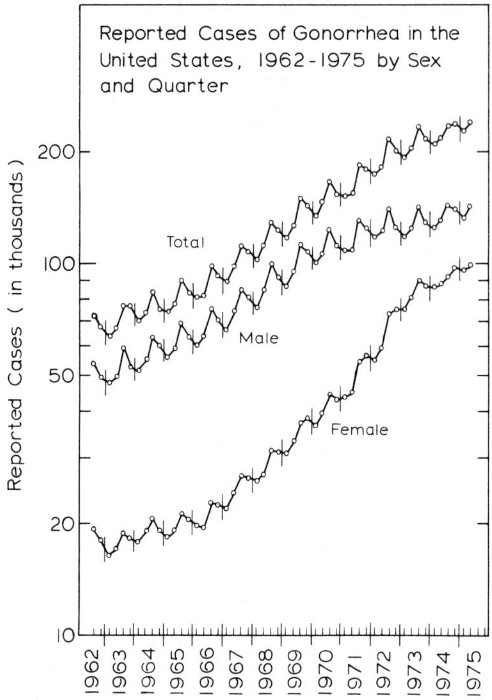

Figure 1

with gonorrhea. Of the approximately 8.0 million females cultured for gonorrhea in fiscal year 1974, 4.3% were positive. These aspects are described in [1].

Gonorrhea is a more serious health problem than many believe. Complications (particularly pelvic inflamatory disease) of gonorrhea in women are estimated to cost the nation $211 million annually, accounting for approximately 175,000 hospital admissions totaling 1.2 million hospital-patient days [2]. Pelvic inflamatory disease is one of the most important causes of sterility in females. The actual incidence of gonorrhea is not known since many cases of gonorrhea are not detected and many which are treated are not reported. The Public Health Service estimates that there were 2.6 million new cases of gonorrhea in the United States during fiscal year 1974.

Our initial attraction to the epidemiology of gonorrhea was the unique role we saw it as having in the mathematical theory of epidemics, namely a disease whose prevalence level is not limited by the accumulation of immunes. No significant immunity has been identified as being conferred by having gonorrhea. In developing models for gonorrhea it became apparent to us that such models were worth little unless they could be developed further, to the point of being medically valuable in reducing incidence and preventing suffering.

To understand our objectives in developing a useful theoretical understanding of the spread of gonorrhea, it is necessary to understand how limited the theoretical basis is for the present control program. The present control screening program results in the treatment of some 300,000 women who are identified as having gonorrhea. The Center for Disease Control has no useful estimates on how high this figure would have to be raised to significantly reduce the incidence of gonorrhea. (We feel such estimates can only be obtained by modeling.) Henderson [3] believes that "gonorrhea epidemiology is our sleeping giant with respect to interrupting the transmission of this disease." While CDC has recently begun to emphasize the preventive nature of their program, they do not speak in terms of how many cases on the average are actually prevented, say for each additional woman positively identified and cured. Modeling can provide such estimates.

MATHEMATICAL EPIDEMIOLOGY

The mathematical modeling process involves formulation of models from observations, deduction of conclusions from the models, comparison of the models and conclusions with reality, and refinement or reformulation of models as necessary. One reason for doing modeling is to take advantage of the economy, clarity, and precision of mathematical notations and techniques. Another reason is to obtain a better understanding of the processes involved by identifying and focusing on the essential aspects or variables. A third reason is to pose sharp questions and to suggest additional areas of study or experimentation. The final and obvious reason is to use the models to predict outcomes. We believe that mathematical modeling can be an effective tool in the study of gonorrhea epidemiology and control. We support this belief with four examples involving modeling of other infectious diseases.

The first example involves recurrent outbreaks of measles, chickenpox, and mumps studied by London and Yorke [4] and Yorke and London [5]. Recurrent outbreaks of measles, chickenpox and mumps in cities are studied with a mathematical model of ordinary differential delay equations. For each calendar month a mean contact rate (fraction of susceptibles contacted per day by an infective) is estimated from the monthly reported cases over a 30- to 35-year period. For each disease the mean monthly contact rate is 1.7 to 2 times higher in the winter months than in the summer months; the seasonal variation is attributed primarily to the gathering of children in school. Computer simulations that use the seasonally varying contact rates reproduce the observed pattern of undamped recurrent outbreaks: annual outbreaks of chickpox and mumps and biennieal outbreaks of measles. The two-year period of measles outbreaks is the signature of an endemic infectious disease that would exhaust itself and become nonendemic if there were a minor increase in infectivity or a decrease in the length of the incubation period. For populations in which most members are vaccinated, simulations show that the persistence of the biennial pattern of measles outbreaks implies that the vaccine is not being used uniformly throughout the population. The mean monthly contact rates for measles, chickenpox and mumps estimated from the monthly reported cases show systematic differences between the years with many cases and the years with few cases. In New York City, the mean contact rates for chickenpox were different during the years 1931-1945 than during 1946-1960. The clustering of cases within social groups is proposed to account for these differences in the contact rates and for other empirical observations. The irregularity of outbreaks of measles in cities of fewer than two million people can be explained by stochastic effects. Outbreaks of measles in distant large metropolitan areas are highly correlated in time, but the reasons for the correlation are not clear.

In the second example, Elveback et. al. [6] study interference and competition between wild enteric viruses and live poliovirus vaccine in a community of families, using two stochastic, discrete-time-interval simulation models. The viruses spread by contact; the live poliovirus vaccine is known to spread from successfully vaccinated children to susceptible children, although not so quickly as the wild poliovirus. An infection with one of these agents confers temporary protection from the other. The rate and pattern of spread and the interference between agents affect the results of vaccination programs. In the first model, intended to simulate a small community in the United States, infections spread by person-to-person contact through a highly structured population of 100 families. There are four exposure groups: the family, the preschool play group, the grade school, and the family cluster comprising two to five families. Each individual is assigned a susceptibility for each agent, and a contact rate for each exposure group. Comparison is made with a simpler model in which the two interfering agents spread through an unstructured, randomly mixing population. Population structure has a marked effect, especially on the choice of an optimal vaccination program; fewer vaccinations are needed for the unstructured population, partly because the vaccine virus spreads more quickly through it. Their second model applies to underdeveloped regions, where polio vaccination failure rates of as high as 50% are attributed to interference by widespread enterovirus infections. For the simulation, all wild nonpolio enteroviruses are lumped into one agent labeled E which confers temporary immunity to contact infection with both wild vaccine polioviruses. There is no permanent immunity for E since there are many types of enteric viruses which can infect a child in succession. For this model, the susceptible population consists of small children, who mix in the family group and play groups. The trial epidemics showed that vaccination programs can be expected to cut polio epidemic size by as much as one half despite the interference of wild viruses.

The third example involves the use by Hethcote and Waltman [7] of the operations research method of dynamic programming to find optimal vaccination schedules in a deterministic epidemic model. A deterministic epidemic model is modified to allow vaccination, and a definition of preventing or controlling an epidemic is given. The question of finding an optimal vaccination strategy (a least-cost vaccination program for a given cost functional) to prevent an epidemic is posed. The use of the dynamic programming technique to construct the optimal vaccination program is illustrated. Optimal vaccination schedules are given which were computed for some specific theoretical epidemics. The ideas in the above two papers are certainly complex and we have tried to follow the authors phrasing wherever possible.

The fourth example involves indirect use of data by Lajmanovich [8] to determine results of possible measles vaccination programs. In the first example above, London and Yorke used over 30 years of data from pre-vaccine periods for New York City and Bultimore to evaluate and determine the coefficients in the modeling equations for measles. Lajmanovich used their results to study the results of measles vaccination programs. It was necessary to use this indirect approach, using models developed with prevaccine data, because the data from later periods is much less useful. For data from vaccine periods to be complete, it would have been necessary to have known not only measles incidence rates but also vaccination rates by age. Realistically it has been impossible to obtain reliable data for the number of children susceptible to measles, while a number of studies from prevaccine periods are useful. Lajmanovich developed a model with a number of distinct groups, categorizing children by age as well as immunization status. It was found that for a vaccination program to be successful in halting measles, at least half the preschool children had to be immune at each point in time, even assuming all school age children were immunized. This is an example of a result which can be obtained even though the model has a number of coefficients which can be only approximately evaluated, namely the interaction rates between children of various age groups. These interaction rates can be chosen from a wide range of possible data, and as long as they are chosen to be consistent with prevaccine incidence experience, the estimate of needing 50% of the preschoolers to be vaccinated remains valid. Her model also indicated how fast measles outbreaks would increase when fewer than 50% of the preschoolers were immunized.

The last three examples have a number of features in common with the problem of modeling the control of gonorrhea. The data available is poor so results obtained must be valid over a number of possible ways of choosing parameters. Experience with the models must be developed to determine which types of models exhibit the least dependence on the unknown societal parameters, and the modeling must directly incorporate clinically realizable control features.

GONORRHEA EPIDEMIOLOGY

Gonorrhea has distinct epidemiological characteristics which must be considered in the construction of models for its spread in a community. First, the model need only consider the sexually active individuals in the community who potentially could be infected by their contacts. Gonorrhea is an essentially nonseasonal disease with less than ten percent variation in incidence [9] so that models can have time-independent parameters. The average incubation period, three to seven days, is short compared to the often quite long period of active infectiousness. An infected individual seems to remain infectious until he or she receives antibiotic treatment. While some infected individuals (especially males) quickly develop painful symptoms and therefore seek prompt medical attention, others develop no symptoms. Infected females may have no easily recognizable symptoms [10], even while the disease does substantial internal damage. A recent study [11] of service men demonstrated that men also can be infected without symptoms.

Homosexual transmission is not as important for gonorrhea as it is for syphilis so we can assume that the mode of transmission is predominantly heterosexual. Related to the inability [12] of the human body to throw off a gonorrheal infection is the important fact that no significant physiological immunity is derived from having previously been infected. There are individuals who have been infected and cured more than ten times a year. As soon as the curing antibodies have left the body, we assume that the individual is again susceptible. Thus a model should consider only susceptibles and infectives with few immunes and negligibly few incubating the disease. We assume that the sexually active population is constant in size and equals the number of infectives plus the number of susceptibles. Unlike many diseases, the contact rates and duration of infection are extremely variable, and this complication should be taken into account in a realistic model.

The first mathematical model specifically for gonorrhea was developed by Cooke and Yorke [13]. They considered a single homogeneous population using time delays to represent variation in the infectious period. Clearly a model for the spread of gonorrhea should divide the population not only by sex, but also into groups with similar ages, social behavior, and risks of infection. In Fiscal year 1974 the ratio of male to female reported cases was 1.5 to 1 whereas it was 1.7 to 1 in fiscal year 1973 and 2.2 to 1 in fiscal year 1972. Reported gonorrhea rates in the 20-24 year age group are almost twice as high as in the 25-29 year age group and in the 15-19 year age group [1]. The difficulty of analyzing differential equations with time delays makes it unlikely that this approach can be extended to a more general model with a nonhomogeneous population divided into subgroups.

A model involving ordinary differential equations without time delays for a host vector infectious disease where recovery does not give immunity was formulated and analyzed by Hethcote [14,15]. Gonorrhea with the sexually active population divided into females and males could be considered a host vector infectious disease.

Reynolds [16] of the Center for Disease Control modeled two interacting populations, males and females, each of which is treated as a homogeneous population. Her work was the first effort to use modeling to gain useful information about how to combat the gonorrhea epidemic.

A general model for gonorrhea in a nonhomogeneous population has been developed and analyzed by Lajmanovich and Yorke [17]. The model involves a system of n ordinary differential equations without time delays which correspond to n groups in the nonhomogeneous population. Each of the n groups is homogeneous in the sense that the individuals belonging to the same group have similar infectious periods and similar contact rates with other groups. Let $x_i(t)$ be the number of susceptibles in the ith group, $y_i(t)$ be the number of infectives, $1/\alpha_i$ be the average period of infectiousness and c_i be the constant total size of the ith group. Let β_{ij} be the contact rate of the ith group susceptibles with the jth group infectives. Thus β_{ij} is the proportion of the ith group susceptible population contacted (infected) by one infective from the jth group per unit time. We assume that any two groups in the population are connected by some chain of transmission. The system of equations analyzed is

$$\frac{dy_i}{dt} = -\alpha_i y_i + \sum_{j=1}^{n} \beta_{ij} c_i y_j - \sum_{j=1}^{n} \beta_{ij} y_i y_j .$$

It is proved in [17] that either the epidemic will die out naturally for every possible initial stage of the epidemic, or when this is not true and the initial number of infectives of at least one group is nonzero, the disease will remain endemic for all future times. Moreover, the number of infectives and susceptibles of each group will approach nonzero constant levels. This behavior of gonorrhea is different from most other infectious diseases, which usually have occasional major outbreaks followed by low disease levels. This different behavior of gonorrhea is attributable to the lack of developed immunity and the basic nonseasonality of the disease.

The important practical implication of the paper of Lajmanovich and Yorke [17] is that the model developed and analyzed there includes all of the factors which are epidemiologically significant for gonorrhea and, consequently, could be used as a basis for a theoretical study of gonorrhea control procedures. Moreover, the analysis shows that a primary object of study should be the equilibrium point, which corresponds to the endemic levels approached asymptotically. The equilibrium point is easier to analyze than the system of differential equations since it is the solution of a system of n simultaneous algebraic equations which involve the contact rates and average lengths of infectious periods for the groups. Thus one can determine how the equilibrium point changes as the parameter values change. It is emphasized that changes in the equilibrium levels (as dependent on parameter changes) require much less precise data than would be required for the model to accurately reflect actual equilibrium levels.

The low quality of data avaliable and more importantly, the lack of understanding of the dynamics of gonorrhea epidemics, makes preliminary heuristic modeling necessary. We studied [18] the record of reported cases to obtain an overview of the chain of infections. We define the infectee number ϕ as the average number of new cases caused by an infected individual during a typical case. In [18] we have found the following result to be of considerable importance for studying a closed population such as that of the United States.

BIOTHEOREM. If an endemic communicable disease is at a constant equilibrium in a population, then the average infected person spreads the disease to exactly one other person during the duration of his or her case, that is, $\phi = 1$.

This result is "obviously" true but we have proofs available only for special differential equations models. The "infective replacement number" defined by Hethcote [15] is the same as our infectee number ϕ.

During the years 1968 to 1972 (before the inception of the screening program for women) the number of reported cases rose steadily by about 13% per year. This implies that the seasonally averaged infectee number ϕ was approximately 1.01. Having ϕ near 1 does not by itself imply that the disease is near equilibrium.

The value of ϕ is certainly less than the number of infections κ which would occur if no one contacted by an infective was already infected; when an infected individual is contacted, no new case results, so as gonorrhea prevalence increases, the infectee number declines toward the equilibrium value 1.0. We call this pre-emption of potential new cases the "saturation effect".

Since the infectee number has hovered just above 1.0, we claim that the prevalence of gonorrhea has been close to equilibrium. It is implausible that κ is so close to 1.0;

if it were, slight improvements in medical care would have lowered incidence substantially and in particular the screening program would have caused incidence to level off, then decline during 1974 and 1975. The observed upward trend in case reports of about 13% per year represents a drift in the equilibrium caused by such agents as social change and changes in bacterial strains.

The core. A simple calculation shows that this present near-equilibrium level is less than 4% of the at-risk populations, that is of the people who are in the interconnected network of sexual contacts and thereby have some real chance becoming infected. The only possible inference seems to be that the saturation effect must be confined to certain highly active subgroups. We show that the persistence of the upward trend in gonorrhea cases despite the screening program indicates that the extra cures have been rapidly offset by reinfections spread from groups in which disease prevalence is high, 15% or 20% or more. We call the union of the high prevalence groups the core. We estimate that the core contains less than 250,000 individuals and would be a small portion of those screened.

Identification of these core individuals is to some extent possible through rescreening of individuals who have previously been identified as having gonorrhea. The potential effectiveness of rescreening programs can be suggested by modeling. Because these individuals play such a major role in the gonorrhea network, the more frequent treatment of core individuals promises significant impact in reducing prevalence in the rest of the population.

We demonstrate that if the prevalence were substantially reduced in the core, the incidence in the rest of the population would decrease proportionately. Further analysis is necessary to identify core composition and to find better methods of identification of core members.

ACKNOWLEDGEMENT

We have been encouraged to develop our ideas beyond those in [17] by R. Henderson's enthusiastic reception and encouragement of the application of mathematical epidemiology to gonorrhea at the 1974 SIAMS SIMS conference on Epidemiology. Since then perhaps a dozen individuals at CDC's VD division have made valuable comments in helping us understand the problems they face. H. Hethcote's comments were also valuable. The data we use has been provided by the Center for Disease Control.

REFERENCES

1. Today's VD Control Problem 1975, American Social Health Association, 1975.

2. Rendtorff, R. C. et al., Economic consequences of gonnorhea in women, *J. Amer. VC Assoc. 1*(1974), 40-47.

3. Henderson, R. H., Director of VD Control at Center for Disease Control, Report to VD Clinic Directors, March 21, 1975.

4. London, W. P., and Yorke, J. A., Recurrent outbreaks of measles, chickenpox, and mumps I: seasonal variation in contact rates, *Amer. J. Epid. 98*(1973), 453-468.

5. Yorke, J. A., and London, W. P., Recurrent outbreaks of measles, chickenpox, and mumps II: systematic differences in contact rates and stochastic effects, *Amer. J. Epid. 98*(1973), 469-482.

6. Elveback. L., Ackerman, E., Gatewood, L., and Fox, J. P., Stochastic two-agent simulation models for a community of families, *Amer. J. Epid. 93*(1971), 267-280.

7. Hethcote, H. W., and Waltman, P., Optimal vaccination schedules in a deterministic epidemic model, *Math. Biosciences 18*(1973), 365-382.

8. Lajmanovich, A., Mathematical models and the control of infections diseases, Ph.D. dissertation, University of Maryland, 1974.

9. Cornelius, C. E. (III), Seasonality of gonorrhea in the United States, *HSMHA Health Reports 86*(1971), 157-160.

10. Johnson, D. et al., An evaluation of gonorrhea case finding in the chronically infected female, *Am. J. of Epid. 90*(1969), 438-448.

11. Hunter Handsfield, H. et al., Asymptomatic gonorrhea in men, *New England J. Med. 290*(1974), 117-123.

12. Kolata, G. B., Research News: Gonorrhea, more of a problem but less of a mystery, *Science 199* No. 4236 (16 April 1976), 244-7.

13. Cooke, K., and Yorke, J., Some equations modeling growth processes and gonorrhea epidemics, *Math. Biosciences 16*(1973), 75-101.

14. Hethcote, H. W., Asympototic behavior and stability in epidemic modes, in Mathematical Problems in Biology, Victoria Conference 1973, P. Van den Driessche, ed., Lecture Notes in Biomathematics 2, Springer Verlag, 1974.

15. Hethcote, H. W., Qualitative analyses of communicable disease models, *Math. Biosciences, 28*(1976), 335-356.

16. Reynolds, G. H., A control model for gonorrhea, Ph.D. dissertation, Emory University, 1973.

17. Lajmanovich, A., and Yorke, J. A., A deterministic model for gonorrhea in a nonhomogeneous population, *Math. Biosciences 28*(1976), 221-236.

18. Yorke, J. A., and Nold, A., The gonorrhea epidemic near equilibrium, to appear.

*This research was supported in part by National Science Foundation Grant MPS 74 24310-A01.

Contributed Papers

ON THE SOLUTIONS OF SECOND AND THIRD ORDER DIFFERENTIAL EQUATIONS

Shoshana Abramovich
Department of Mathematics
University of Haifa

We use the nature of solutions of the second order linear differential equation

(1) $$(ry')' + py = 0,$$

to investigate the nature of the nonhomogeneous second order linear differential equation

(2) $$(ry')' + py = f$$

and of the third order equation

(3) $$(ry'')' + py' = qy.$$

Under these conditions: *when $p(x)$, $f(x)$ and $r(x)$ are positive, having continuous derivative, $p(x)r(x)$ is increasing and $f(x)r(x)$ is decreasing, we get that the zeros of a solution $y_1(x)$ of equation (1) and a solution $y_a(x)$ of equation (2), defined by $y_1(0) = y_a(o) = 0$, $y_a'(0) = a$, seperate each other, moreover, the zeros of $y_a(x)$ for different values of a seperate each other* [1, Th. 1].

By making the same assumptions as before we get that *between two zeros of a solution of (2) there is only one extremum* [1, Th. 2].

Under these restrictions that: $p(x)$, $f(x)$, $r(x)$ are positive, $p(x)r(x)$ is nondecreasing, and $f(x)/p(x)$ is a nonincreasing function we get that the maxima of solutions of (2) are decreasing [1, Th. 3].

Parts of the above theorems about monotone functions can be extended to a larger class of functions, for instance, to a class of functions for which the condition $p(a-x) \geq p(a+x)$, $x > 0$, (such $p(x)$ is called left balanced with respect to a).

As for the third order linear equation (3), *under the conditions when: $r(x) > 0$, $r'(x) \leq 0$, and $q(x) \geq 0$ are continuous functions for $x \geq 0$, and under the condition when $r(x)p(x)$ is a continuous increasing function we get, that a nontrivial solution of (3) defined by $y(0) = \alpha$, $y'(0) = 0$, $y''(0) = \beta$, $\alpha, \beta \geq 0$, is positive and its minima are increasing for $x \geq 0$.* [2, Th. 1].

Some qualities of comparison seperation and oscillation are obtained [2]: For instance, *if we name the values of x in which a solution of (3) has consecutive minimum, maximum and minimum, by b_i, $i = 1,2,3$, then $b_2-b_1 \geq c_3-b_2 \geq b_3-b_2$, where b_2 and c_3 are consecutive zeros of a solution of (1)* [2, Th. 1].

We get similar results for the nonlinear equation:

(3') $\quad y''(x)+p(x)y'(x) = q(x)g(y)+f(x)$, $yg(y)>0$, $y\neq 0$, $g(0)=0$,

see [2, Th. 2].

We present results concerning the n-th eigenfunction of the system

$$(1,\lambda) \qquad (ry')' + \lambda py = 0, \; y(\pm 1) = 0.$$

We name the zeros of the n-th eigenfunction by x_i: $-1 = x_0 < x_1 < \dots < x_n = 1$, and the zeros of its derivative by $\bar{x}_i$: $-1 < \bar{x}_1 < \dots < \bar{x}_n < 1$. *When these conditions exist: $r(x)p(x) \geq 0$ is left balanced with respect to $a = 0$, $1/r(x) > 0$ and $r(x)p(x)$ have one minimum in the interval $[-1,1]$, and*

$$\int_{-1}^{-x} \frac{dt}{r(t)} \geq \int_{x}^{1} \frac{dt}{r(t)}, \quad x \geq 0.$$

(For instance, these conditions hold when $r(x)$ is increasing and $r(x)p(x)$ decreasing.)

Then we get [1, Th. 4]

$$x_i + x_{n-i} \leq x_{i=1} + x_{n-i+1}, \quad 1 \leq i \leq [\tfrac{n}{2}]$$

$$\bar{x}_i + \bar{x}_{n+1-i} \leq x_{i-1} + x_{n-i+1}, \quad 1 \leq i \leq [\tfrac{n+1}{2}]$$

REFERENCES

1. Abramovich, S., On the solutions, eigenfunctions, and eigenvalues of second order linear differential equation, *J. Math. Anal. Appl.*, Vol. 55, No. 3, 531-536.

2. Abramovich, S., Application of the nature of second order equations to third order equations, *J. Math. Anal. Appl.*, to appear, (Vol. 56).

FIXED POINT THEOREMS FOR NONLINEAR FUNCTIONAL OPERATORS

S. R. Bernfeld,
V. Lakshmikantham
Department of Mathematics
University of Texas

and

Y. M. Reddy
Department of Mathematics
Paul Quinn College

The application of Liapunov functions to the study of the qualitative behavior of solutions of functional differential equations has proved to be very successful [3], [4]. Razumikhin [5] developed much of the early work where he utilized particular subsets of the function space in order to develop a comparison between solutions of the functional differential equation and the solutions of an ordinary differential equation. In this note we extend Razumikhin's idea to an analysis of the fixed points of operators whose domain is a function space and range is an appropriate Banach Space. Such operators arise naturally from delay and mixed differential equations.

Let E be any Banach Space, $[a,b]$ any compact interval on the real line, and $E_0 = C[[a,b],E]$, the space of continuous functions on $[a,b]$ with range in E with the sup norm topology. Letting c be any point in the interval $[a,b]$, we say ϕ is a fixed point of a mapping $T:E_0 \to E$ if $T\phi = \phi(c)$. For each $\psi \in E_0$ define $\Omega_\psi = \{f \in E_0 \mid \|f-\psi\|_{E_0} = \|f(c)-\psi(c)\|\}$. When $c = b$, $\psi \equiv 0$ then Ω_ψ is the class used by Razumikhin.

We present some fixed point theorems that are motivated by some of the recent work of Caristi and Kirk [1,2]. Our results, in contrast to the above work, makes strong use of the interplay between E_0, E, c, and Ω_ψ.

1. Theorem. Let $T: E_0 \to E$ and assume there exists a closed linear subspace Ω^L of Ω_0 such that $T(\Omega^L) \subseteq \Omega^L(c)$ where $\Omega^L(c) = \{f(c) | f \in \Omega^L\}$, and c is any point in $[a,b]$. If for all $\phi \in \Omega^L$

$$(1) \qquad ||\phi(c) - T\phi|| \leq \omega(\phi(c)) - \omega(T\phi),$$

where $\omega: E \to [0,\infty)$ is lower semicontinuous, then T has a fixed point in Ω^L. A complementary result is the following:

2. Theorem. Let $T: E_0 \to E$ such that (1) holds for all $\phi \in \Omega_0$. Then T has a fixed point in E_0.

If T satisfies a contraction condition $||T\phi - T\psi||_E \leq K||\phi - \psi||_{E_0}$ $0 \leq K < 1$ then by letting $\omega(\phi(c)) = (1-K)^{-1}||\phi(c) - T\phi||_E$ we obtain from Theorem 2 the following corollary which is a generalization in our setting of the classical contraction mapping principle.

1. Corollary. Assume T is a contraction mapping on E_0. Then T has a fixed point in E_0. If ϕ_1 and ϕ_2 are fixed points of T then $\phi_1 \notin \Omega_{\phi_2}$.

Thus under contraction conditions we cannot assert the uniqueness of fixed points but do obtain uniqueness within certain subsets of E_0.

Now let the mapping T be a contraction on only a subset $K_0 = C([a,b],K)$ of E_0, where K is a closed convex subset of E. Suppose T satisfies the boundary or inward condition

$$(2) \qquad \overline{\lim_{h\to 0^+}} \frac{1}{h} d[(1-h)\phi(c) + hT\phi, K] = 0$$

for all $\phi \in S_0 \equiv K_0 \cap \Omega_0$. (Here d represents the distance function in E). We now have the following result.

3. Theorem. Assume T is a contraction on K_0 and satisfies (2) on S_0. Then T has a fixed point in K_0.

A similar result holds using the idea of Theorem 1.

Further details about these results, as well as applications and other results appear in the forthcoming work.

REFERENCES

1. Caristi, J., Fixed point theorems for mappings satisfying inwardness conditions, *Trans. Amer. Math. Soc.* (to appear).

2. Caristi, J. and Kirk, W. A., Geometric fixed point theory and inwardness conditions, (Conference on Geometry of Metric and Linear Spaces, Michigan State University) June, 1974.

3. Hale, J. K., *Functional Differential Equations*, Springer-Verlag, New York, 1971.

4. Lakshmikantham, V. and Leela, S., *Differential and Integral Inequalities, Vol. 2*, Academic Press, New York, 1969.

5. Razumikhin, B. S., Application of Liapunov's method to problems in the stability of systems with a delay, *Avtomat i. Telemch. 21*(1969), 740-748.

ON THE CONVERGENCE OF SOLUTIONS OF LIÉNARD'S EQUATION WITH A FORCING TERM

Yuri Bibikov
Division of Applied Mathematics
Brown University

Consider a second order o.d.e.

$$\ddot{x} + f(x)\dot{x} + g(x) = p(t) \tag{1}$$

where $f > 0$ and continuous, p is continuous and periodic, g is of class C^1 and $xg(x) > 0$ for $x \neq 0$. Assume that equation (1) is dissipative in the sense of Levinson; i.e., there exist $r > 0$ such that for any solution $x(t)$ of equation (1)

$$\overline{\lim_{t\to+\infty}} |x(t)| < r, \quad \overline{\lim_{t\to+\infty}} |\dot{x}(t)| < r.$$

Under this assumption equation (1) has a periodic solution $\phi(t)$. We are interested in criteria for global asymptotic stability of $\phi(t)$.

1. Theorem. Let for all x

$$g'(x) > 0, \quad f(x) \geq a > 0. \tag{2}$$

We define

$$h(s) = \begin{cases} \inf\limits_{\substack{x\in(-r,r) \\ f(x) < s}} \dfrac{g'(x)}{s-f(x)} & \text{if } s > a \\[2ex] \inf\limits_{x\in(-r,r)} \dfrac{g'(x)}{s-f(x)} & \text{if } s < a. \end{cases}$$

Consider the o.d.e.

$$\frac{dw}{dv} + u\left(\frac{w}{v}\right) = 0, \tag{3}$$

where $u(s)$ $(s\neq a)$ is any continuous function such that $u(s) \leq h(s)$. Consider the integral curve of equation (3) in the right half-plane vw that passes through the point (0,1) and crosses the negative ordinate half-axis at the point $(0,w^-)$.

If

$$-w^- < 1, \tag{4}$$

then for any solution x(t) of equation (1)

$$x(t) - \phi(t) \to 0, \ \dot{x}(t) - \dot{\phi}(t) \to 0 \text{ as } t \to +\infty.$$

This criterion cannot be sharpened by means of a time-independent Liapunov function. It is easily applied to concrete equations (1). It also can be applied to certain classes of such equations. Consider for example a class of equations (1) satisfying the condition

$$g'(x) < \alpha + \beta f(x), \quad \alpha > 0, \ \beta > 0. \tag{5}$$

For $s > a$ we set $u(s) = 0$. For $s < a$ (t) implies

$$u(s) = \inf_f \frac{\alpha+\beta f}{s-f}$$

Hence

$$u(s) = \begin{cases} 0 & \text{for } s > a, \\ \dfrac{\alpha+\beta a}{s-a} & \text{for } a > s > -\dfrac{\alpha}{\beta}, \\ -\beta & \text{for } s \le -\dfrac{\alpha}{\beta}. \end{cases}$$

It is convenient to introduce new parameters

$$\alpha = ma^2, \ \beta = na. \tag{6}$$

Then (4) leads to the following result.

2. Theorem. If (2), (5) and (6) hold, then the periodic solution of equation (1) is globally asymptotically stable if

$$\sqrt{m+n^2} < \exp \frac{1}{\sqrt{4(m+n)-1}} \left[\text{arctg}\frac{1}{\sqrt{4(m+n)-1}} + \text{arctg}\frac{2m+n}{n\sqrt{4(m+n)-1}} \right].$$

For $n = 0$ we obtain the criterion of Z. Opial [1].

In conclusion note that the assumption of periodicity of p(t) in (1) can be replaced by that of almost periodicity.

REFERENCES

1. Opial, Z., Sur un théorème de C. E. Langenhop and G. Seifert, *Ann. Polon. Math.* *9*(1960/61), 145-155.

APPROXIMATE AND COMPLETE CONTROLLABILITY OF NONLINEAR SYSTEMS TO A CONVEX TARGET SET

Ethelbert N. Chukwu
and
Jan M. Gronski
Department of Mathematics
Cleveland State University

Introduction. Consider the nonlinear control system

$$\frac{dx}{dt} = f(t,x,u) \tag{1}$$

where f is a continuous function from $E \times E^n \times E^m$ into E^n. Let $I = [t_0,t_1]$ be a compact subset of E. We say that the system (1) is G-controllable where $G \subseteq E^n$ if for any $x_0 \in R^n$, there exists a bounded measurable function $u\colon I \to E^m$ such that the solution $x(t) = x(t,t_0,x_0,u)$ of

$$\frac{dx}{dt} = f(t,x,u(t)) \qquad x(t_0) = x_0 \tag{2}$$

satisfies $x(t_1) \in G$. The system (1) is approximately G-controllable if for any $x_0 \in E^n$ and for any $\lambda > 0$ there exists a bounded measurable function $u\colon I \to E^m$ such that the solution $x(t)$ of (2) satisfies

$$d(x(t_1),G) < \lambda$$

where $d(x,G) = \inf\{|x-p| : p \in G\}$ and $|\cdot|$ denotes a norm in E^n. Note that G can be taken to be a point, and in particular if $G = \{0\}$ the first definition corresponds to the usual one for null-controllability.

Recall that the system (1) is completely controllable if for any x_0, $x_1 \in E^n$ there exists a bounded measurable function $u\colon I \to E^n$ such that the solution $x(t) = x(t;t_0,x_0,u)$ of (2) satisfies $x(t_1) = x_1$.

This paper deals with G-controllability and approximate G-controllability of systems where G is assumed to be an arbitrary closed and convex target set.

Notations. In what follows $S_N^M(0)$ denotes the M-dimensional ball of radius N center the origin. The set $h(t,S_\rho^M(0))$ is defined by $h(t,S_\rho^M(0)) = \{h(t,u) : u \in S_\rho^M(0)\}$. The Aumann's integral of this set is given by

$\int_I h(t,S_\rho^M(0))\, dt = \{\int_I h(t,u(t))dt : u: I \to S^M(0),$ measurable, $u(t) \in S_\rho^M(0)\}$.

Consider a special case of (1), namely

$$\frac{dx}{dt} = A(t)x + k(t,u) \tag{3}$$

and its perturbation

$$\frac{dx}{dt} = A(t)x + k(t,u) + g(t,x,u) \tag{4}$$

where A is an n-square matrix function, and k and g are continuous n-vector functions.

1. Proposition. A necessary and sufficient condition that (3) be G-Controllable, where G is a convex subset of E^n is that for every $\varepsilon > 0$ there exists a $\delta > 0$ such that

$$\int_I X^{-1}(t)\ (k,(t,S_\delta^M(0))\ dt - X^{-1}(t_1)\ G \supseteq S_\varepsilon^M(0). \tag{5}$$

This result is distantly related to "expanding" characterization of reachable sets in [12, p. 52]. This is equivalent to linear systems being proper in [12, p. 73, 78] and being completely controllable [1, p. 93].

As a corollary we have the following curious result that (3) is completely controllable if and only if it is null controllable. Indeed the following more general result is true.

Corollary. Let $x_1 \in E^n$. The system (3) is x_1-controllable if and only if it is completely controllable.

We now state the main results.

1. Theorem. Assume that g is bounded on $I \times E^n \times E^m$. Further assume that for sufficiently large μ the set

$$k(t,S_\mu{}^m(0)) + g(t,x,S_\mu{}^m(0))$$

is convex for $(t,x) \in I \times E^n$. Let G be a fixed convex subset of E^n. Then the system (4) is G-controllable if and only if (3) is G-controllable.

Next we shall remove the convexity assumption and then deduce an approximate G-controllability result.

2. Theorem. Assume that g is bounded on $I \times E^n \times E^m$ and satisfies a Lipschitz condition

$$|g(t,x,u) - g(t,y,u)| \le \omega(t) \; |x-y|$$

with $\omega \in L'(I)$. Let G be a fixed convex subset of E^n. Then the system (3) is G-controllable if and only if (4) is approximately G-controllable.

The proofs of both results utilize extensions of theorems 1 and 2 of Dauer [1] on the existence of an absolutely continuous function satisfying

$$\dot{x}(t) \in R(t,x(t)) \quad \text{a.e. on } I = [t_0,t_1]$$

$$x(t_0) = x_0$$

$$d(x(t_1),G) \le \varepsilon$$

where $\varepsilon \ge 0$. Here R denotes a set-valued mapping $I \times E^n$ into the set of nonempty closed subsets of E^n which is upper semi-continuous with respect to set inclusion.

1. Example. Consider the system

$$\dot{x} = x + u$$

$$\dot{y} = y + e^{t-y^2} \sin u^2$$

on $I = [0,1]$ $G = \text{span}\{[0,1]^T\}$, where T denotes the transpose.

Here $A = \begin{bmatrix} 1 & 0 \\ 0 & 1 \end{bmatrix}$ $k(t,u) = \begin{bmatrix} 1 \\ 0 \end{bmatrix} u$,

$$g(t,\underline{x},u) = [0, e^{t-y^2} \sin u^2]^T.$$

It is a consequence of [10, Corollary 2.2] that $x = Ax+k(t,u)$ is G-controllable. Obviously g is bounded on $I \times E^2 \times E^2$. The convexity assumption is clearly satisfied. By Theorem 1,

$$\dot{x} = Ax + k(t,u) + g(t,\underline{x},u)$$

is G-controllable.

2. Example. Consider the system

$$\dot{x} = x + y + u + e^t\left(\frac{\sin^2 y + \cos^2 u}{t+1}\right)$$

$$\dot{y} = \frac{t^2 e^{-u^2}}{2+\cos(y-x)}$$

Here $A = \begin{bmatrix} 1 & 1 \\ 0 & 0 \end{bmatrix}$ $k(t,u) = \begin{bmatrix} 1 \\ 0 \end{bmatrix}$

$$g(t,x,y,u) = \begin{pmatrix} \dfrac{e^t(\sin^2 y + \cos^2 u)}{t+1} \\ \\ \dfrac{t^2 e^{-u^2}}{2+\cos(y-x)} \end{pmatrix}$$

Let $G = \text{spac}\,\{[-1,1]^T\}$.

Because the controllability space $\{A|B\} = \text{span}\,\{[1,0]^T\}$ and

$$e^{-At}\, G = \text{span}\,\{[-1,1]^T\}$$

it follows from [10, Corollary 2.2] and the fact that

$$E^2 = \{A|B\} + \bigcup_{t\geq 0} e^{-At} G$$

that the base system is G-controllable even though it is not null-controllable. Since g is bounded and satisfies the Lipschitz condition in x and y, Theorem 2 implies that the system is approximately G-controllable.

REFERENCES

1. Dauer, J. P., A controllability technique for nonlinear systems, *J. Math. Anal. Appl.* *37* (1972), 442-451.

2. Neustadt, L. W., The existence of optimal controls in the absence of convexity conditions, *J. Math. Anal. Appl.* *7* (1963), 110-117.

3. LaSalle, J. P., The time optimal control problem, in *Theory of Nonlinear Oscillations, Vol. 5*, Princeton University Press, Princeton, New Jersey, 1959, 1-24.

4. Fan, K., Fixed-point and minimax theorems in locally convex topological linear spaces, *Proc. Nat. Acad. Sci. U.S.A., 38*(1952), 121-126.

5. Filippov, A. F., Classical solutions of differential equations with multivalued right-hand side, *SIAM J. Control 5*(1967), 609-621.

6. Dauer, J. P., A note on bounded perturbations of controllable systems, *J. Math. Anal. Appl. 42*(1973), 221-225.

7. Dauer, J. P., Approximate controllability of nonlinear systems with restrained controls, *J. Math. Anal. Appl. 46*(1974), 126-131.

8. Filippov, A. F., On certain questions in the theory of optimal control, *SIAM J. Control 1*(1962), 76-84.

9. Richter, H., Verallgemeinerung eines in der statistik berätigten statzes der masstheorie, *Math. Ann. 150*(1963), 85-90, 440-441.

10. Chukwu, E. N., and Silliman, S. D., Complete controllability to a closed target set, *J. Optimization Theory Appl.*, to appear. (Technical report CSU D 42, Department of Mathematics, Clevelend State University, Cleveland, Ohio 44115) August, 1975.

11. Wazewski, T., Sur une generalisation de la notion des solutions d'une equation au contigent, *Bull. Acad. Polon. Sci. sér. Sci. Math. Astronom. Phys. 10*(1962), 11-15.

12. Hermes, H., and LaSalle, J. P., *Functional Analysis and Time Optimal Control*, Academic Press, New York, 1969.

LINEAR NEUTRAL FUNCTIONAL DIFFERENTIAL EQUATIONS ON A BANACH SPACE

Richard Datko
Department of Mathematics
Georgetown University

Consider the linear autonomous neutral differential equation defined on a real Banach space X by the relations:

$$\frac{d}{dt}\left[x(t) - \sum_{j=1}^{m} B_j x(t-h_j)\right] = A\, x\,(t) + \sum_{j=1}^{m} A_j\; x(t-h_j) \qquad 1(a)$$

if $t \geq 0$ and by

$$x(t) = \phi(t) \text{ if } t \in [-h,0]. \qquad 1(b)$$

Here $0 < h_1 < h_2 < \ldots h_m = h$ are constants, $A: X \to X$ is a closed unbounded operator with dense domain, D(A), on X which generates a semi-group of class C_0 (see e.g. [1]), $\{B_j\}$, $1 \leq j \leq m$ are bounded endomorphisms on X such that range $(B_j) \subset \mathcal{D}(A)$ for each j, $\{A_j\}$, $1 \leq j \leq m$ are bounded endomorphisms on X and $\phi:[-h,0] \to X$ is a continuous mapping.

If T(t) denotes the semi-group of operators generated by A then it can be shown that the "integrated" version of (1) is given by the equations

$$x(t) = \sum_{j=1}^{m} B_j\; x(t-h_j) + T(t)\;[\phi(0) - \sum_{j=1}^{m} B_j\phi(-h)] +$$
$$\int_0^t T(t-\sigma)\;[\sum_{j=1}^{m}\; (A_j+AB_j)\; x\;(\sigma-h_j)]d\sigma \qquad 2(a)$$

if $t \geq 0$ and by

$$x(t) = \phi(t) \text{ if } t \in [-h,0] \qquad 2(b)$$

If C[[-h,0],X] denotes the Banach space of continuous mappings from [-h,0] into X which has norm

$$|\phi(\cdot)| = \sup\{|\phi(t)|: t \in [-h,0]\}, \qquad (3)$$

then it can be shown that the equations (2) generate a class C_0 semi-group, $\tau(t)$, on $C[[-h,0],X]$. The infinitesimal generator, $\mathcal{a}$, of $\tau(t)$ satisfies the following theorem:

1. Theorem. The domain of $\mathcal{a}$ consists of those $\phi \in C[[-h,0],X]$ which posses continuous derivatives, for which $\phi(0) \in \mathcal{D}(A)$ and which satisfy the relation

$$\dot{\phi}(0) = \sum_{j=1}^{m} B_j \dot{\phi}(-h_j) + A\phi(0) + \sum_{j=1}^{m} A_j \phi(-h_j). \tag{4}$$

If ϕ is in the domain of $\mathcal{a}$ then the solution of (2) satisfies the differential equation (1).

The next theorem describes a representation for the solutions of (2) which is the analogue of the one found in the finite dimensional case.

2. Theorem. There exists on $[0,\infty)$ a family of bounded linear mappings, $\{S(t)\}$, from X into itself such that (i) $S(t)$ is strongly continuous on $[0,\infty)$ except possibly at points of the form

$$n_1h_1 + \cdots + n_mh_m, \tag{5}$$

where $n_1, \cdots n_m$ are positive integers or zero and $n_1 + \cdots + n_m \neq 0$, (ii) $S(0) = I$ (the identity mapping), (iii) $|S(t)| = 0$ if $t < 0$ and (iv) if ϕ and $\dot{\phi}$ are in $C[[-h,0],X]$ the solution of (2) for $t \geq 0$ can be written uniquely in the form

$$x(t) = [S(t) - \sum_{j=1}^{m} S(t-h_j)B_j]\phi(0) + \sum_{j=1}^{m} \int_{-h_j}^{0} S(t-\sigma-h_j)[A_j\phi(\sigma) + B_j\dot{\phi}(\sigma)]d\sigma. \tag{6}$$

If $\dot{\phi}$ is not continuous or does not exist there exists a further family of bounded linear mappings, $\{R(t)\}$, from X into itself which is strongly continuous, except possibly at points of the type described by (5), and such that for $t \geq 0$

every solution of (2) can be written in the form

$$x(t) = S(t)[\phi(0) - \sum_{j=1}^{m} B_j\phi(-h_j)] + \sum_{j=1}^{m} \int_{-h_j}^{0} S(t-\sigma-h_j)A_j\phi(\sigma)d\sigma + R(t)\phi. \quad (7)$$

The next theorem defined a relationship between T(t) and S(t).

3. Theorem. The mappings {S(t)} described in Theorem 2 satisfy the recursive relation

$$S(t) = T(t) + \sum_{j=1}^{m} \chi(t-h_j)B_jS(t-h_j) + \int_{0}^{t} T(t-\sigma)[\sum_{j=1}^{m} \chi(\sigma-h_j)(A_j+AB_j)S(\sigma-h_j)d\sigma, \quad (8)$$

where $\chi(\sigma) = 1$ if $\sigma > 0$ and $\chi(\sigma) = 0$ if $\sigma < 0$.

REFERENCE

1. Hille, E., and Phillips, R. S., Functional analysis and semi-groups, *A.M.S. Colloquium Publications, Vol. XXXI*, 1957.

PERIODIC SOLUTOINS OF DIFFERENTIAL SYSTEMS OF LIÉNARD AND RAYLEIGH TYPE

Richard I. DeVries
Department of Mathematics
University of Michigan

We consider here the problem of existence of periodic solutions to nonlinear second order differential systems of Liénard and Rayleigh type.

1. We first consider Liénard systems with forcing terms of the type

$$x''(t)=(d/dt)F(x(t))+(d/dt)V(x(t),t)+Ax(t)+g(x(t))=e(t) \qquad (1)$$

$x(t)=(x_1(t),\dots,x_n(t))$, $F(x)=(F_1(x),\dots,F_n(x))$, $A = (a_{ij})$
$V(x,t)=(V_1(x,t),\dots,V_n(x,t))$, $g(x)=(g_1(x),\dots,g_n(x))$, $e = (e_1,\dots,e_n)$

where we assume that $e(t)$ is a 2π-periodic continuous function of mean value zero, that A is a constant $n \times n$ matrix, that $V(x,t)$ is 2π-periodic in t for every x and is such that all V_i, $\partial V_i/\partial x_j$, $i, j = 1,\dots,n$, are of class C^1 in R^{n+1}, that F is the gradient of some function $G(x)$ mapping R^n to R^1 and is such that all F_i, $\partial F_i/\partial x_j$, $i, j = 1,\dots,n$ are of class C^1 in R^n, and that $g(x)$ is continuous in R^n and such that $g(x)/|x| \to 0$ as $|x| \to +\infty$.

Cesari and Kannan [3,4] proved existence theorems for systems of this very general type using only qualitative hypotheses. Their proof was in the framework of Cesari's alternative method [1], using degree theory and a priori bounds. These theorems required that F be of the form $F(x) = \text{grad } G(x)$, $G(x)$ homogeneous of order $2p$, $p > 0$, and of constant sign. However, modifications in the proofs of these theorems allows this condition to be dropped so that the theorems can be restated as below.

(1.i) <u>Theorem</u>. Under the above general assumptions, the system (1) has at least one 2π-periodic solution $x(t)$, $-\infty < t < +\infty$, provided one of the following assumptions holds:

(α) There exist constants c,d,C,D,C',D' with $c > 0$ such that for some integer $p > 1$ $xF(x) = \sum_1^n (x_i F_i(x)) > c|x|^{2p} + d$, $V(x,t) = V_1(x,t) + V_2(x,t)$, where $V_1(x,t) = P(t)\text{grad.}(x)$ for some C^2 $W(x)$ and $P(t)$ a 2π-periodic function which is C^1 from R^1 to R^n, and $|V_1(x,t)| < C'|x|^{2p-2} + D'$, $|V_2(x,t)| < C|x|^p + D$. Finally, we assume that A is a nonsingular $n \times n$ constant matrix.

(α') $g \equiv 0$, F and V as in (α) above, and A an arbitrary constant matrix.

(β) The same as in (α) with $p = 1$ and $c > C + ||A-A_{-1}||/2$.

(β') $g \equiv 0$, F and V as in (β) above, and A an arbitrary constant matrix.

(γ) The same as (α) with $|V_1(x,t)| < C'|x|^{2p-1} + D'$, $c > C'$ for $p > 1$, and $c > C + C' + ||A-A_{-1}||/2$ for $p = 1$.

(γ') $g \equiv 0$, F and V as in (γ) above, and A an arbitrary constant matrix.

The proof (see [5]) is a suitable modification of the proofs contained in the work of Cesari and Kannan (see [3] and [4]).

It is now possible to see the following result of H. Knolle [6] as a special case of the extended version of the Cesari-Kannan Theorem (1.i).

(1.ii) The Liénard type system with 2π-periodic forcing terms $e_i(t)$

$$x_i'' + \sum_{i}^{n}((d/dt)F_{ik}(x_k(t)) + a_{ik}x_k) = e_i(t) \quad i=1,\ldots,n \quad F_{ik}\in C^1 \qquad (2)$$

has at least one 2π-periodic solution if $A = (a_{ik})$ is symmetric nonsingular and

(i) $xF_{ii}(x) > c_{ii}|x|^{2p} + d$ for constants $c_{ii} > 0$, $p > 1$, and d

(ii) $F_{ik}(x) < c_{ik}|x|^p + b_{ik}$ for some constants c_{ik}, b_{ik}

(iii) $e_i(t)$ has mean value zero for $i = 1,\ldots,n$

Setting $F_i(x_1,\ldots,x_n) = F_{ii}(x_i)$, $i = 1,\ldots,n$, $V_i(x_1,\ldots,x_n,t) = \sum_{i\neq k} F_{ik}(x_k)$, $i = 1,\ldots,n$, $F(x) = (F_1,\ldots,F_n)$, and $V(x,t) = (V_1,\ldots,V_n)$, clearly case ($\alpha$) of (1.i) will be satisfied by F and V defined in this way. Knolle's statement (1.ii) becomes a particular case of (1.i).

As we shall see in [5] the existence theorem (1.i) is invariant under transformations of the type $x(t) = y(t)+P(t)$ where $P(t)$ is 2π-periodic and C^1 from R^1 to R^n. Precisely, equation (1) with F,V,A, and g as in (1.i) is transformed by $x = y + P(t)$ into an equation of the form

$$y'' + (d/dt)F(y) + (d/dt)\tilde{V}(y,t) + Ay + \tilde{g}(y) = \tilde{e}(t)$$

where, as we shall see, F, $\tilde{V}$, A, $\tilde{g}$, $\tilde{e}$ still satisfy the conditions of (1.i).

2. The same type analysis as was applied to the Liénard equation above can be used to formulate existence theorems for systems of the form

$$x''(t)+(G_1(x')+G_2(x',t))+H(x,x',t)+(F_1(x)+F_2(x,t))=e(t,x,x') \quad (3)$$

where $x = (x_1,\ldots,x_n)$, $G_1:R^n \to R^n$, $G_2:R^{n+1} \to R^n$, $H:R^{2n+1} \to R^n$, $F_1:R^n \to R^n$, $F_2:R^{n+1} \to R^n$, $e:R^{2n+1} \to R^n$, and e, G_2, F_2, and H are all 2π-periodic in t for each x and x'.

Mawhin [8] has proved existence theorems for Rayleigh type systems which can be reduced to the form (3) with $G_2 = F_2 = H = 0$. Using the same type of analysis as required in Section 1 we have extended the results in [8] to include the case of nontrivial G_2, F_2, and H.

The existence theorems in [8] required that F_1 and G_1 satisfy certain symmetry conditions. In our extension of these results we make use of the following definition due to Hale [6].

(2.i) Definition: The system $x' = \lambda f(x,t,\lambda)$, $\lambda \epsilon (0,1)$, has property E with respect to (Q,ε,τ) is there exists a constant matrix Q and two real numbers ε and τ such that $Q^2 = I$, $\varepsilon^2 = 1$, $Qf(Qx,\varepsilon t+\tau,\lambda) = f(x,t,\lambda)$ for each λ. We shall assume e, G_1, G_2, F_1, F_2, and H are all such that the above system satisfies property E with respect to $(-I,1,\pi)$, that is:

$$e(t+\pi,-x,-x') = -e(t,x,x') \qquad H(t+\pi,-x,-x') = -H(t,x,x')$$

$$F_1(-x) = -F_1(x) \qquad G_1(-x) = -G_1(x)$$

$$F_2(-x,t+\pi) = -F_2(x,t) \qquad G_2(-x',t+\pi) = -G_2(x',t)$$

We will also assume that e, G_1, G_2, H, F_1, and F_2 are all continuous functions.

(2.ii) <u>Theorem</u>. Under the above general assumptions there exists at least one 2π-periodic odd-harmonic solution to the system (3) providing there exists some integer $p \geq 1$ and constants c,C,d,D,e,f,h,H,C',e',h',M such that one of the following conditions is satisfied:

(A) $|x'G_1(x')| = |\sum_j x'_j G_{1j}(x)| > c|x'|^{2p} + d$ with $c > 0$

$|G_2(x',t)| < C|x'|^{2p-2} + D \qquad |e(t,x,x')| < M$

$|F_2(x,t)| < e|x|^{2p-2} + f$

$|H(x,x',t)| < h|x|^q|x'|^1 + H$ with $q + 1 \leq 2p - 2$

(B) $|x'G_1(x')| > c|x'|^{2p} + d$

$|G_2(x',t)| < c'|x'|^{2p-1} + C|x'|^{2p-2} + D$

$|F_2(x,t)| < e'|x|^{2p-1} + e|x|^{2p-2} + f$

$|H(x,x',t)| < h'|x|^{q'}|x'|^{l'} + h|x|^{q}|x'|^{l} + H$ $q' + l' \leq$ $2p-1$ and $q+l \leq qp-2$

$|e(t,x,x')| < M$

where $c > C' + (2\pi)^{2p-1}e' + (2\pi)^{q'}h'$ and $c > 0$

For the proof we refer to [5].

REFERENCES

1. Cesari, L., Alternative methods in nonlinear analysis, (International Conference on Differential Equations, Los Angeles) 1974.

2. Cesari, L., *Asymptotic Behavior and Stability Problems in Ordinary Differential Equations*, Springer-Verlag, 1971.

3. Cesari, L., and Kannan, R., Periodic solutions in the large of nonlinear ordinary differential equations, *Rendiconti di Matematica 8*(1975), (2) Serie VI.

4. Cesari, L., and Kanna, R., Solutions in the large of Liénard systems with forcing terms, to appear in *Annali Matem. pura appl.*

5. DeVries, R., Solutions in the large of nonlinear dynamical systems, University of Michigan Thesis, 1977.

6. Hale, J. K., *Oscillations in Nonlinear Systems*, McGraw-Hill, New York, 1963.

7. Knolle, H., Existenz periodischer Losungen eines heteronomen Lienarschen systems, *Angewandte Analysis and Mathematische Physik, ZAMM 55*(1975).

8. Mawhin, J., Periodic solutions of strongly nonlinear differential systems, (International Conference on Nonlinear Oscillations, Kiev, 1969, Vol. 1), 380-399.

EXISTENCE OF AN OPTIMAL CONTROL FOR STOCHASTIC SYSTEMS GOVERNED BY ITO EQUATIONS

Robert M. Goor
Research Laboratories
Mathematics Department
General Motors Technical Center

In this paper, we state an existence theorem for stochastic control problems with a lower semi-continuous cost functional, and governed by a system of Ito integral equations. The proof makes use of the properties of a weakly convergent sequence of probability measures (see [1] for terminology) and is given in [4]. The approach is similar to that developed in [3]. The result presented here is a generalization of that of Kushner [5] in that we do not require uniform integrability of the covariance terms of admissible trajectories as is done in [5]. Instead, we assume the weaker condition of tightness of the drift and covariance terms.

We take as given a fixed time interval $[0,T]$, and, for given positive integers m and n, a fixed closed subset U of R^m and a fixed element x_0 of R^n. We define C_t^n to be the space of continuous maps from $[0,t]$ into R^n for $0 \le t \le T$, and if $x \in C_T^n$, we define x_t to be the element of C_t^n obtained from x by restriction to $[0,t]$. We let W_n denote Weiner measure on C_T^n and we let w denote an arbitrary R^n-valued standard Weiner process with independent components. That is, w maps some probability space Ω into C_T^n and has W_n as distribution. We let ϕ denote a real-valued functional on C_T^n, that is lower semi-continuous. We let f be a continuous map from $[0,T] \times C_T^n \times U \to R^n$ such that $f(t,x,u) = f(t,x_t,u)$ for all (t,x,u), that is, f is non-anticipative. Finally, we let σ denote a continuous map from $[0,T] \times C_T^n$ into the set of positive semi-definite $n \times n$ matrices such that $\sigma(t,x) = \sigma(t,x_t)$ for all (t,x).

If $z \in R^n$, we denote by $|z|$ the usual Euclidean norm of z. If B is an $n \times n$ matrix, we denote by $||B||$ the expression $(\text{Tr } BB^*)^{1/2}$, where Tr is the trace operator.

We consider the Ito equation

$$x(t) = x_o + \int_0^t f(s,x_s,u(s))ds + \int_0^t \sigma(s,x_s)dw(s), \tag{1}$$

where $u:[0,T] \times \Omega \to U$, is measurable. We let $\mathcal{D}$ be the class of such maps u such that: u yields a non-anticipative response x in C_T^n via (1); u, in turn, is non-anticipative with respect to the x process; and the following three relations hold:

$$\int_0^T E\{|f(t,x_t,u(t))|\}dt < +\infty; \tag{2}$$

$$\int_0^T ||\sigma(t,x_t)||^2 dt < +\infty \quad \text{a.s.}; \tag{3}$$

$$E\{\phi(x)\} < +\infty, \tag{4}$$

where $E\{\cdot\}$ represents the expectation operator. If $u \in \mathcal{D}$ and x is a response via (1), we will say that the pair $[x,u]$ is admissible. It is possible to show that if $[x,u]$ is admissible, there is a measurable map $\hat{u}$: $[0,T] \times C_T^n \to U$ such that $\hat{u}$ in non-anticipative and $\hat{u}(t,x_t) = u(t)$ a.s. That is, $\mathcal{D}$ represents a class of feedback controls.

We define $J[x,u] = E\{\phi(x)\}$ for admissible pairs $[x,u]$. The problem, then, is to minimize $J[x,u]$ for $u \in \mathcal{D}$.

We will require an upper semi-continuity property for the set-valued function $Q(t,x) = f(t,x,U)$. We will say that $Q(t,x)$ satisfies property (Q) (with respect to the x variable only) if, for all (t,x),

$$Q(t,x) = \bigcap_{\delta>0} \overline{co} \bigcup_{y \in N_\delta(x)} Q(t,y)$$

where $\overline{co}$ signifies the closure of the convex hull and $N_\delta(x)$ is the δ-neighborhood of x in C_T^n. Property (Q) was developed by Cesari (see [2] for example) for deterministic control problems with unbounded control spaces.

Theorem. Assume that: $\mathcal{D}$ is non-empty; $f(t,\cdot,u)$ is bounded on bounded C_T^n sets, uniformly for $(t,u) \in [0,T] \times U$; $Q(t,x)$ satisfies property (Q); there is a constant K so that, uniformly in admissible pairs $[x,u]$,

$$E\{\int_0^T |f(s,x_s,u(s))|ds\} \le K \tag{5}$$

and

$$E\{\int_0^T ||\sigma(s,x_s)||^2 ds\} \le K; \tag{6}$$

and, finally, given $\varepsilon > 0$, $\eta > 0$, there is a $\delta > 0$ so that, uniformly in $[x,u]$,

$$\Pr\{\sup_{|s-t|<\delta} |\int_s^t f(\tau,x_\tau,u(\tau))d\tau| > \varepsilon\} < \eta \tag{7}$$

and

$$\Pr\{\sup_{|s-t|<\delta} |\int_s^t \sigma(\tau,x_\tau)dw(\tau)| > \varepsilon\} < \eta. \tag{8}$$

Then, $J[\cdot]$ attains its minimum in the class $\mathcal{D}$.

We note that properties (5) - (8) entail tightness of the set of admissible trajectories, but not uniform integrability. See [4] for details.

REFERENCES

1. Billingsley, P., *Convergence of Probability Measures*, John Wiley, New York, 1968.

2. Cesari, L., Existence theorems for weak and usual solutions in Lagrange problems with unilateral constraints, *Trans. A.M.S. 124*(1966), 369-412, 413-429.

3. Goor, R. M., Existence of an optimal control for systems with jump Markov disturbances, *SIAM J. Control 14*(1976), 899-918.

4. Goor, R. M., Existence of an optimal control for stochastic systems governed by Ito equations, *J.O.T.A.*, to appear.

5. Kushner, H. J., Existence results for optimal stochastic controls, *J.O.T.A. 15*(No. 4, 1975), 347-359.

ASYMPTOTIC BEHAVIOR OF SOLUTIONS OF NONLINEAR FUNCTIONAL DIFFERENTIAL EQUATIONS

John R. Graef and Paul W. Spikes
Department of Mathematics
Mississippi State University

The results announced in this paper are motivated by the recent results of Kartsatos [3], Staikos and Sficas [5], and the present authors [1]. Kartsatos [3] considered the n-th order nonlinear differential equation

$$x^{(n)} + f(t,x,x',\ldots,x^{(n-1)}) = r(t,x,x',\ldots,x^{(n-1)}) \qquad (*)$$

and gave sufficient conditions for any bounded nonoscillatory solution $x(t)$ of (*) to satisfy

$$\liminf_{t\to\infty} |x(t)| = 0. \qquad (**)$$

He then raised the question as to whether conditions could be found which would guarantee that all solutions of (*) satisfy (**). Staikos and Sficas [5] improved Kartsatos' result and at the same time extended it to equations with perturbed arguments, but again this was for bounded solutions. Under more restrictive conditions [5; Theorem 2], they also showed that any solution $x(t)$ satisfying

$$|x(t)| = 0(t^k), \quad t \to \infty$$

satisfied (**) as well.

In Theorem 1 below we answer the question raised by Kartsatos but for the more general equation with perturbed arguments as studied by Staikos and Sficas. As a consequence of this theorem we are able to extend results of Graef and Spikes [1], Kartsatos and Manougian [4] and Staikos and

Sficas [6]. Compelte details of the results presented here will appear in [2].

Consider the equation

$$x^{(n)}+f(t,x,x',\ldots,x^{(n-1)},x(\tau_1(t)),x'(\tau_2(t)),\ldots,x^{(n-1)}(\tau_n(t)))$$
$$=r(t,x,x',\ldots,x^{(n-1)},x(\tau_1(t)),x'(\tau_2(t)),\ldots,x^{(n-1)}(\tau_n(t))) \quad (1)$$

where $f,r\colon [t_0,\infty) \times R^{2n} \to R$, $\tau_i\colon [t_0\infty) \to R$ are continuous and $\tau_i(t) \to \infty$ as $t \to \infty$, $i = 1,2,\ldots,n$.

1. Theorem. If there exists $K \geq 0$ with the property that for any $u \in C^n[t_0,\infty)$ such that $\liminf_{t\to\infty} u(t) > K$ ($\limsup_{t\to\infty} u(t) < -K$) we have

$$\int^{\infty}\{f(t,u(t),\ldots,u^{(n-1)}(t),u(\tau_1(t)),\ldots,u^{(n-1)}(\tau_n(t)))$$
$$- r(t,u(t),\ldots,u^{(n-1)}(t),u(\tau_1(t)),\ldots,u^{(n-1)}(\tau_n(t)))\}dt=$$
$$+\infty \ (-\infty),$$

then every nonoscillatory solution $x(t)$ of (1) satisfies $\liminf_{t\to\infty} |x(t)| \leq K$.

2. Corollary. For the equation

$$x^{(n)}+q(t)f(t,x,x',\ldots,x^{(n-1)}) = r(t,x,x',\ldots,x^{(n-1)}) \quad (2)$$

assume that

(i) $f,r\colon[t_0,\infty) \times R^n \to R$ and $q\colon[t_0,\infty) \to R$ are continuous and $q(t) > 0$,

(ii) $x_1 f(t,x_1,\ldots,x_n) \geq 0$ is $x_1 \neq 0$, and $f(t,x_1,\ldots,x_n)$ is bounded away from zero if x_1 is bounded away from zero,

(iii) $|r(t,x_1,\ldots,x_n)| \leq h(t)$ and $\int_{t_0}^{\infty} q(s)ds = \infty$.

If $h(t)/q(t) \to 0$ as $t \to \infty$, then any nonoscillatory solution $x(t)$ of (2) satisfies $\liminf_{t\to\infty} |x(t)| = 0$.

3. Theorem. Suppose that $r(t,x,\ldots,x^{(n-1)}(\tau_n(t))) \equiv 0$,

$$x_1 f(t,x_1,\ldots,x_n,y_1,\ldots,y_n) \geq 0 \quad (3)$$

if both x_1 and y_1 are nonnegative or nonpositive, and the conditions of Theorem 1 are satisfied with $K = 0$. Then if n is even, all solutions of (1) are oscillatory, while if n is odd, all solutions are oscillatory or tend monotonically to zero together with their first n-1 derivatives.

Remark. When $K = 0$, Theorem 1 answers the question raised by Kartsatos [3]. Corollary 2 generalizes Lemma 10 in [1]; Theorem 3 extends Theorem 2 in [6] and Lemma 9 in [1].

The following theorem generalizes Theorem 2.1 in [4].

4. Theorem. Suppose that $r(t,x,\ldots,x^{(n-1)}(\tau_n(t))) \equiv r(t)$ and condition (3) holds. If for every $T \geq t_0$ and every $k > 0$ we have

$$\limsup_{t\to\infty} \left[\int_T^t (t-s)^{n-1} r(s)ds - kt^{n-1}\right] \geq 0$$

and

$$\liminf_{t\to\infty} \left[\int_T^t (t-s)^{n-1} r(s)ds + kt^{n-1}\right] \leq 0,$$

then every solution of (1) is oscillatory.

REFERENCES

1. Graef, J. R. and Spikes, P. W., Asymptotic behavior of solutions of a second order nonlinear differential equation, *J. Differential Equations* *17*(1975), 461-476.

2. Graef, J. R. and Spikes, P. W., Asymptotic properties of solutions of functional differential equations of arbitrary order, *J. Math. Anal. Appl.*, to appear.

3. Kartsatos, A. G., On the maintenance of oscillations of nth order equations under the effect of a small forcing term, *J. Differential Equations* *10*(1971), 355-363.

4. Kartsatos, A. G. and Manougian, M. N., Further results on oscillation of functional - differential equations, *J. Math. Anal. Appl.* *53*(1976), 28-37.

5. Staikos, V. A. and Sficas, Y. G., Forced oscillations for differential equations of arbitrary order, *J. Differential Equations* *17*(1975), 1-11.

6. Staikos, V. A. and Sficas, Y. G., Oscillatory and asymptotic behavior of functional differential equations, *J. Differential Equations* *12*(1972), 426-437.

BEFORE BIFURCATION AND NONLINEARITY: SOLUTIONS OF THE LINEAR PROBLEMS

John Gregory
Department of Mathematics
Southern Illinois University at Carbondale

I. INTRODUCTION

The purpose of this brief note is to announce a uniform numerical theory to deal with prebifurcation and prenonlinearity problems, that is, self adjoining problems on Hilbert Spaces. To date we have constructed computer programs which solve second order equations such as $L(x) = (rx')' + px = 0$ or the associated eigenvalue problem $L(x) = (rx')' + px - \lambda qx = 0$. Our methods are numerical solutions of the appropriate extremal problems and are being applied to higher order differential equations and partial differential equations. As a biproduct we obtain new theoretical and numerical results for symmetric banded matrices. Trial results indicate that our programs are more accurate, faster, and require significantly less storage than more conventional numerical methods.

II. BASIC IDEAS

We now discuss the correspondence between the differential equations $L(x) = 0$ described above which is the Euler Lagrange equation for the appropriate quadratic functional $J(x)$. For convenience of presentation we assume that $L(x)$ is second order although the reader may picture $L(x)$ as an eigenvalue equation, self adjoint (2nth order) differential equation, or elliptic partial differential equation. A convenient picture is

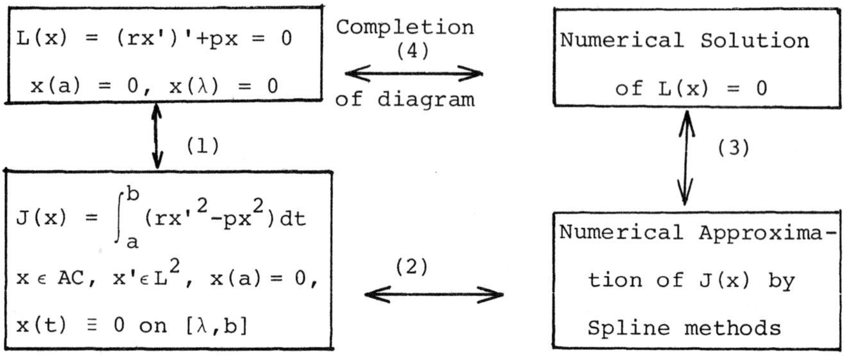

Briefly arrow (1) represents the equivalence between focal (oscillation) point problems and extremal quadratic functionals $J(x)$ initially given by Hestenes and then by the author. Arrow (2) denotes the "equivalence" between $J(x)$ and finite dimensional approximations associated with a symmetric tridiagonal matrix $D(\sigma)$. Arrow (3) represents the "Euler-Lagrange" solution $x_\sigma(t)$ of $D(\sigma)$. Thus if $x_0(t)$ is the solution of $L(x) = 0$, then the integral of the square of $x_0'(t) - x_\sigma'(t)$ goes to 0 as σ does. The proofs of these results are very technical and involve an approximation theory previously given by the author.

III. NUMERICAL APPROXIMATIONS

The main thrust of this paper is the approximation of quadratic form and the associate Euler-Lagrange solution as symbolized by the arrows (2) and (3). We will omit a discussion of negative spaces (indices or signature) and of this unique type of approximation theory which can be found in previous works by the author. For more general problems the ideas of this section are unchanged, but the "tools" (such as spline approximating elements and banded matrix) are generalized in the similar manner.

Thus let A denote the arcs $x(t)$ which are absolutely continuous on $[a,b]$ with $x'(t)$ in $L^2[a,b]$, and such that $x(a) = x(b) = 0$. Let $\Sigma = \{\frac{1}{n} \mid n = 1,2,3,\ldots\} \cup \{0\}$. For each $\sigma = 1/n$ choose a partition $\pi(\sigma) = (a=a_0<a_1<a_2\ldots<a_N\leq b)$ of $[a,b]$ where $a_k = k\sigma + a$. Let (σ) denote the space of broken linear functions with vertices at $\pi(\sigma)$, that is, the spline space of degree two with basis elements $z_k(t) = 1 - |t-a_k|/\sigma$ if $t \in [a_{k-1}, a_{k+1}]$, and $z_k(t) \equiv 0$ otherwise. Finally, choose $r_\sigma(t) = r(a_k)$ and $p_\sigma(t) = p(a_k)$ if $t \in [a_k, a_{k+1}]$. Then our approximating quadratic form defined on (σ) is

$$J(x;\sigma) = \int_a^b [r_\sigma(t)x'^2(t) - p_\sigma(t)x^2(t)]dt$$

$$= b_\alpha b_\beta e_{\alpha\beta}(\sigma) = X^T D(\sigma)X$$

where repeated indices are summed, $x = (b_1,b_2,\ldots)^T = b_\alpha z_\alpha$ is in (σ), and $e_{\alpha\beta}(\sigma) = J(z_\alpha,z_\beta;\sigma)$. We note that $D(\sigma)$ is a symmetric tridiagonal matrix.

<u>Theorem</u>. <u>Let</u> $x_\sigma(t) - c_\alpha z_\alpha$ <u>be normalized so that</u> $x'_\sigma(a) = x'_0(a)$ <u>where</u> $x_0(t)$ <u>is a nontrivial solution to</u> $L(x) = (rx')' + px = 0$, $x(a) = 0$ <u>and</u> c_α <u>satisfies</u> $c_1e_{11} + c_2e_{12} = 0$, $c_{r-1}e_{r,r-1} + c_re_{rr} + c_{r+1}e_{r,r+1} = 0$ $(r = 2,3,4,\ldots)$ <u>then</u>

$$\lim_{\sigma\to 0} \int_a^b (x'_0(t) - x'(t))^2 dt = 0.$$

We note that intuitively the numbers c_α are chosen to satisfy the "Euler-Lagrange equation" for $D(\sigma)$ namely $D(\sigma)C = 0$ where $C = (c_1,c_2,c_3,\ldots)^T$. Similar solutions of appropriate banded symmetric matrices lead to numerical solutions of corresponding equations $L(x) = 0$ and extremals of $J(x)$.

REFERENCES

1. Gregory, John, An approximation theory for elliptic quadratic forms on Hilbert spaces, *Pacific J. of Math.* *37* (No. 2, 1970), 383-395.

2. Gregory, John, and Richards, Franklin, Numerical approximation for 2nth order differential equations via splines, *Rocky Mountain Journal* *5* (No. 1, Winter 1975), 107-116.

SUMS OF RANGES OF OPERATORS AND APPLICATIONS

Chaitan P. Gupta
Department of Mathematical Sciences
Northern Illinois University

Let H denote a Hilbert space and let A, B be two operators on H with ranges R(A), R(B) in H. It is well-known that, in general R(A) + R(B) is much larger than the range R(A+B) of the operator A + B. Recently, Brézis ([1]) observed that R(A) + R(B) is almost equal to R(A+B), in the sense that Int[R(A)+R(B)] = Int R(A+B) and cl[R(A)+R(B)] = cl R(A+B), when both A and B are monotone operators and satisfied some additional conditions. The equality Int [R(A)+R(B)] = Int R(A+B) or, more generally, the inclusion Int[R(A)+R(B)] $\subset$ R(A+B) gives a sufficient condition, on the right hand side f, for the solvability of the equation Au + Bu = f in H. This condition is analogous to the well-known Landesman-Lazer condition for the solvability of a nonlinear elliptic boundary value problem. We observe that a further study of the inclusion Int[R(A)+R(B)] $\subset$ R(A+B), leads to a sufficient condition for the solvability of the perturbed equation Au + Bu + B'u=f, namely, Int[R(A)+R(B)] $\subset$ R(A+B+B'). We shall give here only the statements of some applications of theorems (in [3], [4]) on sums of ranges of operators to nonlinear elliptic boundary value problems, for reasons of lack of space. Details of proofs will appear elsewhere. We also refer to the bibliographies of [1] - [4] for further references.

Let, now, Ω be a bounded domain in an Euclidean space $\mathbb{R}^N (N \geq 1)$ with smooth boundary Γ. Let $g\colon \Omega \times \mathbb{R} \to \mathbb{R}$ be a function satisfying Caratheodory conditions. Suppose that there exist $a(x) \in L^2(\Omega)$ and a constant $b \geq 0$ such that $|g(x,t)| \leq a(x)+b|t|$ for $x \in \Omega$, $t \in \mathbb{R}$ and (ii) there is a $T(x) \in L^2(\Omega)$, $T(x) \geq 0$ a.e. such that $g(x,t)t \geq 0$ for $|t| \geq T(x)$ a.e. For $x \in \Omega$, let $g_+(x) = \liminf_{t\to\infty} g(x,t)$ and $g_-(x) = \limsup_{t\to-\infty} g(x,t)$.

1. Theorem. Let $f \in L^2(\Omega)$. The Neumann-boundary value

$$-\Delta u + g(x,u) = f \qquad \text{a.e. on } \Omega$$

$$\frac{\partial u}{\partial n} = 0 \qquad \text{a.e. on } \Gamma$$

is solvable provided

$$\int_\Omega g_- < \int_\Omega f < \int_\Omega g_+ .$$

Let $A\colon D(A) \subset L^2(\Omega) \to L^2(\Omega)$ be defined by $Au = -\Delta u - \lambda u$, $D(A) \subset H^2(\Omega) \cap H_0^1(\Omega)$, λ on eigen-value of $-\Delta$. Suppose that $|g(x,t)| \leq a(x)$ for $x \in \Omega$ a.e. and $t \in \mathbb{R}$ and $a(x) \in L^2(\Omega)$. Suppose further that $g_+(x) \liminf_{t\to\infty} g(x,t)$ and $g_-(x) = \limsup_{t\to-\infty} g(x,t)$ uniformly for $x \in \Omega$.

2. Theorem. Let $f \in L^2(\Omega)$. The boundary value problem

$$-\Delta u - \lambda u + g(x,u) = f \quad \text{a.e. on } \Omega$$

$$u = 0 \quad \text{a.e. on } \Gamma$$

is solvable, provided

$$\int_\Omega fv\,dx < \int_\Omega (g_+v^+ - g_-v^-)\,dx$$

for $v \in \ker A$, $v \neq 0$. ($v^+ = \max(v,0)$ and $v^- = -\min(v,0)$).

REFERENCES

1. Brézis, H., Monotone operators, nonlinear semi-groups and applications, *Proc. Int. Congress Math. Vancouver, Vol. II*(1974), 249-255.

2. Brézis, H., and Haraux, A., Sur l'image d'une somme d'operateurs monotones, applications, *Israel J. Math.* *23*(1976), 165-186.

3. Gupta, C. P., and Hess, P., Existence theorems for non-linear non-coercive operator equations and nonlinear elliptic boundary value problems, *Jour. Diff. Equations*, to appear.

4. Gupta, C. P., Sums of ranges of operators and nonlinear elliptic boundary value problems, submitted.

RADIATION REACTION IN ELECTRODYNAMICS

Deh-phone K. Hsing
and
R. D. Driver
Department of Mathematics
University of Rhode Island

In 1943 Eliezer [2] studied the Dirac equation for the one-dimensional motion of a classical point electron under the influence of a stationary point charge. He obtained the paradoxical conclusions that if the two charged particles have opposite signs they can never collide, but if they have like signs they may collide.

But really, one cannot consider even a massive point charge to be stationary when another particle comes close to it. And if both particles move, then, due to the finite speed of propagation of electromagnetic effects, one must analyze a system of delay differential equations with state-dependent delays. This has recently been done, and Eliezer's conclusions remain intact [3].

To present the essential ideas in the simplest possible form and to make this announcement self contained, we consider a special case.

Let two particles having identical rest masses and having charges of equal magnitude be moving symmetrically on the x-axis with positions $x(t)$ and $-x(t)$ at time t. Then the equations of motion become

$$c\tau = x + x(t-\tau), \tag{1}$$

$$x' = cv, \tag{2}$$

$$\frac{av'}{(1-v^2)^{3/2}} - \frac{v''}{(1-v^2)^2} - \frac{3vv'^2}{(1-v^2)^3} = \frac{b}{\tau^2}\,\frac{1-v(t-\tau)}{1+v(t-\tau)}, \tag{3}$$

where τ, x, v, v', and v'' stand for $\tau(t)$, $x(t)$, $v(t)$, $v'(t)$, and $v''(t)$. In Eqs. (1) - (3), a, b, and c are constants with a and c (the speed of light) being positive and b being positive (or negative) if both charges have the same (or opposite) signs. The second and third terms on the left hand side of (3) are the "radiation reaction" terms.

Lemma (from [1]). Let $\alpha \le 0 < \beta$ and let $x \in C^1$ with $|x'(t)| = c|v(t)| < c$ on $[\alpha,\beta)$ and $x(t) > 0$ on $[0,\beta)$. Then $\tau(t)$ is a solution of Eq. (1) on $[0,\beta)$ if and only if $\tau(0)$ satisfies (1) at $t = 0$ and

$$\tau' = \frac{v+v(t-\tau)}{1+v(t-\tau)} \text{ on } [0,\beta). \tag{1'}$$

Moreover, in this case $\tau(t)$ is unique and $t-\tau(t)$ is strictly increasing. If $v(t) \ge -M > -1$ on $[\alpha,\beta)$, then $\tau(t) \le 2x(t)/c(1-M)$ on $[0,\beta)$.

Now introduce $z \equiv v/(1-v^2)^{1/2}$. Then $z' = v'/(1-v^2)^{3/2}$, and (3) reduces to

$$z'' - \frac{az'}{(1+z^2)^{1/2}} = -\frac{b}{\tau^2} f(z,z(t-\tau)), \tag{3'}$$

where f is C^1 and $0 < f(\xi,\eta) \le 1 + 2|\eta| + 4\eta^2$ on R^2.

1. Theorem. Let $x(t) = \phi(t)$ for $t \le 0$, $v(0) = \phi'(0-)/c$, anc $v'(0) = w_0$ be given, where $\phi(0) > 0$, ϕ' is Lipschitzian with $|\phi'(t)| < c$, and Eq. (1) has a solution at $t = 0$. Then x has a unique C^1 extension to $\beta > 0$ which satisfies (1), (2), and (3) on $[0,\beta)$. Moreover, either $\beta = \infty$ or $x(t) \to 0$ as $t \to \beta$ -- a collision. (In [3] it is shown that the Lipschitz condition on ϕ' can be relaxed to ordinary continuity.)

The proof uses a "method-of-steps" argument to find τ, x, v, and v' from Eqs. (1'), (2), and (3) for $t > 0$. Assume the solution has been found for $t < \beta < \infty$ and cannot be extended any further; and suppose (for contradiction) that $x(t) \not\to 0$ as $t \to \beta$. Then $x(t) \geq \delta > 0$, and so $\tau(t) \geq \delta/c$ on $[0,\beta)$. This in turn implies $z(t-\tau)$ bounded on $[0,\beta)$. Hence, from (3'),

$$|z'' - a(1+z^2)^{-1/2}z'| \leq A, \text{ some constant.}$$

Now multiply through by an "integrating factor" to find

$$\left|\frac{d}{dt}\{z'(t)e^{-\int_0^t a[1+z^2(s)]^{-1/2}ds}\}\right| \leq Ae^{-\int_0^t \ldots ds} \leq A.$$

Integrating and rearranging terms, one then finds

$$|z'(t)| \leq [|z'(0)| + A\beta]e^{a\beta} \text{ on } [0,\beta).$$

This bound for $z'(t)$ also assures that $z(t)$ is bounded. Therefore $v'(t)$ is bounded and $|v(t)| \leq M_1 < 1$ on $[0,\beta)$. It follows that the solution can be extended beyond β -- a contradiction. □

2. Theorem. Particles of opposite sign cannot collide in a finite time.

Proof. Let $b < 0$, and suppose (for contradiction) that $\beta < \infty$. Then $x(t) \to 0$ as $t \to \beta$.

Case (a). Let $z'(t) < 0$ on $[0,\beta)$. Then (3') gives $z''-az' > 0$, which implies $z'(t) \geq z'(0)e^{at} \geq -|z'(0)|e^{a\beta}$. Thus z' is bounded, and hence z is bounded on $[0,\beta)$. Therefore $|v(t)| \leq M < 1$, and with the aid of the lemma, (3') implies $z'' - az' \geq K/x^2 > -Kx'/cx^2$ for some constant K. Integration gives

$$z'(t) - z'(0) - az(t) + az(0) > (K/c)[1/x(t)-1/x(0)] \to \infty$$

as $t \to \beta$. But, the left hand side is bounded; so Case (a) is impossible.

Case (b). Let $z'(t_1) \geq 0$ for some $t_1 \geq 0$ (so that, by (3'), $z'(t) \geq 0$ for $t_1 \leq t < \beta$), and let $|v(t)| \leq M < 1$ on $[0,\beta)$. Then, from (3'), $z'' \geq K/x^2 > -Kx'/cx^2$ on $[t_1,\beta)$ for some K. Thus $z' \geq z' - z'(t_1) \geq (K/c)[1/x - 1/x(t_1)] \geq -Kx'/c^2x - K/cx(t_1)$. Another integration then gives

$$z(t) - z(t_1) \geq -(K/c^2) \ln [x(t)/x(t_1)] - K\beta/cx(t_1).$$

The left hand side is bounded above, while the right hand side is not; so Case (b) cannot occur. This leaves only

Case (c). $z'(t) \geq 0$ for $t_1 \leq t < \beta$ and $v(t) \to 1$ as $t \to \beta$. Hence $x(t) \not\to 0$ -- which is also a contradiction. □

3. Theorem (Particles of like sign). Let $b > 0$.

(a) If $\phi'(0) \leq 0$ and $v'(0) = w_0 \leq 0$, then $\beta < \infty$ and $x(t) \to 0$ as $t \to \beta$.

(b) If $\phi'(0) > 0$ and $v'(0) = w_0 > 2c^2b[1-\phi'^2(0)/c^2]^{3/2}/(1-M/c)\phi'(0)\phi(0)$, where $M = \max_{t\leq 0} -\phi'(t)$, then $\beta = \infty$ and $x(t) \to \infty$ as $t \to \infty$.

The proof can be found in [3].

REFERENCES

1. Driver, R. D., *Ann. Physics 21*(1963), 122-142.

2. Eliezer, C. J., *Proc. Cambridge Phil. Soc. 39*(1943), 173-180.

3. Hsing, D. K., and Driver, R. D., *Radiation Reaction in the Two-Body Problem of Classical Electrodynamics*, (Technical report No. 61, Department of Mathematics, University of Rhode Island), October, 1975.

INVARIANT SETS FOR CERTAIN LIÉNARD EQUATIONS WITH DELAY

J. Inciura
Department of Applied Mathematics
University of Waterloo

In this paper, we find appropriate invariant sets that enable us to use Schauder's fixed point theorem to prove the existence of periodic solutions for certain delayed Liénard equations of the form

$$\ddot{x}(t) + \mu f(x(t))\dot{x}(t) + g(x(t-r)) = \mu q(t),\ \mu,\ r > o. \tag{1}$$

With $F(x) = \int_o^x f(s)ds$ and $Q(t) = \int^t q(s)ds$, we write (1) as

$$\dot{x}(t) = \mu[y(t) - F(x(t)) + Q(t)]$$
$$\dot{y}(t) = -\frac{1}{\mu} g(x(t-r)). \tag{2}$$

We assume that F and g are continuously differentiable, odd and unbounded on R; $g'(x) > o$ on R with $g(o) = o$, and $xg(x) > o$ for $x \neq o$; that for some $a > o$, $F'(x) < o$ on $|x| < a$, $F'(x) > o$ on $|x| > a$, with $F(a) = -b < o$; that Q is ω-periodic on R with $\max|Q(t)| = \lambda$, and that $o \leq 2\lambda < b$, $o < r \leq \omega$, $\mu r(b-2\lambda) \geq a$.

We first consider the autonomous case $\lambda = o$. For $A > o$ and $B > b$, define the closed bounded convex set $K = K(A,B) \subseteq X$ by $K(A,B) = \{(\phi,y) \mid -A \leq \phi(\theta) \leq o,\ -r \leq \theta \leq o,\ \phi(o) = o, b \leq y \leq B\}$ and define the mapping $P: X \to R^2$ by $P((\phi,y)) = (\phi(o),y)$. In R^2, let Γ be the curve $y = F(x)$ and for arbitrary $d > o$ let $L(d)$ be the line $y = -(b+d)x/a$.

If $z_t(z_o)$ denotes a solution of (2) with initial value $z_o \in K$, at time $t \geq o$, then the point $Pz_t = (x(t),y(t))$ is on the positive y-axis at $t = o$ and cuts Γ, $L(d)$ and the negative y-axis at times t_1, t_2, t_3 respectively. The condition $\mu rb \geq a$ implies $x(r) \geq a$ so that $r < t_1 < t_2 < t_3 < +\infty$ and $y(t_3) < -b$. We define the continuous map $T: K \to X$ by $Tz_o = -z_{t_3}(z_o)$. For $t \in [o,t_3]$ we have $y(t) \leq B + rg(A)/\mu$ so that $x(t) \leq F^{-1}(B+rg(A)/\mu) = A_1$, and for $t \in [t_2,t_3]$ we have $\dot{x}(t) \leq [b-(b+d)x(t_2)/a]$, whence

$$t_3 - t_2 \leq \frac{x(t_2)}{\mu[(b+d)x(t_2)/a-b]} \leq \frac{a}{\mu d}, \text{ if } x(t_2) \geq a.$$

This, together with the equation for $\dot{y}(t)$ in (2) gives

$$-y(t_3) \leq \frac{b+d}{a} A_1 + \frac{a}{\mu^2 d} g(A_1) = B_1.$$

If $o < x(t_2) < a$, then the same bound for $-y(t_3)$ holds and we have $T(K(A,B)) \subseteq K(A_1,B_1)$. If the equation

$$\frac{b+d}{a} u + \left(\frac{r}{\mu} + \frac{a}{\mu^2 d}\right) g(u) = F(u)$$

has a positive root u_o, we can choose $A = A_1 = u_o$ and $B = B_1 = \frac{b+d}{a} u_o + \frac{a}{\mu^2 d} g(u_o)$ and we have the required invariant set K for T. Schauder's theorem applies and the fixed point is the initial value of a nontrivial periodic solution of (2). As an example, if $f(x) = x^2 - 1$ and $g(x) = x + \beta x^3$ then (1) has a periodic solution whenever $\mu r \geq 3/2$ and $o \leq 3r\beta < \mu$.

For the nonautonomous case $o < 2\lambda < b$ it can be shown by arguments similar to those shown above that if $\mu \geq m = a/r(b-2\lambda)$ then some iterate of the mapping $S: X \to X$ defined by $Sz_o = z_\omega(z_o)$ has a fixed point in the set $Y = Y(v_o,v_1)$, $v_o, v_1 > o$, defined by $Y(v_o,v_1) = \{(\phi,y) \mid \|\phi\| \leq v_o, |y| \leq v_1\}$, where we suppose v_o can be chosen sufficiently large so that

$$b + 2\lambda + \frac{b+d+\lambda}{a} v_o + \left(\frac{7r}{m} + \frac{a}{m^2 d}\right) g(v_o) \leq F(v_o), \text{ and where}$$

$$v_1 = b + \lambda + \frac{b+d+\lambda}{a} v_o + \left(\frac{5r}{m} + \frac{a}{m^2 d}\right) g(v_o).$$

REFERENCES

1. Hale, J. K., *Functional Differential Equations*, Springer-Verlag, New York, 1971.

2. Grafton, R. B., Periodic solutions of Liénard equations with delay: some theoretical and numerical results, in *Delay Differential Equations and Their Applications*, (K. Schmitt, Ed.), Academic Press, New York, 1972, 321-334.

OSCILLATION OF A FORCED NON-LINEAR SECOND ORDER DIFFERENTIAL EQUATION

Gary D. Jones
and
Samuel M. Rankin, III
Department of Mathematics
Murray State University

Oscillation criteria are given for the forced non-linear equation

$$(1) \qquad y'' + p(t)g(y) = f(t)$$

where p, $f \in C(0,\infty)$, $p > 0$ on $(0,\infty)$, $g \in C(-\infty,\infty)$, $yg(y) > 0$ for $y \neq 0$ and $g'(y) \geq 0$ for $y \neq 0$. Our criteria differ from known results in that we do not require all solutions of

$$(2) \qquad u'' + p(t)g(u) = 0$$

to be oscillatory. In fact all solutions of equation (2) are permitted to be nonoscillatory. Our main theorem follows.

<u>Theorem</u>. If (i) $\overline{\lim_{t\to\infty}} \int_T^t p(s) + \lambda f(s)ds = \infty$ for each $T > 0$ and for each $\lambda \neq 0$ (ii) there exist points a, b > 0 such that $h_1(t) \equiv \int_a^t (t-s)f(s)ds \geq 0$ for all $t \geq a$ and $h_2(t) \equiv \int_b^t (t-s)f(s)ds \leq 0$ for all $t \geq b$, (iii) $h_1(t)$ and $h_2(t)$ are oscillatory, and (iv) $|\frac{1}{t}\int_T^t (t-s)f(s)ds| < M$ for all $T > 0$ then every solution of equation (1) is oscillatory. The equations $y'' + \frac{1}{4}t^{-2}y = t\cos t$ and $y'' + ty^3 = \sin t$ satisfy the conditions of the theorem.

SOLUTIONS OF ALGEBRAIC MATRIX EQUATIONS RELATED TO OPTIMAL CONTROL THEORY

John Jones, Jr.
Air Force Institute of Technology
Wright-Patterson Air Force Base

Abstract. The purpose of this paper is to obtain necessary conditions and sufficient conditions for the existence of solutions of algebraic matrix equations which occur in optimal control theory and elswhere. Such equations include the Lyapunov equation and the Riccati nonlinear matrix equation. Results include a representation of their solutions.

1. INTRODUCTION

The main purpose of this paper is to obtain necessary conditions and sufficient conditions for the existence of solutions along with a representation of their solutions of algebraic matrix Riccati equations of the type

$$A^tX + XA + C - XDX = 0, \quad t = \text{transpose}, \tag{1.1}$$

which arise in optimal control theory and elsewhere. For the case D = 0 equation (1.1) is the Lyapunov equation which occurs in stability theory of ordinary differential equations.

Capital letters will denote n x n matrices with elements in the field F of complex numbers. The capital letter I will represent unit matrices of order n or 3n to agree with that of other matrices in the same expression. The coefficients of all polynomials which arise will also belong to the field

F. Thus the similarity of matrices and the reducibility of polynomials will remain valid under the rational operations of F. The results obtained in this paper extend those of J. Jones, Jr. [1], J. E. Potter [3], W. E. Roth [4], and others. Further results in this area will appear elsewhere.

2. NECESSARY CONDITIONS

Let the $3n \times 3n$ matrices $\tilde{R}$, $\tilde{\tilde{R}}$ with elements belonging to F be denoted by

$$(2.1)\qquad \tilde{R} = \begin{pmatrix} -A & 0 & -D \\ 0 & I & 0 \\ -C & 0 & A^t \end{pmatrix}, \quad \tilde{\tilde{R}} = \begin{pmatrix} A^t-XD & 0 & 0 \\ 0 & I & 0 \\ -D & 0 & DX-A \end{pmatrix}$$

where X is a solution of (1.1) such that $A - DX + I = 0$ is also satisfied.

The following theorem will be used in obtaining solutions of the given equation (1.1).

1. <u>Theorem</u>. If equation (1.1) has a solution X with elements in F, then there exists at least one triple of polynomials $f_\alpha(\lambda)$, $g_\beta(\lambda)$, and $h_\gamma(\lambda)$ of degree $\alpha \le n$, $\beta \le n$, $\gamma \le n$ respectively with coefficients in F such that $f_\alpha(\tilde{R})g_\beta(\tilde{R})h_\gamma(\tilde{R}) = 0$, where $f_\alpha(\lambda)g_\beta(\lambda)h_\gamma(\lambda)$ is not necessarily the minimum polynomial satisfied by $\tilde{R}$ but is a divisor of $|\tilde{R}-\lambda I|$ and such that $f_\alpha(A^t-XD) = 0$, $g_\beta(I) = 0$, and $h_\gamma(DX-A) = 0$.

<u>Proof</u>. Let X be any solution of (1.1) with elements in F such that $A - DX = -I$, then

$$(2.2)\qquad \begin{pmatrix} X & A-DX+I & I \\ 0 & I & 0 \\ I & 0 & 0 \end{pmatrix} \tilde{R} \begin{pmatrix} 0 & 0 & I \\ 0 & I & 0 \\ I & -A+DX-I & -X \end{pmatrix} = \begin{pmatrix} A^t-XD & 0 & 0 \\ 0 & I & 0 \\ -D & 0 & DX-A \end{pmatrix} = \tilde{\tilde{R}},$$

and the matrices $\tilde{R}$ and $\tilde{\tilde{R}}$ are similar and $|R-\lambda I| = f_\alpha(\lambda)\cdot g_\beta(\lambda)\cdot h_\gamma(\lambda) = |(A^t-XD)-\lambda I|\cdot|I-\lambda I|\cdot|(DX-A)-\lambda I|$, and the theorem follows.

The set of polynomials $f_\alpha(\lambda)$, $g_\beta(\lambda)$, $h_\gamma(\lambda)$ of degree $\alpha \le n$, $\beta \le n$, $\gamma \le n$, respectively, such that $[f_\alpha(\lambda)g_\beta(\lambda)h_\gamma(\lambda)]$ is a divisor $|\tilde{R}-\lambda I|$ and a multiple of the minimum polynomial satisfeid by $\tilde{R}$ will be called a set of admissible polynomials denoted by $\mathcal{A}(\tilde{R})$.

We will use the following $3n \times 3n$ matrices

$$f_\alpha(\tilde{R}) = \begin{pmatrix} U & \overline{0} & M \\ P & Q & R \\ V & S & N \end{pmatrix}, \quad \tilde{R} = \begin{pmatrix} -A & 0 & -D \\ 0 & I & 0 \\ -C & 0 & A^t \end{pmatrix} \tag{2.3}$$

where U, V, M, N, $\overline{0}$, P, Q, R. S are polynomials in the matrices A, A^t, C, D, and $f_\alpha(\tilde{R}) \in \mathcal{A}(\tilde{R})$.

3. SUFFICIENT CONDITIONS

<u>2</u>. <u>Theorem</u>. Let $f_\alpha(\lambda) \in \mathcal{A}(\tilde{R})$ be a polynomial of degree $\alpha \le n$ with coefficients in $\mathcal{F}$ such that U^{-1} or M^{-1} and $\overline{0}^{-1}$ exists then a common solution X of $XU + V = 0$, $X\overline{0} + S = 0, XM + N = 0$ is a solution of (1.1) and $A - DX + I = 0$, where $\tilde{R}$, $f_\alpha(\tilde{R})$ are given in (2.3).

<u>Proof</u>. Let X be a common solution of $XU + V = 0$, $X\overline{0} + S = 0$, $XM + N = 0$ and U^{-1} or M^{-1} and $[\overline{0}]^{-1}$ exist then

$$0 = -(XM+N)C = -XUA-XMC+XUA-NC = -X(UA+MC)+XUA-NC. \tag{3.1}$$

Now $\tilde{R}$ and f $(\tilde{R})$ commute and making use of these identities we have

$$\begin{aligned} 0 &= -X(AU+DV) + XUA - NC = -XAU + XDXU + XUA - NC \\ &= XDXU - A^tXU - XAU + A^tXU + XUA - NC \\ &= XDXU - A^tXU - XAU - A^tV - VA - NC \qquad (3.2) \\ &= XDXU - XAU - A^tXU - CU \\ &= -(XDX + A^tX + XA + C)U \end{aligned}$$

Since U^{-1} exists X is a solution of (1.1). A similar type argument holds in case M^{-1} exists by starting with $0 = (XM+N)A^t$ and using the identities mentioned above. If X is also a solution of $X\overline{0} + S = 0$ we have the matrix identity $\overline{0} = -A\overline{0} - DS$ from $\tilde{R}f_\alpha(\tilde{R}) = f_\alpha(\tilde{R})\tilde{R}$, and

$$(3.3) \qquad 0 = \overline{0} + A\overline{0} + DS = \overline{0} + A\overline{0} - DX\overline{0} = (I+A-DX)\overline{0}.$$

Since $[\overline{0}]^{-1}$ exists X is also a solution of $A - DX = -I$. Making use of S. K. Mitra [2] we have $X = D^-(A+I) + W - D^-DW$, (W is arbitrary). If D^{-1} exists, such a solution is unique. For $X = X^t$, $D = D^t$ such a positive semi-definite solution can be shown to be unique.

RERERENCES

1. Jones, J., Jr., Solutions of certain matrix equations, *Proc. AMS, 31* (No. 2, 1972). 333-339.

2. Mitra, S. K., Common solutions to a pair of linear matrix equations $A_1XB_1 = C_1$ and $A_2XB_2 = C_2$, *Proc. Camb. Phil. Soc. 74* (1973), 213-216.

3. Potter, J. E., Matrix quadratic solutions, *SIAM J. Math. 14* (1966), 496-501, MR 34 #1341.

4. Roth, W. E., On the matrix equation $X^2 + AX + XB + C = 0$, *Proc. AMS, 1* (1950), 586-589, MR 12 #471.

RAZUMIKHIN TYPE THEOREM FOR DIFFERENTIAL EQUATIONS WITH INFINITE DELAY*

Junji Kato
Mathematical Institute
*Tohoku University***

Our concern is on the stability problem for functional differential equations with infinite delay

(1) $$\dot{x}(t) = f(t, x_t),$$

where $x_t \in C$ is given by

$$x_t(s) = x(t+s) \text{ for } s \in (-\infty, 0]$$

and C denotes a function space of $(-\infty, 0]$ into R^n with a norm $||\cdot||$ such that if $x(s)$ satisfies $x_\tau \in C$ and is continuous on $[\tau, t]$, then $x_t \in C$, it is continuous in $t \geqq \tau$ and

$$|x(t)| \leqq ||x_t|| \leqq K_1(t-\tau)||x_\tau|| + K_2(t-\tau) \sup_{\tau \le s \le t} |x(s)|$$

for continuous functions $K_1(s)$ and $K_2(s)$.

The following theorem is a simple version of the Liapunov-Krasovski's theorem (see [1] also [2]).

<u>Theorem A</u>. Suppose that there exists a continuous function $V(t,\phi)$ defined on $(-\infty,\infty) \times C$ such that

$$a(|\phi(0)|) \leqq V(t,\phi) \leqq b(||\phi||)$$

for continuous positive-definite functions $a(r)$ and $b(r)$ and that for a continuous function $c(t,r) \geqq 0$

(2) $$\dot{V}(t, x_t) \leqq -c(t, V(t, x_t))$$

along any solution $x(t)$ of (1). Then, the zero solution of (1) is uniformly asymptotically stable, if for any $r > 0$

$$\int_t^{t+T} c(s,r)ds \to \infty \text{ as } T \to \infty \text{ uniformly in } t \geqq 0.$$

Since the solutions become more restrictive as the time elapses, the following theorem is expected to be more effectively (refer [3] and [4]).

Theorem B. Suppose that for a constant $L > 0$

$$|f(t,\phi)| \leqq L||\phi||.$$

Then, in Theorem A, it is sufficient for $V(t,\phi)$ to satisfy (2) under the case (*) $x(s)$ is a solution in (1) at least on the interval $[t-p(V(t,x_t)),t]$, where $p(r) \geqq 0$ is a continuous function for $r > 0$.

Generally speaking, a Razumikhin type Liapunov function can be constructed easier than a usual Liapunov function. Razumikhin type theorems for infinite delay cases are given in [2], [5], [6].

Here, by extending the idea in [4] (also [7]) we shall state the following theorem.

Theorem C. In Theorem B, we can assume that the condition (2) holds when, in addition to (*), $x(s)$ satisfies

$$V(s,x_s) \leqq f(V(t,x_t)) \text{ for } s \in [t-q(V(t,x_t)),t]$$

where $f(r)$ and $q(r)$ are continuous functions such that $f(r) > r$ and

$$0 \leqq p(r) \leqq q(r)$$

for $r > 0$. (Here we may assume $\frac{f(r)}{r}$ is non-decreading and $q(r) > 0$.)

To prove Theorem C, it is sufficient to note that the function $W(t,\phi)$ defined by

$$W(t,x_t) = \sup_{s \leqq t} V(s,x_s)e^{\alpha(V(s,x_s))(s-t)},$$

where $\alpha(r) = \{\log(r/f^{-1}(r))\}/q(f^{-1}(r/2))$, satisfies the conditions endowed for V in Theorem B.

Example. By applying Theorem C with $V(\phi) = \phi(0)^2$ we can show that the zero solution of the equation

$$\dot{x}(t) = -a(x(t))x(t-p(x(t)))$$

is uniformly asymptotically stable under the conditions

$$a(r) > 0, \; a(r)p(r) \leqq \alpha < 1 \text{ for } r \neq 0$$

with continuity of $a(r)$ (all r) and $p(r)$ ($r \neq 0$).

I wish to thank Professor Driver for his comments.

REFERENCES

1. Krasovskii, N. N., *Stability of Motion*, Stanford Univ. Press, 1963.

2. Driver, R. D., *Arch. Rat. Mech. Anal. 10*(1962), 401-426.

3. Barnea, D. I., *SIAM J. Appl. Math. 17*(1969), 681-697.

4. Kato, J., *Springer Lect. Notes in Math. 243*(1971), 54-65.

5. Seifert, G., *J. Differential Eqs. 16*(1974), 289-297.

6. Grimmer, R. and Seifert, G., ibd. *19*(1975), 142-166.

7. Kato, J., *Funkcial. Ekvac. 16*(1973), 225-239.

---

*The detail will appear somewhere else.

**Visiting Professor, Department of Mathematics, Michigan State University.

PERIODIC AND NONPERIODIC MOTIONS OF HOMOGENEOUS TURBULENCE

Jon Lee
Flight Dynamics Laboratory
Wright-Patterson Air Force Base

In the absence of mean flow, the Navier-Stokes equations can be reduced to an infinite set of ordinary differential equations by Fourier analyzing the velocity field. As in classical statistical mechanics, the theory of homogeneous turbulence is based on the assumption that the Navier-Stokes dynamical system would exhibit the stochastic behavior necessary for ergodicity and mixing. Although this assumption has not yet been examined in detail, the predictability problem in numerical weather forecasting[(1)] has a direct bearing on it. Inevitably, the investigation of stochasticity takes us into numerical work, hence the details are not appropriate for this note. However, one can briefly outline the general framework of analysis, thereby indicating how periodic and nonperiodic motions can emerge from the Navier-Stokes equations.

For the homogeneous velocity field in a cyclic box of side L, the incompressible Navier-Stokes equations can be put in the form[(2)]

$$(\partial/\partial t+\nu k^2)u^{\mu}(\underset{\sim}{k},t) = -i(2\pi/L)^{3/2}\sum_{\lambda,\rho}\sum_{\underset{\sim}{k}=\underset{\sim}{p}+\underset{\sim}{q}}\bar{\phi}^{\mu|\lambda\ \rho}_{\underset{\sim}{k}|\underset{\sim}{p},\underset{\sim}{q}}(\xi)\ u^{\lambda}{}^{*}(\underset{\sim}{p},t)u^{\rho}{}^{*}(\underset{\sim}{q},t), \qquad (1)$$

where ν is the kinematic viscosity, and $\mu,\ \lambda,\rho = 1,2$. The coupling coefficient $\bar{\phi}^{\mu|\lambda,\rho}_{\underset{\sim}{k}|\underset{\sim}{p},\underset{\sim}{q}}(\xi)$ is defined in terms of the wavevectors $\underset{\sim}{k}$, $\underset{\sim}{p}$, and $\underset{\sim}{q}$, and the associated polarization vectors, in which the parameter ξ can take any value in $[0,2\pi]$. Certain constraints are obeyed by the coupling coefficients to assure the energy and helicity conservations in the inviscid limit.

To exhibit the fundamental triad-interaction, one may single out a typical nonlinear term over the triad wavevector $\underset{\sim}{K} + \underset{\sim}{P} + \underset{\sim}{Q} = 0$, and write it in detail (by denoting $\partial/\partial t$ by a dot, and setting $\nu = 0$ and $L = 2\pi$)

$$
\begin{aligned}
\dot{u}^{\mu}(\underset{\sim}{K}) &= -i \sum_{\lambda,\rho} \bar{\phi}^{\mu|\lambda,\rho}_{\underset{\sim}{K}|\underset{\sim}{P},\underset{\sim}{Q}}(\xi)\; u^{\lambda}{}^{*}(P)u^{\rho}{}^{*}(Q),\\
\dot{u}^{\mu}(\underset{\sim}{P}) &= -i \sum_{\lambda,\rho} \bar{\phi}^{\mu|\lambda,\rho}_{\underset{\sim}{P}|\underset{\sim}{Q},\underset{\sim}{K}}(\xi)\; u^{\lambda}{}^{*}(\underset{\sim}{Q})u^{\rho}{}^{*}(\underset{\sim}{K}),\\
\dot{u}^{\mu}(\underset{\sim}{Q}) &= -i \sum_{\lambda,\rho} \bar{\phi}^{\mu|\lambda,\rho}_{\underset{\sim}{Q}|\underset{\sim}{K},\underset{\sim}{P}}(\xi)\; u^{\lambda}{}^{*}(\underset{\sim}{K})u^{\rho}{}^{*}(\underset{\sim}{P}).
\end{aligned}
\tag{2}
$$

Split (2) into the real and imaginary parts by $u^{\mu}(\underset{\sim}{K}) = u^{\mu}_{r}(\underset{\sim}{K}) + iu^{\mu}_{i}(\underset{\sim}{K})$ with similar expressions for $u^{\mu}(\underset{\sim}{P})$ and $u^{\mu}(\underset{\sim}{Q})$. Because $\Sigma_{\mu}\,\partial\dot{u}^{\mu}_{r}(\underset{\sim}{K})/\partial u^{\mu}_{r}(\underset{\sim}{K}) + \ldots + \partial\dot{u}^{\mu}_{i}(\underset{\sim}{Q})/\partial u^{\mu}_{i}(\underset{\sim}{Q}) = 0$, the volume element $du^{1}_{r}(\underset{\sim}{K}), \ldots, du^{2}_{i}(\underset{\sim}{Q})$ in the 12-D space is an integral invariant.[3]

Now consider a subsystem of (2) for $\mu = \lambda = \rho = 1$. After dropping the superfluous indices and denoting $\bar{\phi}^{1|1,1}_{\underset{\sim}{K}|\underset{\sim}{P},\underset{\sim}{Q}}(\xi) = \phi_{\underset{\sim}{K}}$, $\bar{\phi}^{1|1,1}_{\underset{\sim}{P}|\underset{\sim}{Q},\underset{\sim}{K}}(\xi) = \phi_{\underset{\sim}{P}}$, and $\bar{\phi}^{1|1,1}_{\underset{\sim}{Q}|\underset{\sim}{K},\underset{\sim}{P}}(\xi) = \phi_{\underset{\sim}{Q}}$, we have

$$
\dot{u}(\underset{\sim}{K}) = -i\phi_{\underset{\sim}{K}}\; u^{*}(\underset{\sim}{P})u^{*}(\underset{\sim}{Q}),\quad \dot{u}(\underset{\sim}{P}) = -i\phi_{\underset{\sim}{P}}\; u^{*}(\underset{\sim}{Q})u^{*}(\underset{\sim}{K}),
$$
$$
\dot{u}(Q) = -i\phi_{\underset{\sim}{Q}}\; u^{*}(\underset{\sim}{K})u^{*}(\underset{\sim}{P}),
\tag{3}
$$

which is the basic triad-interaction in 2-D flow.[4] The energy is conserved for $\phi_{\underset{\sim}{K}} + \phi_{\underset{\sim}{P}} + \phi_{\underset{\sim}{Q}} = 0$. Since the u's are complex, (3) contains in all 6 real equations, each of which consists of the 3-mode interactions of the form

$$\dot{x}(\underset{\sim}{K}) = \phi_{\underset{\sim}{K}} x(\underset{\sim}{P})x(\underset{\sim}{Q}), \quad \dot{x}(\underset{\sim}{P}) = \phi_{\underset{\sim}{P}} x(\underset{\sim}{Q})x(\underset{\sim}{K}), \quad \dot{x}(\underset{\sim}{Q}) = \phi_{\underset{\sim}{Q}} x(\underset{\sim}{K})x(\underset{\sim}{P}) \tag{4}$$

where x's are real. This is identical to Euler's equation of a rigid body moving with one point fixed under no external forces.[5] Its periodic motion is known in terms of the Jacobian elliptic functions. It is, however, more illuminating to note that the trajectory of (4) lies in the intersection of the energy sphere $\frac{1}{2}(x^2(\underset{\sim}{K}) + x^2(\underset{\sim}{P}) + x^2(\underset{\sim}{Q})) =$ E (=const) and the ellipsoid $a_{\underset{\sim}{K}} x^2(\underset{\sim}{K}) + a_{\underset{\sim}{P}} x^2(\underset{\sim}{P}) + a_{\underset{\sim}{Q}} x^2(\underset{\sim}{Q}) =$ C (= const), where $(a_{\underset{\sim}{K}}, a_{\underset{\sim}{P}}, a_{\underset{\sim}{Q}}) = (\phi_{\underset{\sim}{Q}}-\phi_{\underset{\sim}{P}}, \phi_{\underset{\sim}{K}}-\phi_{\underset{\sim}{Q}}, \phi_{\underset{\sim}{P}}-\phi_{\underset{\sim}{K}})$.

Since (3) is made up of the 3-mode interactions, the question is whether the periodic motion of (4) can survive in (3). To this end, let $u(\underset{\sim}{K}) = r(\underset{\sim}{K})e^{i2\pi\omega(\underset{\sim}{K})}$, $u(\underset{\sim}{P}) = r(\underset{\sim}{P})e^{i2\pi\omega(\underset{\sim}{P})}$, and $u(\underset{\sim}{Q}) = r(\underset{\sim}{Q})e^{i2\pi\omega(\underset{\sim}{Q})}$. Since $r(\underset{\sim}{K})r(\underset{\sim}{P})r(\underset{\sim}{Q}) \cos 2\pi\Omega \equiv A$, where $\Omega = \omega(\underset{\sim}{K}) + \omega(\underset{\sim}{P}) + \omega(\underset{\sim}{Q})$, is another constant of motion, the polar representation of (3) becomes

$$\dot{r}(\underset{\sim}{K}) = -\phi_{\underset{\sim}{K}}\, r(\underset{\sim}{P})r(\underset{\sim}{Q}) \sin 2\pi\Omega, \quad \dot{r}(\underset{\sim}{P}) = -\phi_{\underset{\sim}{P}}\, r(\underset{\sim}{Q})r(\underset{\sim}{K}) \sin 2\pi\Omega,$$
$$\dot{r}(\underset{\sim}{Q}) = -\phi_{\underset{\sim}{Q}}\, r(\underset{\sim}{K})r(\underset{\sim}{P}) \sin 2\pi\Omega, \tag{5a}$$

$$\dot{\omega}(\underset{\sim}{K}) = -\phi_{\underset{\sim}{K}} A/2\pi\; r^2(\underset{\sim}{K}), \quad \dot{\omega}(\underset{\sim}{P}) = -\phi_{\underset{\sim}{P}} A/2\pi\; r^2(\underset{\sim}{P}), \quad \dot{\omega}(\underset{\sim}{Q}) =$$
$$-\phi_{\underset{\sim}{Q}} A/2\pi\; r^2(\underset{\sim}{Q}). \tag{5b}$$

Regardless of the factor $\sin 2\pi\Omega$, (5a) has the same type of solution as (4); hence the amplitude r's will have periodic trajectories. For the emergence of almost periodic motion, let us consider the invariant set = $\{r(\underset{\sim}{K}) = r(\underset{\sim}{P}) = r(\underset{\sim}{Q}) =$ const., and $\Omega = 0\}$. Then, expressing the angle variable in terms of the frequency, $\omega(\underset{\sim}{K}) = f(\underset{\sim}{K})t$, $\omega(\underset{\sim}{P}) = f(\underset{\sim}{P})t$, and $\omega(\underset{\sim}{Q}) = f(\underset{\sim}{Q})t$, one finds that the frequencies $f(\underset{\sim}{K})$, $f(\underset{\sim}{P})$, and $f(\underset{\sim}{Q})$ are rationally independent because of $\phi_{\underset{\sim}{K}} + \phi_{\underset{\sim}{P}} + \phi_{\underset{\sim}{Q}} = 0$.

Hence, the invariant set corresponds to an everywhere dense trajectory on a 3-D torus. In general, the r's are periodic so that one may visualize the trajectory of (5) to lie on a 3-D torus with the periodically varying configuration, i.e., the meridian and parallel in the case of 2-D. Finally, when the almost periodic motions of (3) are combined into the form of (2), the emerging trajectory is a recurrence motion.

REFERENCES

1. Lorenz, E. H., *Tellus 21*(1969), 289.

2. Lee, J., *J. Math. Phys. 16*(1975), 1359.

3. Nemytskii, V. V., and Stepanov, V. V., *Qualitative Theory of Differential Equations*, Princeton Univ. Press, Princeton, New Jersey, 1960.

4. Kraichnan, R. H., *Phys. Fluids 6*(1963), 1603.

5. Lamb, H., *Higher Mechanics*, Cambridge Univ. Press, Cambridge, 1943.

BUCKLING OF CYLINDRICAL SHELLS WITH SMALL CURVATURE*

John Mallet-Paret
Lefschetz Center for Dynamical Systems
Division of Applied Mathematics
Brown University

Consider the bifurcation buckling of a rectangular plate $\Omega = (0,\ell) \times (0,1)$ under a lateral force of magnitude λ applied to $x = 0,\ell$. In the absence of forces, the plate has an imperfection of magnitude α and assumes a form $z = \alpha w_0(x,y)$ for some known $w_0:\bar{\Omega} \to R$; the buckled state of the plate is given by $z = \alpha w_0(x,y) + w(x,y)$ where w is the solution to be determined. The boundary of the plate is simply supported (hinged). We consider this problem, described by the von Kármán equations, for parameters (λ,α) varying independently near $(\lambda_0,0)$ and for w near the unbuckled state $w = 0$, where λ_0 represents the critical buckling load.

The differential equations describing w can be rewritten as an operator equation

$$(I-\lambda L+\alpha^2\Lambda^2)w + \alpha Q(w) + C(w) = \alpha\lambda p \tag{1}$$

where $w \in X =$ the Hilbert space $H_0^1(\Omega) \cap H^2(\Omega)$. Here L and Λ are compact self adjoint linear operators on X, Q and C are compact, nonlinear, and homogeneous of orders two and three respectively, and $p \in X$ is fixed. The smallest characteristic value of L is λ_0.

If the kernel of $I - \lambda_0 L$ is one-dimensional, the analysis can be carried out using ideas motivated by the theory of unfoldings of functions [1]. We therefore consider the case where $\ker(I-\lambda_0 L)$ is two-dimensional; in particular, this occurs when $\ell = \sqrt{2}$. Some techniques for studying this case are presented in [2], where an equation similar to (1) is considered with p satisfying a certain generic condition, essentially that p have non-zero projection on $\ker(I-\lambda_0 L)$.

Here we suppose $p = 0$; this is equivalent to assuming the imperfection is of the form $w_0(x,y) = \frac{1}{2} y^2(\sigma x+\tau)$ for constants σ and τ. This is motivated by work of Knightly and Sather [3] who study the buckling of a cylindrical surface with curvature α in the y-direction, where α may be large. In our case, fixing $\sigma = 0$, $\tau = 1$ would therefore correspond to the situation in [3] for α near zero, and λ in a uniform neighborhood of λ_0.

Two cases are considered: first, when $\sigma = 0$, $\tau = 1$, giving rise to an imperfection with curvature α in y-direction, and second, when $0 < |\sigma| << 1$, $\tau = 1$, giving rise to an imperfection with curvature roughly $\alpha(\sigma x+1)$ varying slightly with x. An important feature here is the presence of a high degree of symmetry in the first case which is lost as σ becomes non-zero in the second case. In particular, for case one, equation (1) takes the form

$$(I-\lambda L+\alpha^2 L^2)w + \alpha Q_0(w) + C(w) = 0 \tag{2}$$

while in case two it becomes

$$\begin{gathered}(I-\lambda L+\alpha^2 \Lambda^2)w + \alpha Q(w) + C(w) = 0 \\ \Lambda = L + O(\sigma) \;, \; Q = Q_0 + O(\sigma).\end{gathered} \tag{3}$$

In addition (2) possesses a left-right symmetry which is lost in (3).

This implies that the double eigenvalue in (2) persists for small imperfections α, while in (3) it splits into distinct simple eigenvalues. One also sees the appearance of secondary solutions bifurcating from a primary solution in (2), while in (3) the secondary solutions do not connect up to the primary solution.

REFERENCES

1. Chow, S. N., Hale, J. K., and Mallet-Paret, J., Applications of generic bifurcation. I, *Arch. Rat. Mech. Anal.* *59* (1975), 159-188.

2. Chow, S. N., Hale, J. K., and Mallet-Paret, J., Applications of generic bifurcation. II., *Arch. Rat. Mech. Anal.* *62* (1976), 209-235.

3. Knightly, G. H., Some mathematical problems from plate and shell theory, *Nonlinear Functional Analysis and Differential Equations*, (Proceedings of the Michigan State University Conference), Marcel Dekker, Inc., New York, 1976.

---

*This research was supported in part by the National Science Foundation under Grants GP-28931X3 and MPS 71-02923 and in part by U. S. Army Research Office under Grants AROD DAH CO4-75-G-007 and AROD AAG 29-76-6-0052.

FUNCTION SPACE CONTROLLABILITY OF RETARDED SYSTEMS: A DERIVATION FROM ABSTRACT OPERATOR CONDITIONS

A. Manitius
Centre de Recherches Mathématiques
Université de Montréal

and

R. Triggiani
Department of Mathematics
Iowa State University

In this note we announce some of our recent results on function space controllability of retarded systems that were obtained by using a representation of such systems by an abstract differential equation in Banach space $R^n \times L_2([-h,0];R^n)$ (denoted shortly as M_2). The proofs and a more comprehensive treatment of these results are given in [1,2,3].

We investigate the controllability of systems

$$\dot{y}(t) = A_0 y(t) + A_1 y(t-h) + Bu(t) \tag{1}$$

where $y \in R^n$, $u \in R^m$, A_0, A_1, B are matrices of appropriate dimensions. The system (1) has an equivalent representation

$$\dot{x}(t) = \tilde{A}x(t) + \tilde{B}u(t) \tag{2}$$

where $x \in M_2$, $x = (x^0, x^1)$, $x^0 \in R^n$, $x^1 \in L_2([-h,0],R^n)$;

$x^0(t) = y(t)$, $x^1(t) = y_t = y_t(\theta) = y(t+\theta)$, $\theta \in [-h,0]$;

$\tilde{A}$ is the infinitesimal generator of a strongly continuous semigroup of bounded linear operators, given by $\tilde{A}x = (A_0x^0 + A_1x^1(-h), dx^1/d\theta)$, and $\tilde{B}u = (Bu,0)$. We define the resolvent $R(\lambda,\tilde{A}) = (I\lambda - \tilde{A})^{-1}$ and the resolvent set $\rho(\tilde{A}) = \{\lambda \mid \det \Delta(\lambda) \neq 0\}$, where $\Delta(\lambda)$ is an $n \times n$ matrix given by $\Delta(\lambda) = I\lambda - A_0 - A_1e^{-\lambda h}$, and we compute the operator $R(\lambda,\tilde{A})\tilde{B}$

(3) $$R(\lambda,\tilde{A})\tilde{B}u = (\Delta^{-1}(\lambda)Bu,\ e^{\lambda\theta}\Delta^{-1}(\lambda)Bu),\ \lambda \in \rho(\tilde{A}).$$

For a fixed t, let C_t denote the set of attainable functions $y_t \in L_2([-h,0],R^n)$, and let K_t denote the set of attainable pairs $(y(t),y_t) \in R^n \times L_2([-h,0],R^n)$. The system (1) will be called L_2-approximately controllable if $\overline{UC_t} = L_2([-h,0],R^n)$, and M_2-approximately controllable if, $\overline{UK_t} = M_2$, where the union is taken over all $t > 0$, and the bar denotes closure in L_2 or M_2, respectively.

The general characterization of approximate controllability for abstract systems (2) given in earlier works by Fattorini and by Triggiani, when specialized to the space M_2, gives that the system is M_2-approximately controllable if and only if

(4) $$\forall\eta \in M_2\ \langle \eta, R(\lambda,\tilde{A})\tilde{B}u\rangle_{M_2} = 0\ \forall\lambda \in \rho(\tilde{A})$$
$$\forall u \in R^m \text{ implies } \eta = 0$$

For any $\xi \in L_2([-h,0],R^n)$ define

(5) $$q(\lambda) = \int_{-h}^{0} \xi(\theta)e^{\lambda\theta}d\theta$$

The function $q(\lambda)$ is a finite Laplace transform of an R^n-valued L_2-function with support on $[-h,0]$. A class of such functions will be denoted by $FLT_2([-h,0],R^n)$; let $c \in R^n$, and let T denote transposition. By using (3) (4) (5) we have proved the following result.

1. Theorem. The system (2), i.e. (1) is (i) L_2-approximately controllable if and only if $\forall q(\cdot) \in FLT_2([-h,0],R^n)$ $q^T(\lambda)\Delta^{-1}(\lambda)B \equiv 0\ \forall\lambda \in \rho(\tilde{A})$ implies $q(\lambda) \equiv 0$. (ii) M_2-approximately controllable if and only if $\forall c \in R^n$, $\forall q(\cdot) \in FLT_2([-h,0],R^n)$ $[c^T+q^T(\lambda)]\Delta^{-1}(\lambda)B \equiv 0\ \forall\lambda \in \rho(\tilde{A})$ implies $c = 0$, $q(\lambda) \equiv 0$. (iii) R^n-controllable if and only if $\forall c \in R^n$ $c^T\Delta^{-1}(\lambda)B \equiv 0\ \forall\lambda \in \rho(\tilde{A})$ implies $c = 0$.

This theorem is general in the sense that it applies to any retarded system (with an appropriate modification of the matrix $\Delta(\lambda)$), not just to the simple case of system (1).

Since the matrix adj $\Delta(\lambda)$ is a matric polynomial in two variables: λ and $e^{-\lambda h}$, one can write

(6) $\Delta^{-1}(\lambda)B=[\det \Delta(\lambda)]^{-1}[\text{adj } \Delta(\lambda)]B=[\det \Delta(\lambda)]^{-1}P(\lambda)v(e^{-\lambda h})$

where $P(\lambda)$ is an $n \times nm$ polynomial matrix in λ, and $v(e^{-\lambda h}) = [I_m, I_m e^{-\lambda h}, \ldots, I_m e^{-\lambda(n-1)h}]^T$, I_m being an $n \times m$ identity matrix. The matrix $P(\lambda)$ plays an essential note in the controllability conditions. Its columns can be computed via a recursive algorithm [2] using directly the matrices A_0, A_1 and B.

<u>2</u>. <u>Theorem</u>. A necessary condition for L_2-(hence also for M_2-) approximate controllability of (1) is rank $P(\lambda) = n$, or, equivalently rank $H(\lambda) = n$, where

(7) $$H(\lambda) = [G(\lambda), F(\lambda)G(\lambda), \ldots, F^{n-1}(\lambda)G(\lambda)]$$
$$F(\lambda) = (I\lambda - A_0)^{-1}A_1, \quad G(\lambda) = (I\lambda - A_0)^{-1}B.$$

(for $H(\lambda)$ the values of λ are restricted to $\{\lambda | \det(I\lambda - A_0) \neq 0\}$).

This condition is necessary, and it becomes sufficient under some additional hypotheses. For example, for $m = 1$, if $P(\lambda)$ can be transformed, via elementary constant row operations, to a right (resp. left)-triangular matrix, then the condition of Theorem 2 becomes sufficient for M_2 (resp. L_2) approximate controllability. By using such ideas we have proved several algebraic sufficient conditions, one of which is

<u>3</u>. <u>Theorem</u>. Suppose that

(i) rank $[B, A_1B, \ldots, A_1^{n-1}B] = n$

(ii) $A_0 \text{Im} A_1^i B \subset \sum_{j=0}^{i} \text{Im} A_1^j B \quad i = 0, \ldots, n-1$

Then the system (1) is M_2-approximately controllable, for any value of $h > 0$. (Im denotes "image.")

By using spectral analysis of the infinitesimal generator one can prove that a necessary condition for M_2-approximate controllability is that the system (2) be controllable on all its eigenmanifolds. This leads to the following result.

4. Theorem. Let $m = 1$. A necessary condition for M_2 approximate controllability is

(i) $\det P(\lambda) \neq 0$

and

(ii) $P(\lambda)v(e^{-\lambda h}) \neq 0$ for all complex λ.

We note that the condition (ii) restricted to $\{\lambda \mid \text{Re } \lambda \geq 0\}$ is necessary and sufficient for controllability on "unstable" eigenmanifolds, hence for feedback stabilizability of system (1).

Let $p_k(\lambda)$ denote an arbitrary $1 \times m$ row vector valued polynomial in λ of degree at most k. Let $Q(\lambda)$ be a $1 \times (n-1)m$ polynomial matrix of the form

$$Q(\lambda) = [p_{n-2}(\lambda), p_{n-3}(\lambda), \ldots, p_0] \tag{8}$$

and let $[0_m, Q(\lambda)]$ and $[Q(\lambda), 0_m]$ denote $1 \times nm$ matrices, where 0_m denotes a row of m zeros.

5. Theorem. A necessary and sufficient condition for M_2-approximate controllability is that the equation

$$[c^T + q^T(\lambda)]P(\lambda) = [0_m, Q(\lambda)]e^{-\lambda h} - [Q(\lambda), 0_m] \tag{9}$$

has no nonzero solution $\{c, q(\cdot), Q(\cdot)\}$ in the class $c \in R^n$, $q(\cdot) \in FLT_2([-h,0], R^n)$, $Q(\lambda)$ given by (8).

Since one can prove that $q(\lambda) \in FLT_2$ satisfying (9) is of the form $[g_1(\lambda) + g_2(\lambda)e^{-\lambda h}]/\det P(\lambda)$ where $g_1(\lambda)$ and $g_2(\lambda)$

are n-vector polynomials of degree at most N - 1, where $N = \deg \det P(\lambda)$, testing the conditions of Theorem 5 reduces to testing for the existence of nonzero solutions of a system of linear homogeneous algebraic equations with respect to coefficients of $g_1(\lambda)$, $g_2(\lambda)$, c and of all the $p_k(\lambda)$ appearing in $Q(\lambda)$. Solved examples show that this test is practically feasible.

Number of other facts have been established in the detailed report [1].

REFERENCES

1. Manitius, A., and Triggiani, R., *Function Space Controllability of Linear Retarded Systems: A Derivation from Abstract Operator Conditions*, (report CRM-605, Centre de Recherches Mathématiques, Université de Montréal, 1976).

2. Manitius, A., and Triggiani, R., Sufficient conditions for function space controllability and feedback stabilizability of linear retarded systems, (Proc. 1976 IEEE Conference on Decision and Control, Clearwater, Florida, 1-3 Dec. 76) to appear.

3. Manitius, A., and Triggiani, R., New results on functional controllability of time-delay systems, (Proc. 1976 Conference on Information Sciences and Systems) The John Hopkins University, Baltimore, Maryland (1976), 401-405.

4. Delfour, M. C., and Mitter, S. K., Controllability, observability and optimal feedback control of affine hereditary differential systems, *SIAM J. Control 10* (No. 2, 1972), 298-328.

5. Jacobs, M. Q., and Langenhop, C. E., Criteria for function space controllability of linear neutral systems, *SIAM J. Control and Optimization*, to appear.

6. Mantitius, A., *Optimal Control of Hereditary Systems*, Lecture notes for the course "Control Theory and Topics in Functional Analysis," *IAEA, Vol. II*, Vienna 1976, 43-178. (Also report CRM-472, Centre de Recherches Mathématiques, Université de Montréal, 1975.)

7. Triggiani, R., Extensions of rank conditions for controllability and observability to Banach spaces and unbounded operators, *SIAM J. Control 14*(1976), 313-338.

PASSIVITY AND EVENTUAL PASSIVITY OF ELECTRICAL NETWORKS

T. Matsumoto
Department of Electrical Engineering
Waseda University

Consider a network containing ρ resistors (depedent sources are allowed), γ capacitors and λ inductors. Let Λ be the manifold of the branch characteristics [1] and let K be the Kirchhoff space. Then $\Sigma = \Lambda \cap K$ is the state space where the dynamics takes place [1]. In [1] it is shown that the vector field X describing the dynamics of the network is given by the formula

$$\pi^*G(X,\cdot) = \omega(\cdot) \tag{1}$$

where the notation is as in [1]. Here we will try to relate the circuit theoretic concepts passivity and eventual passivity to stability and boundedness of a network. These physical quantities play important roles.

Let v_C and i_L be the capacitor voltages and inductor currents, respectively. Let Γ_C(resp.Γ_L) be a smooth curve joining the origin of $\mathbb{R}^\lambda$ (resp. $\mathbb{R}^\lambda$) with a point v_C(resp. i_L) of $\mathbb{R}^\gamma$ (resp. $\mathbb{R}^\lambda$). Let $C_{mn}(v_C)$ and $L_{mn}(i_L)$ be the incremental capacitance matrix and the incremental inductance matrix, respectively. They are symmetric and positive definite. Then the integral

$$E_1(v_C, i_L) = \int_{\Gamma_C}^{v_C} \sum_{m,n=1}^{\lambda} C_{mn}(v_C')v_{C_m}'dv_{C_m}' + \int_{\Gamma_L}^{i_L} \sum_{m,n=1}^{\lambda} L_{mn}(i_L')i_{L_m}'di_{L_m}' \quad (2)$$

does not depend on the particular choices of Γ_C and Γ_L. Let E be the restriction of E_1 to Σ. It is clear that E is the <u>energy stored in capacitors and inductors</u>. Let $W_1 : \mathbb{R}^\rho \times \mathbb{R}^\rho \to R$ be defined by

$$W_1(v_R, i_R) = \sum_{n=1}^{\rho} v_{R_n} i_{R_n} \quad (3)$$

and let W be its restriction to Σ. W is the <u>power dissipated by resistors</u>.

<u>1</u>. <u>Result</u>. Let $\sigma(t)$ be the flow generated by the network. Then

$$\frac{dE(\sigma(t))}{dt} = -W(\sigma(t)). \quad (4)$$

<u>Proof</u>. Let $(v_C(t), i_C(t), v_L(t), i_L(t))$ be the projection of $\sigma(t)$ onto $\mathbb{R}^\gamma \times \mathbb{R}^\gamma \times \mathbb{R}^\lambda \times \mathbb{R}^\lambda$, where i_C and v_L are capacitor currents and inductor voltages, respectively. It follows from Maxwell's equations that

$$\frac{d}{dt}\int_{\Gamma_C}^{v_C(t)} \sum_{m,n=1}^{\gamma} C_{mn}(v_C')v_{C_m}'dv_{C_m}' + \frac{d}{dt}\int_{\Gamma_L}^{i_L(t)} \sum_{m,n=1}^{\lambda} L_{mn}(i_L')i_{L_m}'di_{L_m}'$$

$$= \sum_{m,n=1}^{\gamma} v_{C_m}(t)C_{mn}(v_C(t))\frac{dv_{C_n}(t)}{dt} + \sum_{m,n=1}^{\lambda} i_{L_m}(t)L_{mn}(i_L(t))\frac{di_{L_n}(t)}{dt}$$

$$= \sum_{n=1}^{\lambda} v_{C_m}(t)i_{C_m}(t) + \sum_{m=1}^{\lambda} i_{L_m}(t)v_{L_m}(t).$$

The last quantity is the power at capacitors and inductors. The result follows from Tellegen's theorem:

$$\sum_{n=1}^{\rho} v_{R_n} i_{R_n} + \sum_{n=1}^{\gamma} v_{C_n} i_{C_n} + \sum_{n=1}^{\lambda} v_{L_n} i_{L_n} = 0 \text{ on } \Sigma.$$

Formula (4) relates the stability and boundedness of a network to passivity and eventual passivity of networks. Passivity has a long history in circuit theory.

Definition. A network is called passive if $W \geq 0$.

It is clear that if each of the resistors is passive so is the network.

2. Result. Let σ_0 be an equilibrium of the dynamics. If there is a neighborhood U of σ_0 such that

(1) $E(\sigma_0) = 0$, $E(\sigma) > 0$ on $U - \{\sigma_0\}$

(2) the network is passive on U, i.e., $W(\sigma) \geq 0$, $\sigma \in U$

then σ_0 is a stable equilibrium.

3. Result. Under the same situation as in Result 2, let condition (2) be replaced by (2)' the network is strictly passive on U, i.e., $W(\sigma) > 0$, $\sigma \in U-\{\sigma_0\}$.

Then σ_0 is an asymptotically stable equilibrium.

Definition. A network is called eventually passive if there is a compact subset Ω of Σ such that

$$W \geq 0 \quad \text{on} \quad \Sigma - \Omega.$$

4. Result. If

(1) E is nonnegative,

(2) E is proper, i.e., for any nonnegative number α, the set $\{\sigma\in\Sigma \mid E(\sigma)\leq\alpha\}$ is bounded

(3) the network is eventually passive,

then the flow $\sigma(t)$ generated by the network is uniformly bounded on $[0,\infty)$.

5. Result. Under the same setting as of Result 4, let

$$\alpha_0 = \max_{\sigma\in\Omega} E(\sigma).$$

Then the set

$$\mathcal{E} = \{\sigma \in \Sigma \mid E(\sigma) \le \alpha_0\}$$

is an invariant set.

Since E is proper, $\mathcal{E}$ is a compact invariant set. This result is convenient for it reduces the analyses of networks on Σ, which is in general unbounded, to the analyses of networks on compact sets. It should also be noted that eventual passivity has a clear physical meaning and it is satisfied by a large class of electrical networks. Results related to the above are found in [2]-[7].

Finally let us cite some of the problems mentioned in [7].

1. Problem. Suppose that Λ and K are not transversal. How can one make an effective perturbation of Λ or K such that the perturbed Λ and K are transversal? This will be a difficult problem since one can perturb Λ only by adding another element to the network.

2. Problem. In [1] it is shown that one cannot synthesize an R-C reciprocal network (with simply connected Σ) with periodic orbits provided that π is nonsingular. Are there examples that admit periodic orbits because of the singularity of π? There are networks with periodic orbits because of the fact that Σ is not simply connected, while all other conditions are satisfied [7].

REFERENCES

1. Matsumoto, T., On the dynamics of electrical networks, *J. Differential Equations, 21* (1976), 179-196.

2. Smale, S., On the mathematical foundations of electrical circuit theory, *J. Differential Geometry 7* (1972), 193-210.

3. Brayton, R., and Moser, J., A theory of nonlinear networks, *Quart. Appl. Math. 22* (1964), 1-33, 81-104.

4. Matsumoto, T., On several geometric aspects of nonlinear networks, J. Franklin Institute, Special Issue on Systems Theory, *301* (1976), 203-225.

5. Matsumoto, T., Dynamical systems arising from electrical networks, in *Dynamical Systems*, Academic Press, 1976.

6. Matsumoto, T., On a class of nonlinear networks, *Int. J. Circuit Theory and Appl. 4* (1976), 55-73.

7. Matsumoto, T., *Dynamics of Electrical Networks*, Research Institute of Mathematical Sciences, Kyoto University, 1975.

INEQUALITIES FOR THE ZEROS OF BESSEL FUNCTIONS

Roger C. McCann
Department of Mathematics
Case Western Reserve University

Let $j_{p,n}$ denote the n-th positive zero of J_p. If we hold n fixed and treat $j_{p,n}$ as a function of p it is known that

(i) $j_{p,n}$ is a strictly increasing function of p, [3],

(ii) $j_{p,n} \sim p + a_n p^{1/3} + b_n p^{1/3} + \theta(p^{-1})$, [2], where the coefficients a_n and b_n are independent of p.

In light of (ii) it would not be surprising if $p^{-1} j_{p,n}$ were a strictly decreasing function of p. In fact, this is what we have proved.

Instead of considering Bessel's equation in its usual form, we will consider the eigenvalue problem

$$-(xy')' + x^{-1}y = \lambda x^{2p-1}y \qquad p > 0 \tag{1}$$

$$y(a) = y(1) = 0 \qquad 0 < a < 1 \tag{2}$$

It is easily verified that $y(x) = c_1 J_{1/p}(\lambda^{1/2}x^p/p) + c_2 Y_{1/p}(\lambda^{1/2}x^p/p)$ is the general solution of (1) and that the eigenvalues λ are solutions of

$$J_{1/p}(\lambda^{1/2}/p) - \frac{J_{1/p}(\lambda^{1/2}a^p/p)}{Y_{1/p}(\lambda^{1/2}a^p/p)} Y_{1/p}(\lambda^{1/2}/p) = 0 \tag{3}$$

Let $z_n(a,1/p)$ denote the n-th positive solution of (3). Then the n-th eigenvalue $\lambda_n(1/p,a)$ of (1), (2) satisfied

$$\lambda_n(1/p,a) = (pz_n(a,1/p))^2$$

Using the asymptotic expansions for $J_{1/p}$ and $Y_{1/p}$ near zero and a simple counting argument it is possible to show that $z_n(a,1/p) \to j_{1/p,n}$ as $a \to 0^+$.

Let R[p.y] denote the Rayleigh quotient

$$R[p,y] = \frac{\int_a^1 (-(xy')' + x^{-1}y)y\,dx}{\int_a^1 x^{2p-1}y^2 dx}$$

It is well known, [1, sections 31 & 35], that the eigenvalue of (1), (2) can be obtained from the Rayleigh quotient by maximizing and minimizing it in an appropriate manner.

Whenever $p \geq q$ we have that $x^{2p-1} \leq x^{2q-1}$ for all $x \in [a,1]$. Hence, $R[p,y] \geq R[q,y]$. It follows that $\lambda_n(1/p,a) \geq \lambda_n(1/q,a)$ whenever $p \geq q$ or, equivalently,

$$pz_n(a,1/p) \geq q\, z_n(a,1/q)$$

whenever $p \geq q$. Setting $t = q^{-1}$ and $s = p^{-1}$ we have that

$$s^{-1}z_n(a,s) \geq t^{-1}z_n(a,t)$$

whenever $t \geq s$. If we now let $a \to 0^+$ we obtain

$$s^{-1}j_{s,n} \geq t^{-1}j_{t,n}$$

whenever $t \geq s > 0$. In fact, it is possible to strengthen the inequality.

Theorem. $s^{-1}j_{s,n} > t^{-1}j_{t,n}$ when $t > s > 0$.

REFERENCES

1. Mikhlin, S. G., *Variational Methods of Mathematical Physics*, Macmillan, New York, 1964.

2. Tricomi, F., Sulle funzioni de Bessel di ordine e argomenta pressoche uguali, *Atti. Accad. Sci. Torimo Cl. Sci. Fis. Math. Nat. 83*(1949), 3-30.

3. Watson, G. N., *A Treatize on the Theory of Bessel Functions*, Cambridge Univ. Press, Cambridge (Eng.), 1958.

A HOMOGENEOUS, NONADIABATIC MODEL OF DELTA CEPHEI

Peter J. Melvin
Liberal Arts and Sciences Administration
University of Illinois at Urbana-Champaign

1. INTRODUCTION

In the pulsation theory of variable stars, it has long been noted [1,3] that the radial velocity curve of the fundamental mode of a homogeneous star seems to fit the observations of Cepheid variable stars. The intention of this paper is to examine just how close this correspondence is for the radial velocity curve of δ Cephei [4].

2. THE INTERNAL STRUCTURE OF A HOMOGENEOUS STAR

In Table 1 the notation of [2] is specialized to this case. The function w(t) is the scale factor. It can be independent of the Lagrangian or comoving coordinate ξ only for this particular mode and star. The position of the spherical shells of the star is given by (1.1). The internal pressure, density and temperature are computed from (1.2), (1.3), and (1.4). Hydrostatic equilibrium is defined to occur when x = y = z = 0.

Table 1. Hydrodynamical Fields of a Homogeneous Star	
$r(\xi,t) = R_e \xi w(t), \quad w(t) = 1 + x(t) + \sigma z(t).$	(1.1)
$P(\xi,t) = w^{-5} P_c (1-\xi^2)[1+\sigma y(t)].$	(1.2)
$\rho(\xi,t) = w^{-3} \rho_e.$	(1.3)
$\Theta(\xi,t) = w^{-2} \Theta_c (1-\xi^2)(1+\sigma y).$	(1.4)

3. INTEGRATION OF THE DYNAMICAL AMPLITUDES

In Table 2 the equations for the dynamical amplitudes of [2] are written for the case of the nonadiabatic, fundamental mode of the homogeneous star.

The simplest means of integrating (2.1) is by use of an elementary transformation. Let

$$u = (1+x)^{-1}, \qquad \alpha'(\tau) = u^2/2\pi, \tag{1}$$

Table 2. Time Separation Equations	
$x''(\tau) = (1+x)^{-3} - (1+x)^{-2} = F(x) - x.$	(2.1)
$y'(\tau) = H(x,0).$	(2.2)
$z''(\tau) + [3(1+x)^{-4} - 2(1+x)^{-3}]z = (1+x)^{-3}y.$	(2.3)

to obtain for (2.1)

$$u''(\alpha) + 4\pi^2 u = 4\pi^2, \tag{2}$$

and the solution is immediately written down as

$$u(\alpha) = 1 + e\cos 2\pi\alpha. \tag{3}$$

The constant of integration e is called the skewness parameter, and the independent variable α is called the rectified phase. The second equation of (1) is inverted parametrically as

$$\tan \pi\alpha = [(1+e)/(1-e)]^{1/2} \tan \pi\beta,$$

$$\phi = \beta - (e/2\pi)\sin 2\pi\beta. \tag{4}$$

The second equation here is familiar as Kepler's equation.

In Table 3 the period of the star is Π, and the constant of gravitation is G. The period-mean density relation is

Table 3. Time Variables and Constants			
$\phi = \text{phase} = t/\Pi,$	(3.1),	$\tau = t/T_H,$	(3.4)
$2\pi\phi = \omega\tau,$	(3.2),	$\omega = (1-e^2)^{3/2},$	(3.5)
$\Pi = 2\pi T_H/\omega,$	(3.3),	$T_H = (R_e^3/GM)^{1/2},$	(3.6)

(3.3). The frequency (3.5) and the hydrodynamic time scale (3.6) are appropriate for the homogeneous star. It must be emphasized that e is a constant of integration, and once its value is known, the adiabatic part of the oscillation of a homogeneous star is completely determined.

The solutions to (2.2) and (2.3) are ninety degrees out of phase from (3) and are found in [2] as

$$y(\phi) = 2 \sum_{j=1}^{\infty} y_j \sin 2\pi j\phi, \quad z(\phi) = 2 \sum_{j=1}^{\infty} z_j \sin 2\pi j\phi.$$

The stability coefficient is an average over the adiabatic cycle of the heat function in (2.2), and it is assumed to vanish so that a limit cycle can exist.

4. A FITTING THEORY

According to [2] the velocity is a superposition of two waves which are ninety degrees out of phase. The antisymmetrical portion is due to adiabatic processes and the symmetrical to nonadiabatic. In (4.1) it is necessary [3] to increase the observed velocity by a factor of 24/17 to compensate for foreshortening and limb darkening. So far I have been unable to find a mathematical means of determining the phase shift ϕ_o in (4.1), and the value of -0.01 is used for δ Cephei. The approximation made in writting (4.4) as (4.5) is that the star is homogeneous.

The last expression in (4.6) is the two parameter (v_o and e) family of curves which seem to fit so many of the Cepheid velocities. One property of (4.6) which has not been realized in the literature is that the skewness parameter can be determined by the shape of the adiabatic velocity curve. It can be shown by (4.6) and (4) that

$$e = (\Sigma_2 - \Sigma_1)/(\Sigma_2 + \Sigma_1),$$

Table 4. Theory of Radial Velocity Curves

$$v(\phi) \equiv \partial r(1,t)/\partial t = \frac{24}{17} v_{ob}(\phi-\phi_o) = v_{ad}(\phi)+v_{na}(\phi). \quad (4.1)$$

$$v_{ad}(\phi) = \frac{1}{2}[v(\phi)-v(1-\phi)], \quad (4.2)$$

$$= -2(\omega R_e/T_H) \sum_{j=1}^{\infty} j(\sum_{\ell=1}^{\infty} \psi_\ell(1)x_{\ell,j})\sin 2\pi j\phi, \quad (4.3)$$

$$= 2(\omega R_e/T_H) \sum_{j=1}^{\infty} j\ x_j \sin 2\pi j\phi, \quad (4.4)$$

$$\simeq 2(R_e/T_H)e(1-e^2)^{1/2} \sum_{j=1}^{\infty} J_j'(je)\sin 2\pi j\phi, \quad (4.5)$$

$$= (R_e/T_H)e \sin 2\pi\alpha = v_o \sin 2\pi\alpha \equiv v_h(\phi). \quad (4.6)$$

$$v_{na}(\phi) = \frac{1}{2}[v(\phi)+v(1-\phi)], \quad (4.7)$$

$$= 2\sigma(\omega R_e/T_H) \sum_{j=1}^{\infty} j(\sum_{\ell=1}^{\infty} \psi_\ell(1)z_{\ell,j})\cos 2\pi j\phi, \quad (4.8)$$

$$\simeq (R_e/T_H)(1-e^2)^{3/2} \sum_{j=1}^{\infty} j(2\sigma z_j)\cos 2\pi j\phi. \quad (4.9)$$

where the areas Σ_1 and Σ_2 are identified in the figure. Once the values of v_o and e are known, it is a simple matter to compute the physical quantities for a homogeneous star which are given in Table 5.

5. DISCUSSION OF THE MODEL

In the figure the adiabatic velocity and (4.6) are plotted as functions of the rectified phase. The maximum of the absolute value of the residuals $v_h - v_{ad}$ is less than 1 km/sec. Thus, a relative error of about 5 percent in the adiabatic velocity is made by the assumption that δ Cephei is homogeneous. The residuals for the nonadiabatic velocity are within the scatter of Shane's data.

The values of the physical parameters in Table 5 are appropriate for a small star. The size of the physical constants is determined primarily by the knowledge of the

Table 5. Homogeneous, Nonadiabatic Model of δ Cephei
$\Pi=5.366296$ days, $e=0.490$, $v_o=\frac{24}{17}\times 19.63=27.71$ km/sec.
$T_H=(1-e^2)^{3/2}\Pi/2\pi=0.566$ days$=4.89\times10^4$ sec.
$R_e=v_oT_H/e=2.77\times10^6$ km$=3.97R_\odot$
$M=R_e^3/GT_H^2=5.56\times10^{32}$ gm$=0.066M_\odot$
$\rho_e=3M/4\pi R_e^3=6.28\times10^{-3}$ gm/cm^3, $\rho_{max}=2.09\times10^{-2}$ gm/cm^3
$P_c=G\rho_eM/2R_e=4.21\times10^{11}$ dynes/cm^2, $P_{max}=3.14\times10^{12}$ dynes/cm^2
$\Theta_c=\mu P_c/\mathcal{R}\rho_e=606\mu(v_o/e)^2\ ^\circ K=1.94\times10^6\mu\ ^\circ K$, $\Theta_{max}=4.5\times10^6\mu\ ^\circ K$
$2\sigma z_j=0.0128,\ 0.0053,\ 0.0089,\ 0.0046,\ 0.0030;\ j=1,\ldots,5.$
$2\sigma z_j=0.0016,\ 0.0010,\ 0.0008,\ 0.0006,\ 0.0005;\ j=5,\ldots,10.$

equilibrium radius R_e which is much smaller than that normally assumed in the literature. In spite of the small residuals, the homogeneous model is usually rejected on philosophical grounds because instead of being a supergiant, δ Cephei is found to be (figuratively speaking) a pulsating Jupiter.

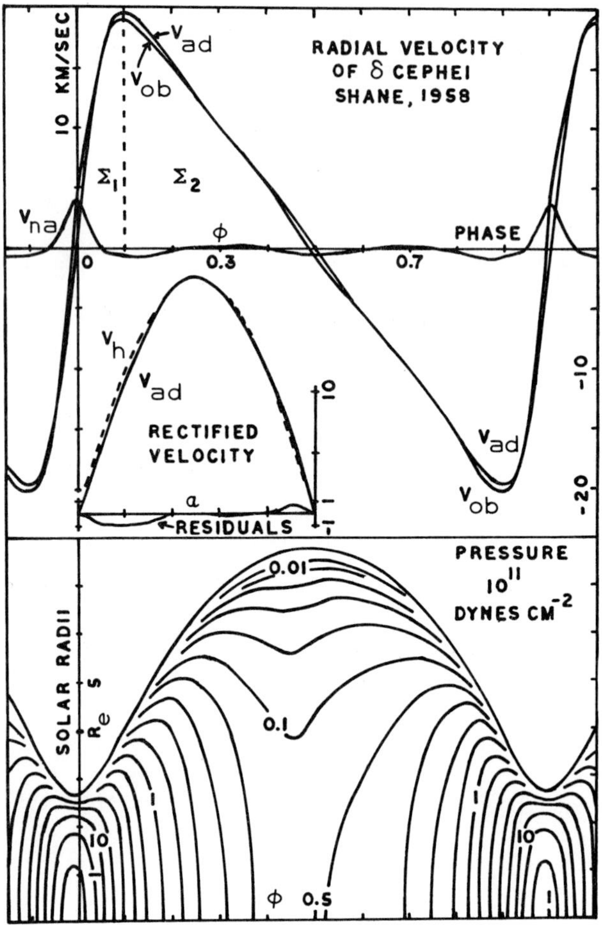

REFERENCES

1. Cox, J. P., *Rept. Prog. Phys.* *37*(1974), 563.

2. Melvin, P. J., An analytical theory of periodic variable stars, unpublished (1975).

3. Rosseland, S., *The Pulsation Theory of Variable Stars*, Dover, 1949.

4. Shane, W. W., *Astrophys. J.*, *127*(1958), 573.

EXPLOSIONS IN COMPLETELY UNSTABLE DYNAMICAL SYSTEMS

Zbigniew Nitecki
Department of Mathematics
Tufts University

A dynamical system X is said to have no C^0 Ω-explosions if for any neighborhood U of its non-wandering set, $\Omega(X)$, all dynamical systems Y sufficiently near X (in the C^0 topology) have $\Omega(Y) \subset U$. When X is a flow on a closed manifold, it was shown in [5] that the existence of Ω-explosions is related to certain kinds of generalized homoclinic phenomena ("cycles") relative to $\Omega(X)$, and that such explosions do not occur in the presence of a fine sequence of filtrations for Ω, or equivalently, in the presence of a Lyapunov function for Ω.

In trying to extend this theory to flows on open (i.e., non-compact) manifolds, one runs into the phenomenon of "completely unstable" systems, where Ω is empty. The results of [5] are formulated in terms of phenomena occurring "at Ω" and hence suggest that systems without Ω might have no C^0 Ω-explosions; however, Krych [3] and Mendes [4] have both observed that, by removing the points of $\Omega(X)$ from the (closed) phase space of a system which has Ω-explosions, one obtains a completely unstable system (on an open phase space) with Ω-explosions. Mendes [4] also noted that, as a consequence of the Brouwer translation theorem, Ω-explosions cannot occur for completely unstable systems in the plane, R^2. On the other hand, he constructed a completely unstable difeomorphism of R^3 with Ω-explosions.

1. Theorem. If M is an open manifold not homeomorphic to R^1 or R^2, then there exists a flow X on M with $\Omega(X) = \phi$, but with arbitrarily small C^0-perturbations Y for which $\Omega(Y) \neq \phi$.

In the examples constructed to prove theorem 1, Y can be taken to agree with X off a (specified) compact set. The simplest such example is the flow in the punctured plane $R^2 - 0$ given in polar coordinates by the system of differential equations

$$\dot{r} = r \cos\theta \quad (r>0)$$
$$\dot{\theta} = \sin^2\theta.$$

This system can be embedded in a flow on any open manifold of dimension 3 or more; the construction of general 2-dimensional examples requires somewhat different techniques.

The Ω-explosion in the above example is possible because the two rays $\theta = 0$ and $\theta = \pi$ are orbits belonging to each other's (first) positive prolongational limit sets. To prevent explosions, we need to deal with higher-order prolongational limit sets, and are led to the generalized limit set $R_A(X)$ of J. Auslander [1] (a point belongs to R_A iff it belongs to one of its positive prolongational limit sets), or to the more general chain-recurrent set $R_C(X)$ of Conley [2]: define $R_C(X)$ to be the set of points x such that , for any positive function $\varepsilon: M \to R^+$ and any positive number T, there exists a sequence of points $x_0 = x, x_1, \ldots, x_{n-1}, x_n = x$ and times $t_0, t_1, \ldots, t_{n-1}$ such that $t_i \geq T$ $(i=0,\ldots, n-1)$ and $d(x_i t_i, x_{i+1}) < \varepsilon(x_i)$ $(i=0,\ldots,n-1)$. (Here, d(x,y) is the distance in some metric on M, and xt is the position of x after time t.) Then, we can formulate the following conditions to prevent C^0 Ω-explosions:

2. Theorem. Suppose X is a flow on an open manifold M, and $\Omega(X) = \phi$. Then the following conditions are equivalent:

(i) For any flow Y sufficiently near X in the strong (Whitney) C^0 topology, $\Omega(Y) = \phi$;

(ii) $R_A(X) = \phi$;

(iii) $R_C(X) = \phi$;

(iv) There exists a function $f:M \to R$ and a Riemann metric on M such that X is the gradient flow of f.

(v) There exists a fine sequence of filtrations for X.

For the definitions of (v), see [5]; here, we allow each filtration to have a bi-sequence of closed pieces, and "fine" means $\cap_i K(M_i) = \phi$.

Details of these results, and a discussion of their relation to the results of Krych [3] and Mendes [4], will be published elsewhere.

REFERENCES

1. Auslander, J., Generalized recurrence in dynamical systems, *Contributions to Diff. Eq., III*(1963), 65-74.

2. Conley, C., The gradient structure of a flow: I, (unpubl. IBM Research Report), 1972.

3. Krych, M., Note on structural stability of open manifolds, *Bull. Acad. Pol. Sci. XXII*(1974), 1033-1038.

4. Mendes, P., On stability of dynamical systems on open manifolds, *J. Diff. Eq. 16*(1974), 144-167.

5. Nitecki, Z., and Shub, M., Filtrations, decompositions, and explosions, *Am. J. Math. 97*(1976), 1029-1047.

AN ALTERNATIVE PROBLEM WITH AN ASYMPTOTICALLY LINEAR NONLINEARITY

J. D. Schuur
Department of Mathematics
Michigan State University

Following the theme of [1] - [4] we study the problem

(1) $$Lx + Nx = f$$

where L is a linear, selfadjoint operator from $\mathcal{D}_L \subset S$ into S (S a real Hilbert space with norm $||\cdot||$) which satisfies: i) S_1, the range of L, is closed; ii) S_0, the kernel of L, satisfies $0 < \dim S_0 < \infty$; and iii) $H = [L|S_1]^{-1}$ is completely continuous and $||H|| \le K$ for some positive constant K; where N is a (nonlinear) operator which satisfied iv) $||Nx|| \le a + b\,||x||$ for $||x_1|| = R$, $||x_0|| \le R$ (for constants $a,b \ge 0$, $r > 0$ satisfying $K(a+b\sqrt{2}R) < R$ and $x = x_1 + x_0$ with $x_1 \in S_1$, $x_0 \in S_0$); and v) $(N(x_1+Ru_0),u_0) > 0$ for $||x_1|| \le R$, $u_0 \in S_0$, $||u_0|| = 1$ ($\cdot,\cdot$ the inner product in S); and where $f \in S$.

With these assumptions (1) has a solution.

Our intent is to bring out the degree theoretical significance of (v).

Under hypotheses (i) - (v), an alternative problem to (1) is

(2) $$x = Px + H(\tilde{I}-P)Nx$$

(3) $$0 = PNx$$

where $\tilde{I}$ is the identity map and $P: S \to S_0$, $(\tilde{I}-P): S \to S_1$ are projections. Further, S may be written as the direct sum $S = S_1 \oplus S_0$. We shall find it convenient to use the norm, $|x| = \max(||x_1||, ||x_0||)$ for $x = x_1 + x_0 \in S_1 \oplus S_0$ as well as the original norm and inner product. ($(S_1 \oplus S_0, |\cdot|)$ is a Banach space.)

Also, we shall let I_i be the identity on S_i $(i = 0,1)$ and $I = (I_1, I_0)$ be the identity on $S_1 \oplus S_0$. Then problem (2), (3) can be written as

(4) $$0 = x_1 + H(I-P)N(x_1,x_0) = I_1 x_1 - T_1(x_1,x_0)$$

(5) $$0 = x_0 - (x_0 - PN(x_1,x_0)) = I_0 x_0 - T_0(x_1,x_0)$$

where $N(x_1,x_0) = N(x_1+x_0)$ and these equations are the definition of $T_i: S_1 \oplus S_0 \to S_i$ $(i=0,1)$.

Or

(6) $0 = (I-T)(x_1,x_0)$ where $T = (T_1,T_0): S_1 \oplus S_0 \to S_1 \oplus S_0$ is, from our hypotheses, continuous and compact.

This use of the direct product space in alternative problems can be found in Cesari [5].

In this setting we may apply the Borsuk theorem to (6) as follows: If

(7) $$(I-T)(x_1,x_0)/|x_1+x_0| \neq (I-T)(-x_1,-x_0)/|-x_1-x_0|$$

on $\partial B_R = \{(x_1,x_0) \in S_1 \oplus S_0 : |x_1 + x_0| = R\}$,

then (6) has a solution in B_R.

Proof that (7) holds - and that (1) has a solution.

a) If equality exists in (7) on $\{(x_1,x_0): ||x_1|| = R, ||x_0|| \leq R\}$, then equality holds in the first entry and

$$2 = \frac{||2x_1||}{|x_1+x_0|} = \frac{||H(I-P)(N(-x_1,-x_0)-N(x_1,x_0))||}{|x_1 + x_0|}$$

$$\le \frac{2K(a+b\sqrt{2}R}{R} < 2$$

since $||x_1|| = |x_1+x_0|$ and $||x_1+x_0||^2 = ||x_1||^2 + ||x_0||^2 \le 2R^2$ - a contradiction.

b) If equality exists in (7) on $\{(x_1,x_0): ||x_1|| \le R, ||x_0|| = R\}$, then equality holds in the second entry and

$$PN(x_1,x_0) = PN(-x_1,-x_0). \tag{8}$$

If we let $u_0 = R^{-1}x_0$ and if we note that $(y,x_0) = (Py,x_0)$ for $y \in S$, then (8) implies that

$$\langle N(x_1+Ru_0), u_0\rangle = -\langle N(-x_1-Ru_1\rangle, -u_0\rangle$$

which contradicts (v).

REFERENCES

1. DeFigueiredo, D. G., On the range of nonlinear operators with linear asymptotes which are not invertable, *Comment. Math. Univ. Carolinae*, *15*(1974), 415-428.

2. Fučík, S., Kučera, M., Nečas, J., Ranges of nonlinear asymptotically linear operators, *J. Differential Equations*, *17*(1975), 375-394.

3. Nečas, J., On the range of nonlinear operators with linear asymptotes which are not invertible, *Comment. Math. Univ. Carolinae*, *14*(1973), 63-72.

4. Fučík, S., Surjectivity of operators involving linear noninvertible part and nonlinear compact perturbation, *Funkcial. Ekvac.* *17*(1974), 73-83.

5. Cesari, L., Functional analysis and Galerkin's method, *Mich. Math. J.* *11*(1964), 385-414.

POSITIVE INVARIANCE OF CLOSED SETS FOR SYSTEMS OF DELAY-DIFFERENTIAL EQUATIONS

George Seifert
Department of Mathematics
Iowa State University

Let $\{X,\langle\,,\rangle\}$ be a complete real inner produce space, and R the real line. We denote by CB the set of functions continuous and bounded on $(-\infty,0]$ to X, and for $\phi \in$ CB we define $||\phi|| = \sup\{\phi(s)\,|\,s\leq 0\}$. Then $\{CB,||\ \ ||\}$ is a Banach space over the reals. Let $f(t,\phi)$ be a function on $R \times CB$ to X, $x(t)$ be a function continuous on $(-\infty,T)$ to X where $T \leq \infty$. For fixed $t < T$ we denote by x_t the function $x(t+s)$, $s \leq 0$. If $t_0 < T$, we say that $x(t)$ is a solution of

(1) $$x'(t) = f(t,x_t)$$

for $t_0 \leq t < T$ if $x_{t_0} \in CB$, and the derivative $x'(t)$ exists on $t_0 \leq t < T$ and satisfies (1). We call x_{t_0} the initial function for this solution.

Let M be a closed subset of X; we denote by CB(M) the subset of CB consisting of ϕ with $\phi(s) \in M$ for $s \leq 0$. We say that M is positively invariant for (1) if for each $(t_0,\phi) \in R\times CB(M)$, every solution $x(t)$ of (1) on $t_0 \leq t < T$ such that $x_{t_0} \in CB(M)$ satisfies $x(t) \in M$ on this interval.

The problem of establishing necessary and sufficient conditions on M and f that M be positively invariant for (1) for the so-called non-delay case where $f(t,\phi) = F(t,\phi(0))$; i.e., the ordinary differential equation $x'(t) = F(t,x(t))$, and for $X = R^n$, real Euclidean n-space, has been studied by by many authors: [1]-[6]. Here, in the definition of M positively invariant we require only that $x(t_0) \in M$. Such equations have also been studied in case where X is a Banach space; [7]. However, even in the cases $X = R^n$ and the delay in (1) is fixed and finite, the necessary and sufficient conditions known for the non delay case cannot be extended in a natural way to necessary and sufficient conditions for the delay case (1). Simple examples show that for example the so-called subtangential condition

$$\lim_{h\to 0+} h^{-1} \operatorname{dist}(\phi(0) + hf(t,\phi),M) = 0 \tag{2}$$

for all $(t,\phi) \in R\times CB(M)$ is not sufficient for the positive invariance of M for (1) even if f is continuous on $R \times CB$ and locally Lipschitz in ϕ there. For the non delay case, (2) turns out to be necessary and sufficient if f has these properties and $X = R^n$. However, under suitably stronger conditions on f and M, in particular that M be convex, (2) has been shown to be necessary and sufficient for the case $X = R^n$; [8].

Another condition introduced in [2] for non-delay equations in R^n, a so-called inner product condition, can also be adapted to the delay case (1). We say that $n \in X$ is an outer normal to M at $x_0 \in \partial M$, the boundary of M, if $n \neq 0$, and $x \in M$ and $|x-x_0-n| \leq |n|$ imply $x = x_0$. Thus for example if $y_0 \notin M$ and there exists a $x_0 \in \partial M$ such that $\operatorname{dist}(y_0,M) = |x_0-y_0|$, then any $\lambda(y_0-x_0)$ with $0 < \lambda < 1$ is an outer normal M at x_0.

If M is such that for each $x_0 \in \partial M$ there exists a $n = n(x_0) \in \partial M$, $n \neq 0$, which is in a sense the unique limit of outer normals to M, then under suitable conditions of f, the condition

(3) $$\langle f(t,\phi), n(x_0)\rangle \le 0$$

for each $(t,\phi) \in R\times CB(M)$ for which $\phi(0) \in x_0$, and each $x_0 \in \partial M$, is necessary and sufficient that M be positively invariant for (1); [9]. This condition is stronger than the analogous one for the non-delay case, where it is only required that (3) hold for only those $x_0 \in \partial M$ at which such outer normals exist. Simple examples again show that such a weakening of (3) for the delay case is impossible.

Roughly speaking, (2) and (3) are to some extent equivalent. However, using (2) requires strong conditions on f, while using (3) requires stong conditions on M. Using (2) seems to involve the construction of a solution with the desired properties as a limit of approximate solutions. For even continuous and Lipschitzian functions f, this presents difficult problems for the infinite delay case of (1). On the other hand, using (3) no such problems arise; however, ∂M must now be pretty "nice."

REFERENCES

1. Nagumo, M., Über die Lage der Integralkurven gewöhnlicher Differentialgleichungen, *Proc. Physico-Math. Soc. Japan, Sec. 3, 24*(1942), 550-559.

2. Bony, J. M., Principe du maximum, inequalite de Harnack et unicite du problemes de Cauchy pour les operateurs elliptiques degeneres, *Ann. Inst. Fourier Grenoble 19* (1969), 227-304.

3. Yorke, J. A., Invariance for ordinary differential equations, *Math. Systems Theory 1*(4)(1967), 353-372.

4. Brezis, H., On a characterization of flow-invariant sets, *Comm. Pure Appl. Math. 23*(1970), 261-263.

5. Crandall, M. G., A generalization of Peano's existence theorem and flow invariance, *Proc. Am. Math. Soc. 36*(1) (1972), 151-155.

6. Hartman, P. On invariant sets and on a theorem of Wazewski, *Proc. Am. Math. Soc.* *32*(1972), 511-520.

7. Martin, R. H., Approximation and existence of solutions to ordinary differential equations in Banach spaces, *Funkc. Ekv.* *16*(1973), 195-211.

8. Seifert, G., Positively invariant closed sets for systems of delay equations, *J. Diff. Eqs.*, to appear.

9. Seifert, G., Positive invariant closed sets for delay-differential equations in inner product spaces, (unpublished).

NONEXISTENCE OF ALMOST PERIODIC SOLUTION*

Yasutaka Sibuya
Department of Mathematics
University of Minnesota

and

Junji Kato
Mathematical Institute
*Tohoku University***

There are several results concerning the existence of an almost periodic solution under a stability property of a preassigned bounded solution (see, for example, [1] and [2]). One of the fundamental theorems is the following.

Theorem. If an almost periodic system has a bounded solution which is totally stable, then it has an almost periodic solution.

This theorem has been proved by R. K. Miller [3] under the uniqueness of the initial value problem for every system in the hull (regularity) and, in the latter, by W. A. Coppel [4] without assuming the regularity.

It is also known that for an almost periodic system the uniformly asymptotic stability does not necessarily imply the total stability ([5] and [6]), while it does when the regularity is assumed [7] or when the system is periodic [5]. However, it was conjectured that in the above theorem the total stability can be replaced by the uniformly asymptotic stability.

Recently, we can get an example which denies this conjecture.

<u>Example</u>. Let $F(\xi,\eta,x)$ be a continuous function on $T \times (-\infty,\infty)$, where $T = [0,1] \times [0,1]$ is a torus. Then for an irrational number α, $F(t,\alpha t,x)$ is an almost periodic function.

Suppose that the system

$$\text{(E)} \qquad \dot{x} = F(t,\alpha t,x)$$

has a bounded solution which is globally uniformly asymptotically stable. Then, the almost periodic solution is unique, if any, and hence, it is of module containment [1, Th. 17.2], and hence it can be represented in the form

$$\text{(A)} \qquad x(t) = X(t,\alpha t)$$

for a continuous function $X(\xi,\eta)$ on T [8].

Therefore, to deny the conjecture, it is sufficient to construct a function $F(\xi,\eta,x)$ for which (E) has the stability property but no solutions of the form (A). Such a function is given in the following way:

Choose two small parallelograms

$$P_i = \{(\xi,\alpha\xi+s);|\xi-\xi_i| < \varepsilon,\ |s-s_i| < \varepsilon\} \quad (i=1,2)$$

in T such that $\bar{P}_1 \cap \bar{P}_2 = \emptyset$ and that there is no real t with

$$t \equiv \xi_2 \pmod 1 \text{ and } \alpha t \equiv \alpha\xi_2 + s_2 \pmod 1.$$

Set

$$F(\xi,\alpha\xi+s,x) = \begin{cases} 0 & T - P_1 \cup P_2 \\ -f(\xi,s)x & \text{on } P_1 \\ g(\xi,s)H(x,s) & P_2 \end{cases}$$

where by setting $\delta(t) = \varepsilon^2 - t^2$, $f(\xi,s) = \delta(\xi-\xi_1)\delta(s-s_1)$, $g(\xi,s) = (\xi_2-\xi)\delta(\xi-\xi_2)\delta(s-s_2)$,

$$H(x,s) = \begin{cases} \dfrac{x^2|\sin\frac{\pi}{x}|}{x^2|\sin\frac{\pi}{x}|+(s-s_2)^2}\,H(x) & s > s_2 \\ H(s) + (s-s_2)^2 & s \leqq s_2, \end{cases}$$

$$H(x) = \begin{cases} \left(\sum_{n=1}^{\infty} \frac{1}{2n} (n|x|-1)^{-1/2} \right)^{-1} & x \neq \frac{1}{n} \\ 0 & x = \frac{1}{n} (n=\pm 1, \pm 2, \ldots), \\ & |x| \geqq 1. \end{cases}$$

Then, we can see that the solution $x(t)$ such that

$$x(t) = 0 \text{ if } (t, \alpha t) \notin P_2 \pmod 1$$

is bounded and uniformly asymptotically stable and that in the expression (A) X should have a discontinuity at $(\xi, \alpha\xi + s_2)$, $|\xi - \xi_2| < \varepsilon$.

REFERENCES

1. Yoshizawa, T., Springer-Verlag, *Appl. Math. Sci. 14*(1975).

2. Fink, A. M., Springer-Verlag, *Lect. Note in Math. 377* (1975).

3. Miller, R. K., *J. Differential Eqs. 1*(1965), 337-345.

4. Coppel, W. A., *Ann. Mat. Pura Appl. 76*(1967), 27-49.

5. Kato, J., *Tohoku Math. J. 22*(1970), 254-269.

6. Sibuya, Y., (International Symposium on Dynamical Systems, Brown University) August, 1974, Academic Press, 1976.

7. Kato, J. and Yoshizawa, T., *Funkcial. Ekvac. 12*(1970), 233-238.

8. Nakajima, F., *Funkcial. Ekvac. 15*(1972), 61-73.

---

*The detail will appear somewhere else.

**Visiting Professor, Department of Mathematics, Michigan State University.

A GENERALIZATION OF THE PROBLEM OF LURIE TO FUNCTIONAL EQUATIONS

Alfredo Somolinos
Division of Applied Mathematics
Brown University

In the problem of Lurie

$$\dot{x} = Ax + b\phi(\sigma), \quad \dot{\sigma} = c'x - \rho\phi(\sigma)$$

where A is a $n \times n$ matrix, x,b,c are n-vectors and σ,ϕ,ρ are scalars, one assumes that the trivial solution of the plant equation

$$\dot{x} = Ax$$

is exponentially stable. This ensures the existence of a Liapunov function which is then used to study the stability of the trivial solution of the controlled system.

If one considers plant equations of a more general type, say non-linear functional equations, under the condition that the trivial solution is exponentially stable one can use inverse theorems ensuring the existence of Liapunov functionals for the plant equation and obtain conditions for the absolute stability of the feedback system.

1. NON-LINEAR PLANT EQUATION

Consider the equation:

$$(1) \qquad \dot{x} = g(t,x_t)$$

where $g(t,\psi)$ is a n-vector function, continuous for $t \geq 0$, and $\psi \in C \equiv C_n[-r,0]$, and Lipschitzian in ψ with Lipschitz constant L. $x_t \equiv x(t+\theta)$, $-r \leq \theta \leq 0$.

Assume that the solutions of (1) satisfy

(2) $$|x(t,t_0,\psi)| \le K||\psi||\exp\{-\alpha(t-t_0)\}.$$

The following Lemma, Hale [1], guarantees the existence of a Liapunov functional.

Lemma. Consider (1) as above and let (2) be satisfied. Then for any choice of q, $0 < q < 1$, there exists a functional $V(t,\psi)$ continuous on (t,ψ) for all $t \ge 0$, $\psi \in C$, such that

(3) (i) $$||\psi|| \le V(t,\psi) \le K||\psi||$$

(ii) $$|V(t,\psi_1) - V(t,\psi_2)| \le M||\psi_1-\psi_2||$$

(iii) $$\dot{V}(t,\psi) \le -\beta V(t,\psi)$$

where $\beta = (1-q)\alpha$, $M = K^{[L+(1-q)\alpha]/q\alpha}$ and $\dot{V}(t,\psi)$ is the usual upper right-hand derivative (see Hale [2]).

Let us consider first the indirect control case

(4) $$\dot{x} = g(t,x_t) + b\phi(\sigma)$$
$$\dot{\sigma} = h(t,x_t) - \rho\phi(\sigma)$$

where g is as above $h(t,\psi)$ is a scalar function continuous in $t \ge 0$, $\psi \in C$, such that

(5) $$|h(t,x_t)| \le c||x_t||,$$

$\rho > 0$ is a scalar and $\phi(\sigma)$ belongs to the class S_∞ of all continuous functions such that $\phi(0) = 0$ and $\sigma\phi(\sigma) > 0$ if $\sigma \ne 0$. We want to impose conditions on b,c,ρ such that the trivial solutions of (4) is absolutely stable, i.e.: it is asymptotically stable in the large for any $\phi \in S_\infty$.

1. Theorem. Consider (4) as above and assume that (2) and (5) are satisfied. Let $\int_0^\sigma \phi(s)ds \to \infty$ as $|\sigma| \to \infty$. Then if

(6) $$4\rho\beta > (M|b|+c)^2$$

the system is absolutely stable.

Proof. Let $V(t,\psi)$ be the Liapunov functional for (1) given by the Lemma. Denote by $\dot{V}_*$ the derivative of V along the solutions of (4). We have

(7) $$\dot{V}_* \le -\beta V + M|b\phi(\sigma)|.$$

Define $W = V^2/2 + \int_0^\sigma \phi(s)ds$. The derivative of W along the solutions of (4) satisfies

(8) $$\dot{W} \le -\beta V^2 + V(M|b|+c)|\phi(\sigma)| - \rho|\phi(\sigma)|^2.$$

(Here, we used (7) and (5).)

The condition (6) makes this quadratic form in V and $|\phi(\sigma)|$ negative definite.

It is easy to prove then that the conditions in Theorem 11.1 in Hale [2] are satisfied. That proves the theorem.

Using the same method we obtain a theorem for the direct control case:

(9) $$\dot{x} = g(t,x_t) + b\phi(\sigma), \quad \sigma = c'x$$

where the letters have the same meaning as above and $c'b = -\rho < 0$.

2. Theorem. Consider the system (9) as above with $\phi \in S_\infty$ and let (2) be satisfied. If either one of the following conditions are satisfied

(i) $\phi(\sigma)/\sigma < \beta/M|b||c|$

(ii) $4\beta\rho > (M|b|+L|c|)^2$

then the system is absolutely stable.

I wish to thank Dr. Hack Hale for his help in the preparation of this paper.

REFERENCES

1. Aizerman, M. A., and Gantmacher, F. R., [1], *Absolute Stability of Regulator Systems*, Holden-Day, San Francisco, 1964.

2. Hale, J., [1], Asymptotic behaviour of the solutions of differential-difference equations, *Proceedings of the International Symposium for Non-linear Oscillations, Vol. II*, Kiev, 1963.

3. Hale, J., [2], *Functional Differential Equations*, Springer-Verlag, New York, 1971.

4. Lefschetz, S., [1], Stability of non-linear control systems, *Math. in Sci. and Engineering*, Vol. 13, Academic Press, New York, 1965.

5. Somolinos, A., [1], Stability of Lurie-type functional equations, to appear in *Journal of Differential Equations*.

ON THE INTERNAL REALIZATION OF NONLINEAR BEHAVIORS

Eduardo D. Sontag*
Center for Mathematical System Theory
Department of Mathematics
University of Florida

The algebraic realization theory of linear input/output (i/o) maps is by now well developed and understood; see for instance KALMAN, FALB, and ARBIB [1969]. Linear mehtods are still useful when treating certain special types of i/o-maps; namely, bilinear i/o-maps (KALMAN [1968, 1976] and internally-bilinear i/o-maps (BROCKETT [1972], ISIDORI [1973], FLIESS [1973]). We outline here an approach designed to deal with a new class of nonlinear (shift-invariant, causal) discrete-time i/o-maps, polynomial maps. This class is large enough to include a wide range of practical examples while at the same time restricted enough to permit the application of useful algebraic tools (commutative algebra, algebraic geometry). Space limitations allow only for an informal exposition; rigorous definitions, proofs of statements and more details can be found in SONTAG [1976a]. The language used is in part that of modern algebraic geometry (MUMFORD [1968], SHAFAREVICH [1975]).

Let us fix an infinite field k and positive integers m, p. The space of input values is U: = k^m and the space of output values is Y: = k^p. We take here p = 1, for notational simplicity; the extension to p > 1 is trivial. An i/o map operates on sequences of input values to produce corresponding sequences of output values. The output values calculated by a polynomial i/o-map f are sums of products of previous input values. A polynomial i/o-map f is specified through a formal power series called the (discrete) Volterra series of f. A particular case is that of bounded i/o map f, when no input value is raised to a power higher than a certain bound; examples of bounded maps include linear, bilinear, and internally-bilinear i/o-maps.

Let U* be the set of all sequences $w = (w_1, \ldots, w_t)$, $t \geq 0$, $w_i \in U$. A polynomial i/o-map f is uniquely determined by its response map $\underline{f}: U^* \to k: w = (w_1, \ldots, w_t) \mapsto$ output at time t + 1. The concatenation of w, $\hat{w} \in U^*$ is $w\hat{w} := (w_1, \ldots w_t, \hat{w}_1, \ldots, \hat{w}_{\hat{t}})$. The elementary observable of f corresponding to a $w \in U^*$ is the map $\underline{f}^w: U^* \to k: v \mapsto \underline{f}(vw)$; this corresponds to the action of w on the states which resulted from previous inputs. The set of all maps $U^* \to k$ is a k-algebra under the pointwise operations. Thus we introduce the observation space $L_f :=$ linear span of $\{\underline{f}^w, w \in U^*\}$ and the observation algebra $A_f :=$ k-algebra generated by L_f. It can be seen that A_f is an integral domain, so A_f has a well-defined quotient field, the observation field Q_f.

The polynomial systems introduced by SONTAG and ROUCHALEAU [1975] constitute an appealing and useful class of state-space realizations of polynomial i/o-maps. Polynomial systems have their one-step transition maps given by systems of first-order polynomial difference equations and have polynomial read-out maps. It turns out, however, that a larger class, "k-systems," must be considered in order to obtain a satisfactory theory of canonical realizations. Therefore we make k-systems our primary object of study, viewing polynomial systems as a particular case.

A k-<u>system</u> Σ has an affine k-scheme as state-space X_Σ and scheme morphisms as one-step transition and read-out maps. The i/o-map of a k-system is a polynomial map. Conversely, we prove that every polynomial i/o map admits at least one k-system internal realization. A <u>canonical</u> k-system is one which is both quasi-reachable (i.e. the reachable states form a Zariski-dense subset of the state-space) and algebraically observable (i.e., the states of the system can be distinguished by means of algebraic operations on i/o data; this is the generalization to k-systems of the definition in SONTAG and ROUCHALEAU [1975]). We prove that <u>any i/o-map</u> f <u>admits a canonical realization</u> Σ_f, <u>unique up to isomorphism</u> (= change of coordinates in the state-space).

The above result is rather abstract, in that no finite-dimensionality conditions are involved. We consider three kinds of finiteness: A k-system Σ is (a) <u>almost-polynomial</u> iff X_Σ = Spec A and A can be embedded in a finitely generated k-algebra; (b) <u>polynomial</u> when X_Σ is an algebraic variety, i.e. X_Σ = Spec A and A is finitely generated; (c) <u>state-affine</u> when polynomial and given by a set of equations linear in the state (but arbitrarily polynomial in the inputs).

Let f be a polynomial i/o-map. We prove that <u>the canonical realization</u> Σ_f <u>is an almost-polynomial system iff</u> Q_f <u>has a finite transcendence degree over</u> k, <u>iff</u> Q_f <u>is a field of algebraic functions over</u> k. <u>Also</u>, Σ_f <u>is a polynomial system iff</u> A_f <u>is a finitely generated</u> k-<u>algebra, and</u>

Σ_f is a state-affine system iff L_f is a finite-dimensional k-vector space.

When f is a bounded i/o-map, all the above finiteness conditions are equivalent. Thus, for instance, a linear i/o-map realizable by a polynomial system is already realizable by a linear system.

An algebraic difference equation for f (of order $r \geq 1$) is an equation of the type $E(y(t), y(t-1),\ldots,y(t-r),u(t-1),\ldots, u(t-r)) = 0$ for all i/o pairs $\{u(\cdot),y(\cdot)\}$ for which $f: u(\cdot) \mapsto y(\cdot)$, where E is a polynomial in $(m+1)r + 1$ variables (each u(j) is an m-vector) and E is nontrivial in y(t). The equation is output-affine iff $y(t),\ldots,y(t-r)$ appear linearly. We prove that Σ_f is an almost-polynomial system iff f satisfies some algebraic difference equation, and Σ_f is a state-affine system iff f satisfies an output-affine equation.

Realization algorithms applicable in the case of bounded i/o-maps can be obtained through the notion of rational power series in noncommutative variables; this is detailed in SONTAG [1976b].

ACKNOWLEDGEMENT: I would like to express my gratitude to Professor R. E. KALMAN for providing the stimulating atmosphere of the Center for Mathematical System Theory and for arranging the long-term support which made possible, among other work, the research reported herein.

REFERENCES

1. Brockett, R. W., On the algebraic structure of bilinear systems, in *Theory and Applications of Variable Structure Systems*, (R. Mohler and A. Ruberti, eds.), A. Press, 1972.

2. Fliess, M., Sur la réalisation des systèmes dynamiques bilinéaires, *C. R. Acad. Sc. Paris, t. 277*(Series A) (1973), 243-247.

3. Isidori, A., Direct construction of minimal bilinear realizations from nonlinear input-output maps, *IEEE Trans. Automatic Control, AC-18*(1973), 626-631.

4. Kalman, R. E., Pattern recognition properties of multilinear machines, IFAC Symposium, Yerevan, Armenian SSR, 1968.

5. Kalman, R. E., On the realization of multilinear input/output maps, *Ricerche de Automatica* (1976), to appear.

6. Kalman, R. E., Falb, P. L., and Arbib, M. A., *Topics in Mathematical System Theory*, McGraw-Hill, 1969.

7. Mumford, D., *Introduction to Algebraic Geometry*, Harvard Lecture Notes, 1968.

8. Shafarevich, I., *Basic Algebraic Geometry*, Springer, 1975.

9. Sontag, E. D., On the realization of polynomial input/output maps, Ph.D. Dissertation, University of Florida, 1976a.

10. Sontag, E. D., Realization theory of discrete nonlinear systems: the bounded case, Proc. IEEE Conference on Decision and Control, December, 1976.

11. Sontag, E. D., Rouchaleau, Y., On discrete-time polynomail systems, CNR-CISM Symposium on Algebraic System Theory, Udine, Italy. To appear in revised form in *J. Nonlinear Analysis, Methods, Theory and Application, 1, 1*(1976).

*This research was supported in part by US Army Research Grant DAAG29-76-G-0203 through the Center for Mathematical System Theory, University of Florida, Gainesville, FL 32611, USA.

INVERSE PROBLEMS FOR DYNAMICAL SYSTEMS IN THE PLANE

Ronald Sverdlove
Department of Mathematics
Stanford University

By an inverse problem, we mean: given certain topological properties of a partition into curves of a phase space (here, R^2), to find a system of differential equations whose solution curves have those properties, locally or globally.

Given a set of closed curves in the plane, C_i: $f_i(x,y) = 0$, $i = 1,\ldots,n$, let us first find a system which has exactly these curves as its set of limit cycles. To do this, we form the function $F(x,y) = \prod_{i=1}^{n} f_i(x,y)$ and consider the system:

(1) $\frac{dx}{dt} = P(x,y) = F_y + GFF_x \qquad \frac{dy}{dt} = Q(x,y) = -F_x + GFF_y$

where G is an arbitrary function which is first taken to be positive. Such systems, with G = 1, have been studied in [1] and [2]. Along an orbit of (1), $\frac{dF}{dt} = FG(F_x{}^2+F_y{}^2)$ so the curves defined by F = 0, namely the C_i's, are closed orbits. Since F changes monotonically along orbits for $F \neq 0$, there are no other closed orbits, so the C_i's must be limit cycles. By changing the sign of G locally, we can achieve a prescribed stability type for each C_i independently of the others.

We can also include prescribed critical points in our system. Using a factor $f_i = (x-a)^2 + (y-b)^2$ in F makes (a,b) a source or sink. $f_i = (x-a)^2 - (y-b)^2 + u(x,y)$ makes (a,b) a saddle point of (1) where u contains higher order terms which insure that the separatrices go off to infinity without running into any of the other prescribed orbits.

While the system (1) has all the prescribed critical orbits desired, it also has additional critical points of two types. First, there must be a critical point inside every closed orbit. But there is also a set of saddle points which result from the topology of the level curves of the function F and which are not essential to the satisfaction of the other prescriptions. These saddle points can be eliminated by "pushing them off to infinity" as follows. Let C_i^*: $w_i(x,y) = 0$, $i = 1,\dots,n-1$ be a curve which is unbounded and separates C_i from $C_{i+1},\dots,C_n$. If two C_i's are nested, then the corresponding C_j^* will be a closed curve between them. Now we let $W(x,y) = \prod_{i=1}^{n-1} w_i(x,y)$ and define $F^*(x,y) = F/W^2$. Then the desired system is:

$$(2)\quad \frac{dx}{dt} = P^*(x,y) = W^5(F^*_y + F^*F^*_x) \quad \frac{dy}{dt} = Q^*(x,y) = W^5(-F^*_x + F^*F^*_y).$$

Although F^* is infinite on the C_i^*'s, the factor W^5 in (2) makes P^* and Q^* well-defined everywhere. It is the presence of these infinity-level curves that prevents the existence of the saddle points since F^* can now change sign by crossing one of them without ever passing through the value zero. A direct calculation shows that the field (2) is transverse to each of the curves C_i^*. We have thus constructed a system which has prescribed local behavior of its critical set and only such additional critical points as are required by index considerations.

Now suppose we are given a structurally stable global topological type of dynamical system in the plane according to one of the classification schemes of [3] or [4]. To construct a system of equations whose solution-flow has the given type, we first arrange for the proper local critical behavior and then insure that the critical orbits are connected to each other properly by choosing the relative positions of the curves C_i^* defined above. The correct positions are found by reducing the flow to a schematic graph in which each vertex represents a generalized source or sink and each edge represents a family of orbits connecting its positive and negative limit sets. A vertex may represent a single critical point, a limit cycle, or a configuration consisting of a saddle point, one of its invariant manifolds (stable or unstable) and the opposite limit sets of the two separatrices comprising that manifold. Since we are dealing with structurally stable systems, the separatrices cannot go to other saddle points. Now, we place the C_i^*'s so that there is one crossing each edge of the graph and they do not intersect each other. We must also choose them in such a way that the other separatrices of the saddle points (the ones not contained in the manifold chosen) are properly connected to their opposite limit sets. When this is done, the system (2) has the prescribed global type. Note that if we choose the C_i's and C_i^*'s to be polynomial curves (and in the global case we can put the C_i's wherever we like), the system (2) will be a polynomial system.

<u>Theorem</u>: Every global topological equivalence class of dynamical systems in the plane, having a finite number of

critical orbits and of structurally stable type, contains the solution-flow of a system of ordinary differential equations. These equations may be taken to be polynomial equations.

Details of the proof will appear elsewhere.

REFERENCES

1. Al'mukhamedov, M. I., On the construction of a differential equation having given curves as limit cycles (Russian), *Izvestia VUZ Mat. 44* (No. 1) (1965), 12-16; MR 30, No. 3254.

2. Valeeva, R. T., The construction of a differential equation with prescribed limit cycles and critical points, (Russian), *Volzh. Mat. Sb.*, Vyp. 5, (1966), 83-85; MR 35, No. 5701.

3. Peixoto, M. M., On the classification of flows on 2-manifolds, *Dynamical Systems:* Proceedings of the Bahia Symposium, Salvador, 1971. M. M. Peixoto, ed., Academic Press, New York, 1973, 389-419; MR 48, No. 12608.

4. Andronov, A. A., Leontovich, E. A., Gordon, I. I., and Maier, A. G., *Qualitative Theory of Second-Order Dynamical Systems*, John Wiley and Sons, New York, 1973; MR 50, No. 2619.

ON THE ALGEBRAIC CRITERIA FOR LOCAL PARETO OPTIMA

Y. H. Wan
Department of Mathematics
State University of New York at Buffalo

Motivated by mathematical economics, we consider the problem of optimizing several functions at once (see [4],[6]). More precisely, let $f = (f_1, \ldots, f_p)$ be a C^μ mapping from a given manifold M^n into R^p. A point $x \in M^n$ is said to be a <u>local Pareto Optimum</u> for f if and only if there exists a neighborhood N of x such that $y \in N$ and $f_i(y) \geq f_i(x)$ for $i = 1, \ldots, p$, imply $f_i(y) = f_i(x)$, for $i = 1, \ldots, p$. The problem is to find conditions for a given point $x \in M^n$ to be a local Pareto Optimum for f.

The first order (necessary) conditions for a local Pareto optimum are well-known (cf. [4]). The second order (sufficient conditions have been obtained recently (cf. [4], [6]). In [7], we show that the best second order conditions in [6] are still insufficient for the characterization of local Pareto optima for any generic class of mappings from M^n into R^p with $n \geq p \geq 3$. Thus, it is natural to ask the question: can one find certain high order sufficient conditions for local Pareto optima which are also necessary for some generic class of mappings from M^n into R^p?

In order to handle those high order criteria in a neat way, one is led to the notion of jets (cf. [2]). Denote by $J^r(n,p)$ the vector space of all r-jets $j^r f$ if local

C^r mappings f from R^n into R^p with $f(0) = 0$.

1. Definition. A set C in $J^k(n,p)$ is called a kth order condition provided the set C is invariant under the group of C^k diffemorphisms around $0 \in R^n$. For a given point $x \in M^n$, a C^μ $(\mu \geq k)$ map f from M^n into R^p is said to satisfy the kth order condition C at $x \in M^n$, if and only if there exists a chart (U,ϕ) around x with $\phi(x) = 0$ such that $j^r(f\circ\phi^{-1}-f\circ\phi^{-1}(0)) \in C$. It is clear that whether f satisfies the kth order condition C is independent of the choice of charts (U,ϕ) around x with $\phi(x) = 0$.

Now, we are ready to state the main results in this paper.

Main Theorem. There exists a necessary and sufficient $\min(n,p) + 1$ order algebraic condition for local Pareto optima for a generic class of $C\mu$ $(\mu>\min(n,p)+1)$ mappings from M^n into R^p. Furthermore, this class of generic mappings can itself be described by a $\min(n,p) + 1$ order algebraic condition.

Here a kth order condition C in $J^k(n,p)$ is said to be algebraic if and only if the set C is semialgebraic.

2. Definition. A subset C in R^ℓ is said to be semialgebraic if and only if the set C can be defined as a finite union of sets K_j, each being defined by a finite set of polynomial equations $P_\ell = 0$ and inequalities $Q_m \geq 0$.

Using standard theorems on stratification theory (cf. [8]) and Thom's transversality theorem, we obtain the following result as a corollary of our main theorem.

Structure theorem. Let $\mu > \min(n,p) + 2$. Then, there exists a generic class of C^μ mappings from M^n into R^p such that the set of local Pareto optima admits a Whitney (pre)-stratification.

The details of these results will be published in Topology (1976).

REFERENCES

1. Kuo, T. C., Characterizations of v-sufficiency of jets, *Topology 11* (1972), 115-131.

2. Levine, H. I., Singularities of differentiable mappings, *Proceedings of the Liverpool Singularities Symposium I, Springer Lecture Note Series no. 192* (1971).

3. Seidenberg, A., A new decision method for elementary algebra, *Annals of Math. 60* (1954), 365-374.

4. Smale, S., Optimizing several functions, *Proceedings of the Tokyo Manifolds Conference* (1973).

5. Thom, R., Local topological properties of differentiable mappings, *Differential Analysis*, Oxford University Press (1964), 191-202.

6. Wan, Y. H., On local Pareto optima, *Journal of Mathematical Economics 2* (1975), 35-42.

7. Wan, Y. H., Remarks on local Pareto optimum (Preprint).

8. Mather, J. N., Stratifications and mappings, *Salvador Symposium on Dynamical Systems* (1971).

STABILITY OF AN INFINITE SYSTEM OF DIFFERENTIAL EQUATIONS FOR THE KINETICS OF POLYMER DEGRADATION

Yong J. Yoon
Department of Mathematics
University of Florida

This paper discusses some qualitative properties (boundedness, stability, etc.) of the solutions of dynamical systems, described by infinite systems of differential equations. Our main purpose is to obtain explicit results for a special class of such systems introduced and studied by Bellman [1] and Oguztorelli [8]. These systems describe the process of dissociation of the polymer molecules into smaller units (i.e. the degradation of polymers). This process, like its opposite, the polymerization, are customarily described by finite systems of ordinary differential equations. Following [8], we consider the "limit-case" of an infinite system of such equations. It is hoped that this approach would lead to a better understanding of the limit behavior of very large systems.

Oguztorelli's system is described by the differential equations

$$(S)\quad \begin{cases} \dot{x}_1 = \sum_{j>1} m_{1j}x_j \\ \dot{x}_i = -\alpha_i x_i + \sum_{j>i} m_{ij}x_j, \quad i = 2,3,\ldots \end{cases}$$

with the initial condition $x_i(0) = c_i$, $i = 1,2,\ldots$. Here the α_i's, the m_{ij}'s and the c_i's are nonnegative constants.

The system can be written in the form $\dot{x} = Ax$ where $x = x(t) = \{x_i(t)\}$. One assumes that

1. $\sum_i c_i < \infty$.
2. $0 < \alpha_i < A$, $i = 1,2,\ldots$ for some positive number A.
3. $\alpha_j = \sum_{i=1}^{j-1} m_{ij}$, $j = 2,3,\ldots$

Guided by the physical interpretation of the equations, we choose to study this system in the ℓ_1 space. Then A becomes a continuous linear operator on ℓ_1 into ℓ_1 (p. 183 in [10]), and therefore the system has a unique solution $x(t) = e^{At}x_o$ (p. 24 in [5] or p. 289 in [2]).

Theorem. If every component of the initial condition is positive, the solution of the system (S) remains positive componentwise.

This property is suggested by the physical interpretation and is very easy to prove in the finite-dimensional case. However, in our infinite-dimensional system the problem is much more delicate. From the point of view of dynamical systems this result means that the set

$$P = \{x \text{ in } \ell_1 \colon x_i > 0,\ i = 1,2,\ldots\}$$

is positively invariant with respect to our system (S). An interesting feature is that the set P is nowhere dense in ℓ_1, and yet P and $\overline{P}$, the closure of P in ℓ_1, are the only subsets of ℓ_1 which are involved in the interpretation of the problem. The fact that in our application we only need to consider the solutions which belong to a nowhere dense subset of our Banach space seems to suggest that the Banach space is "too large" for some applications and that, at least for some applications, it is necessary to develop a special theory of dynamical systems on meager sets in Banach spaces. The concepts of dynamical systems and stability in subsets of a Banach space appear in LaSalle [6], Lefschetz [7] and Hale [3].

Theorem. The solution of the system (S) starting from the initial point x_0 in P approaches the stationary point $(c,0,0\ldots)$ where $c = \sum_i c_i$.

This convergence in ℓ_1-norm is done by a direct argument, without applying any standard stability theory. The main tool we use to prove this theorem is Barbalat's lemma (p. 211 in [9]).

We investigate our system (S) from a different point of view. Let c_0 be the space of all sequences $\{s_i\}$ such that $s_i \to 0$ as $i \to \infty$. The following lemma is important.

1. Lemma. If $x(t) = \{x_i(t)\}$ is the solution of the system (S) starting in P, then the sum $\sum_i x_i(t)$ remains constant for $t \geq 0$ and there is a sequence $\{s_i\}$ from c_0 such that $x_i(t) \leq s_i$ for $t \geq 0$ and every i.

Definition. If $\{s_i\}$ is a sequence from c_0, the set

$$Q = \{x \text{ in } \ell_\infty : x_i \leq s_i \text{ for each } i\}$$

is called a cube in ℓ_∞.

Under the inclusion map $i:\ell_1 \to \ell_\infty$, the range of the orbit $\gamma^+(x_0)$ is contained in a cube Q in ℓ_∞ by Lemma 1. Since Q is a compact subset of ℓ_∞, the set

$$\gamma_\infty^+(x_0) = \{i \circ x(t) \text{ in } \ell_\infty : t \geq 0,\ x(0) = x_0\}$$

has nonempty ω-limit set in ℓ_∞.

2. Lemma. Suppose $\{x^n\}$ is a sequence in ℓ_∞ converging to x^0 and for a positive number c $\sum_i |x_i^n| = c$ for every n. Then $\sum_i |x_i^0| \leq c$.

Proof: By the Fatou's lemma (p. 172 in [4])

$$\sum_i |x_i^0| \le \varliminf_{n\to\infty} \sum_i |x_i^n| = c.$$

This lemma and the fact that $\sum_i x_i(t) = \text{constant}$ for $t \ge 0$ from Lemma 1 imply that the ω-limit set of $\gamma_\infty^+(x_o)$ belongs to $i(\ell_1)$. Suppose $y(t,y_o)$ is the solution of the system (S) considered in the space ℓ_∞. Let V be a scalar-function on ℓ_∞ defined by $V(y) = -y_1^{\,2}$. Then V is continuous, $-c^2 \le V(y) \le 0$ and $\dot V(y) = -2y_1\dot y_1 \le 0$ if $c_i \ge 0$ for every i, where $c_i = y_i(0)$ and $c = \sum_i c_i$. Thus, V is a Liapunov function (see Definition in [6]) on the set

$$P_\infty = \{x \text{ in } \ell_\infty : x_i \ge 0\}.$$

Let S be the set defined by

$$S = \{x \text{ in } P_\infty : \dot V(x)=0\} = \{x \text{ in } \ell_\infty : x_1 \ge 0 \text{ and } x_i=0 \text{ for } i\ne 1\}$$

and $S_c = \{x \text{ in } S : x_1 \le c\}$.

Since every point in S is an equilibrium point, the largest invariant set in S is S itself. By Theorem 1 in Hale [3] and Lemma 2 in this paper we have the following theorem.

Theorem. If $y(t,y_o)$ is the solution of the system (S) considered in the ℓ_∞-space starting in P_∞, then

$$y(t) \to \text{a point in } S_c \text{ as } t\to\infty \text{ where } c = \sum_i y_i(0).$$

It is worth mentioning that the stationary solution of our system is not asymptotically stable in the classical sense, but they are asymptotically stable in the extended sense and in the practical sense.

REFERENCES

1. Bellman, R., The boundedness of solutions of infinte systems of linear differential equations, *Duke Math. J.* *14*(1947), 695-706.

2. Dieudonne, *Foundations of Modern Analysis*, Academic Press, 1960.

3. Hale, J. K., Dynamical systems and stability, *J. Math. Anal. Appl.* *26* (1969), 39-59.

4. Hewitt, E., and Stromberg, K., *Real and Abstract Analysis*, Springer-Verlag, 1969.

5. Ladas, G., and Lakshmikantham, V., *Differential Equations in Abstract Spaces*, Academic Press, New York, 1972.

6. LaSalle, J. P., An invariance principle in the theory of stability, *Int. Sym. Diff. Eqs. Dyn. Sys.*, Academic Press, New York, 1967, p. 277.

7. Lefschetz, Solomon, *Differential Equations: Geometric Theory*, Interscience, 1962.

8. Oguztorelli, On an infinite system of differential equations occurring in the degradation of polymers, *Utilitas Mathematica, Vol. 1* (1972), 141-155.

9. Popov, V. M., *Hyperstability of control systems*, Springer-Verlag, 1973.

10. Taylor, A. E., *Introduction to Functional Analysis*, John Wiley & Sons, Inc., 1958.

BILINEAR INTEGRAL EQUATIONS IN BANACH SPACE

P. F. Zweifel,
R. L. Bowden
Department of Physics
Virginia Polytechnic Institute
and State University

and

R. Menikoff
Theoretical Division
Los Alamos Scientific Laboratory

We consider the Chandrasekhar H-equation

$$H(z) = 1 + H(z) \int_0^1 \psi(t)\ H(t) \frac{x}{x+t}\, dt \tag{1}$$

and its generalizations. In Eq. (1) $\psi(t)$ is the so-called "Phase function." Bittoni, Casadei, and Lorenzutta[1] have studied Eq. (1) by showing that in the Banach space $L_1(0,1)$ the bilinear equation

$$v = f + A(v,v) \tag{2}$$

has a unique solution in the ball

$$S_1 = \{v \in L_1;\ ||v-f||_1 < 1/2\} \tag{3}$$

if $f(x) \geq 0$ on $[0,1]$ and $||f||_1 < 1/2$. Here A the bilinear operation

$$A(v_1,v_2)(x) = v_1(x) \int_0^1 v_2(t) \frac{dt}{x+t}, \tag{4}$$

Furthermore, this solution is obtained by iteration.

However it is well known[2] Eq. (1) has more than one solution. None the less the solution of Eq. (1) which obeys the constraints

$$1 = \nu_j \int_0^1 \psi(t)\ \frac{dt}{\nu_j - t},\ \mathrm{Re}\nu_j > 0,\ j = 0,\ldots,d-1 \tag{5}$$

is unique. Here $\pm\ \nu_j$, $j = 0,\ldots,d-1$, are the 2d zeros of the dispersion function

$$\Lambda(z) = 1 - z \int_{-1}^{1} \frac{\psi(t)}{t-z}\, dt \qquad (6)$$

By considering the fact that every function which satisfies Eq. (1) also factors $\Lambda(z)$ in the form $H(z)H(-z)\Lambda(z) = 1$[3], we can show

I. Theorem. If $\psi(x)$ is continuous on $[0,1]$, $||\psi||_1 < 1/2$, and v is the unique solution to Eq. (2) in S_1 for $f = \psi$, then

$$H(z) = [1 - z \int_0^1 v(t) \frac{dt}{t+z}]^{-1} \qquad (7)$$

satisfies Eq. (1) and the constraints (5).

The details of the proof of this theorem can be found in Reference 4.

For $\psi(x) \geq 0$, the dispersion function has only one pair of zeros, say $\pm\nu_0$. The case $\psi(x) \geq 0$, $||\psi||_1 \neq 1/2$ (and thus $||\psi||_1 > 1/2$) can be treated by considering the Siewert transformation[5]

$$H(z) = \nu_o^{(1+z)}\, L(z)/\ (\nu_o+z) \qquad (8)$$

and the constraint (5). This leads us to consider Eq. (2) with

$$f(x) = F(x) \equiv \nu_0^2(1-x^2)\ \psi(x)/\ (\nu_0^2-x^2). \qquad (9)$$

We can show that $||F||_1 < 1/2$ for $\psi(x)$ non-negative and even on $[-1,1]$. We then have

II. Theorem. If $\psi(x) \geq 0$ on $[0,1]$, $\psi(x)$ even on $[-1,1]$, $||\psi||_1 \neq 1/2$, and v is the unique solution to Eq. (2) with $f = F$, then H(z) given by Eq. (8) with

$$L(z) = [1-z \int_0^1 v(t)\ \frac{dt}{t+z}]^{-1}, \qquad (10)$$

satisfies Eq. (1) and the contraint (5) for $j = 0$.

The details of the proof of this theorem as well as a similar treatment of the case $||\psi||_1 = 1/2$ can also be found in Reference 4.

We can generalize our results to a system of coupled matrix bilinear equations such as arise in multigroup transport theory:[6]

$$U_1(z) = I + z \int_0^1 U_1(z)\ U_2(s)\ R(s)\frac{ds}{s+z} \tag{11a}$$

and

$$U_2(z) = I + z \int_0^1 R(s)\ U_1(s)\ U_2(z)\frac{ds}{s+z}, \tag{11b}$$

where R is the product of the non-negative matrix $\Sigma^{-1}C$ and a diagonal matrix of step functions. We deal with the Banach space X of ordered pairs of matrices with typical elements $\mathcal{U} = [U_1 U_2]$ and norm

$$||\mathcal{U}||_X = ||[U_1,U_2]||_X = \int_0^1 \max_i\ [||U_i||_M]\ (s)ds, \tag{12}$$

where $||\ \ ||_M$ is the "matrix norm." We define the bilinear operator $\mathcal{A}$ by

$$\mathcal{A}(\mathcal{U},\mathcal{V})(z) = [z \int_0^1 V_1(z)\ U_2(s)\frac{ds}{s+z},\ z \int_0^1 U_1(s)V_2(z)\frac{ds}{s+z}]. \tag{13}$$

and consider the equation

$$\mathcal{U} = \mathcal{F} + \mathcal{A}\ (\mathcal{U},\mathcal{U}), \tag{14}$$

where $\mathcal{F} = [R,R]$. Using a result of Bittoni et al, we can show that if $||F||_X < 1/2$, then Eq. (14) has a unique solution in the ball

$$S_2 = \{\mathcal{U}\in X; ||\mathcal{U}-\mathcal{F}||_X < 1/2\}. \tag{15}$$

We then write

III. Theorem. If $\int_0^1 ||R||_M\ (s)\ ds < 1/2$ and $\hat{U}$ is the unique solution of Eq. (14) in S_2

$$U_1(z) = [I - \int_0^1 \hat{U}_2(s)\frac{z}{z+s}\ ds]^{-1} \tag{16a}$$

and

$$U_2(z) = [I - \int_0^1 \hat{U}_1(s)\frac{z}{z+s}\ ds]^{-1} \tag{16b}$$

then satisfy Eq. (11) and factor the dispersion matrix in the form

$$U_2(z)U_1(-z)\Lambda(z) = U_2(z)U_1(-z)[I - z\int_{-1}^{1} R(s)\frac{ds}{s-z}] \qquad (17)$$

and obeys the constraints

$$\det U_1(-\nu_j) = \det U_2(-\nu_j) = 0 \quad \mathrm{Re}\ \nu_j > 0 \quad j=0,\ldots,d-1 \qquad (18)$$

where here $\pm\nu_j$ are the 2d zeros of det $\Lambda(z)$. Again the details of proof can be found elsewhere.[7]

REFERENCES

1. Bittoni, E., Casadei, G., and Lorenzutta, S., *Boll. U.M.I.* *4*(1969), 435.

2. Mulikin, T. W., *Trans. Am. Math. Soc.* *113*(1964), 316.

3. Chandrasekhar, S., *Radiative Transfer*, Oxford University Press, London, 1960.

4. Bowden, R. L., and Zweifel, P. F., A Banach space analysis of the Chandrasekhar H-equation, (1976) to be published in *Ap. J.*

5. Siewert, C. E., *J. Quant. Spectrosc. Radiat. Transfer 15* (1975), 385.

6. Bowden, R. L., Sancaktar, S., and Zweifel, P. F., *J. Math. Phys.* *17*(1976), 76, 82.

7. Bowden, R. L., Zweifel, P. F., and Menikoff, R., *J. Math. Phys.* *17*(1976), 1722.

A
B 7
C 8
D 9
E 0
F 1
G 2
H 3
I 4
J 5